CISM Courses and Lectures

The series presents lecture notes, monographs, edited works and proceedings in the field of Mechanics, Engineering, Computer Science and Applied Mathematics.
Purpose of the series is to make known in the international scientific and technical community results obtained in some of the activities organized by CISM, the International Centre for Mechanical Sciences.

International Centre for Mechanical Sciences

Courses and Lectures Vol. 555

For further volumes:
www.springer.com/series/76

Gaetan Kerschen
Editor

Modal Analysis of Nonlinear Mechanical Systems

Springer

Editors

Gaetan Kerschen
University of Liège, Belgium

ISSN 0254-1971
ISBN 978-3-7091-1790-3 ISBN 978-3-7091-1791-0 (eBook)
DOI 10.1007/ 978-3-7091-1791-0
Springer Wien Heidelberg New York Dordrecht London

All contributions have been typeset by the authors
Printed in Italy

Printed on acid-free paper

Springer is part of Springer Science+Business Media (www.springer.com)

PREFACE

This book contains lecture material from CISM course on Modal Analysis of Nonlinear Mechanical Systems, delivered in June 25-29, 2012. The concept of normal modes is central in the theory of linear vibrating systems. Besides their obvious physical interpretation, the linear normal modes (LNMs) have interesting mathematical properties. LNMs are therefore exploited routinely for various purposes including experimental modal analysis, model reduction and structural health monitoring. Clearly, nonlinearity is a frequent occurrence in engineering applications and can take its origin from different sources including contact, friction and large displacements. This is why the concept of nonlinear normal modes (NNMs) was introduced in the 1960s by Rosenberg. The course gives a complete and thorough overview of NNMs, going from their definition and fundamental properties to their applications in structural dynamics, which include model reduction, system identification and acoustic and vibration mitigation.

The first two chapters, by G. Kerschen and S.W. Shaw, introduce the two different definitions of NNMs. Even though NNMs have a clear conceptual relation to LNMs, it will be shown that they have properties that are fundamentally different than those of linear modes. Chapter 3, by C. Touzé, details a constructive technique for computing the NNMs of mechanical systems based on the theory of normal forms. The usefulness of NNMs for the interpretation of nonlinear phenomena and for model reduction of nonlinear systems is discussed. Chapter 4, by O. Gendelman, analyses thoroughly the effect of damping on NNMs, which, in turn, allows to study targeted energy transfer in coupled nonlinear oscillators. Chapters 5 and 6, by G. Kerschen and B. Cochelin, address the numerical computation of NNMs using algorithms that have the capability to deal with systems of high dimensionality in strongly nonlinear regimes of motion. Chapter 7, by A.F. Vakakis, discusses the elements of a nonlinear system identification methodology which integrates the concept of nonlinear modes with slow flows and signal processing. Finally, Chapter 8, by B. Cochelin, revisits targeted energy transfer with application to acoustic mitigation.

Gaëtan Kerschen, Liège, 2014

CONTENTS

Definition and Fundamental Properties of Nonlinear Normal Modes

G. Kerschen
Department of Aerospace and Mechanical Engineering
University of Liège
Liège, Belgium

1 A Brief Historical Perspective

The concept of a normal mode is central in the theory of linear vibrating systems. Besides their obvious physical interpretation, the linear normal modes (LNMs) have interesting mathematical properties. They can be used to decouple the governing equations of motion; i.e., a linear system vibrates as if it were made of independent oscillators governed by the eigensolutions. Two important properties that directly result from this decoupling are:

1. *Invariance*: if the motion is initiated on one specific LNM, the remaining LNMs remain quiescent for all time.
2. *Modal superposition*: free and forced oscillations can conveniently be expressed as linear combinations of individual LNM motions.

In addition, LNMs are relevant dynamical features that can be exploited for various purposes including model reduction (e.g., substructuring techniques, experimental modal analysis, finite element model updating and structural health monitoring.

Clearly, though, linearity is an idealization, an exception to the rule; nonlinearity is a frequent occurrence in real-life applications. Any attempt to apply traditional linear analysis to nonlinear systems results, at best, in a suboptimal design. Thus, there is a need for efficient, analytically rigorous, broadly applicable analysis techniques for nonlinear structural dynamics. In this context, nonlinear normal modes (NNMs) offer a solid theoretical and mathematical tool for interpreting a wide class of nonlinear dynamical phenomena, yet they have a clear and simple conceptual relation to the LNMs, with which practicing structural engineers are familiar. Other appealing features of the NNMs are that they are capable of handling strong structural nonlinearity and that they have the potential to address the individualistic nature of nonlinear systems.

G. Kerschen (Ed.), *Modal Analysis of Nonlinear Mechanical Systems*, CISM International Centre for Mechanical Sciences DOI 10.1007/ 978-3-7091-1791-0_1

The most straightforward definition of an NNM is a *vibration in unison* of the system (i.e., a synchronous oscillation). NNMs were pioneered in the 1960s thanks to the seminal work of Rosenberg (1). They were further studied in the 1970s by Rand (2; 3; 4) and Manevitch and Mikhlin (5). They were given a new impetus in the 1990s through the efforts of Vakakis et al. (6; 7; 8; 9; 10; 11) and Shaw and Pierre (12; 13; 14; 15). Since then, a large body of literature has addressed, with notable success, the qualitative and quantitative analysis of nonlinear phenomena using NNMs (see, e.g., (16; 17; 18; 19; 20; 21; 22; 23; 24; 25; 26; 27; 28; 29; 30; 31; 32; 33; 34; 35; 36; 37)). For a few typical applications and comprehensive reviews, the reader can refer to (10; 11; 38; 39; 40; 41).

The objective of the present chapter is to describe and to illustrate in a simple manner the fundamental properties of NNMs. The chapter is organized as follows. In the next section, the two main definitions of NNMs are provided. In addition, their fundamental properties are described, and their representation in a frequency-energy plot is introduced. In Section 3, the different means of computing the NNMs are reviewed and assessed. The application of NNMs to 'linear' and nonlinear modal analysis is then briefly discussed in Section 4.

2 Nonlinear Normal Modes: What Are They ?

The free response of discrete mechanical systems is considered herein, assuming that continuous systems have been spatially discretized using, e.g., the finite element method. The equations of motion are

$$\mathbf{M}\ddot{\mathbf{x}}(t) + \mathbf{K}\mathbf{x}(t) + \mathbf{f}_{nl}\left\{\mathbf{x}(t), \dot{\mathbf{x}}(t)\right\} = 0 \tag{1}$$

where $\mathbf{M}$ is the mass matrix; $\mathbf{K}$ is the stiffness matrix; $\mathbf{x}$, $\dot{\mathbf{x}}$ and $\ddot{\mathbf{x}}$ are the displacement, velocity and acceleration vectors, respectively; $\mathbf{f}_{nl}$ is the nonlinear restoring force vector, assumed to be regular.

To illustrate the different concepts, a two-degree-of-freedom (2DOF) system with a cubic stiffness is chosen. The system is depicted in Figure 1, and its motion is governed by the equations

$$\begin{aligned} \ddot{x}_1 + (2x_1 - x_2) + 0.5\,x_1^3 &= 0 \\ \ddot{x}_2 + (2x_2 - x_1) &= 0 \end{aligned} \tag{2}$$

For comparison purposes, the underlying linear system

$$\begin{aligned} \ddot{x}_1 + (2x_1 - x_2) &= 0 \\ \ddot{x}_2 + (2x_2 - x_1) &= 0 \end{aligned} \tag{3}$$

is also considered in this study. The time series corresponding to in-phase and out-of-phase normal mode motions of the linear system (3) are depicted in Figure 2. Motion in the configuration space (i.e., in the plane of the displacements $x_1(t)$ and $x_2(t)$) is given in Figure 3. Obviously, LNM motions correspond to straight lines in this plane.

2.1 Definition of a Nonlinear Normal Mode

There exist two main definitions of the NNMs in the literature, due to Rosenberg (1) and Shaw and Pierre (12; 13; 14; 15). There have been additional definitions, which include a complex-valued invariant manifold formulation (42; 43) and group theoretic definitions (10), but they are not described herein.

Rosenberg's Definition

During the normal mode motion of a linear conservative system, each system component moves with the same frequency and with a fixed ratio amongst the displacements of the components. Targeting a straightforward nonlinear extension of the LNM concept, Rosenberg defined an NNM as a *vibration in unison* of the system (i.e., a synchronous oscillation). This definition requires that all material points of the system reach their extreme values and pass through zero simultaneously and allows all displacements to be expressed in terms of a single reference displacement.

For illustration, the time series and the configuration space of in-phase and out-of-phase NNM motions during the free response of system (2) are depicted in Figures 4 and 5, respectively. The modal lines of this nonlinear system are curves, resulting from the nonlinear relationship between the coordinates during the periodic motion. These curved NNMs, termed nonsimilar NNMs by Rosenberg, are generic in nonlinear systems, and their existence certainly complicates the concept of orthogonality between modes (44; 45). As discussed in Section 2.2, a particular feature of these NNMs is that their shape depends on the total energy present in the system. When special spatial symmetries exist, the NNMs may degenerate into (energy-invariant) straight modal lines, as in the linear case (see (7) and Section 2.2).

Rosenberg's definition may appear restrictive in two cases:

1. This definition, as such, cannot be easily extended to nonconservative systems.
2. In the presence of internal resonances (i.e., when two or more NNMs interact), some coordinates may have a dominant frequency component different than that of the other coordinates (e.g., some coordi-

nates may vibrate faster than others). In this case, the system no longer vibrates in unison. This is illustrated in Figure 6 for an internally resonant NNM (3:1 internal resonance) of system (2).

However, these two limitations can be circumvented. Firstly, as shown in Section 4.2 and in (26; 33), the damped dynamics can often be interpreted based on the topological structure and bifurcations of the NNMs of the underlying undamped system. Secondly, realizing that the motion is still periodic in the presence of internal resonances, Rosenberg's definition of an NNM can be extended to *a (non-necessarily synchronous) periodic motion of the system*. This extended definition is particularly attractive when targeting a numerical computation of the NNMs. It enables the nonlinear modes to be effectively computed using algorithms for the continuation of periodic solutions, which are really quite sophisticated and advanced.

The Invariant Manifold Approach

Shaw and Pierre proposed a generalization of Rosenberg's definition that provides a direct and elegant extension of the NNM concept to damped systems. Based on geometric arguments and inspired by the center manifold technique, they defined an NNM as a two-dimensional invariant manifold in phase space. Such a manifold is invariant under the flow (i.e., orbits that start out in the manifold remain in it for all time), which extends the invariance property of LNMs to nonlinear systems. In order to parameterize the manifold, a single pair of state variables (i.e., both the displacement and the velocity) are chosen as master coordinates, the remaining variables being functionally related to the chosen pair. Therefore, the system behaves like a nonlinear single-DOF system on the manifold.

Geometrically, LNMs are represented by planes in phase space, and NNMs are two-dimensional surfaces that are tangent to them at the equilibrium point. For illustration, the manifolds corresponding to in-phase and out-of-phase NNMs motions of system (2) are given in Figure 7.

2.2 Fundamental Properties

NNMs have intrinsic properties that are fundamentally different from those of LNMs. They are reviewed and illustrated in what follows.

Frequency-Energy Dependence

One typical dynamical feature of nonlinear systems is the frequency-energy dependence of their oscillations. One important consequence is that the frequency response functions (FRFs) of nonlinear systems are no longer invariant. For illustration, the FRFs of system

$$\begin{aligned} \ddot{x}_1 + (0.02\dot{x}_1 - 0.01\dot{x}_2) + (2x_1 - x_2) + 0.5\,x_1^3 &= F\cos\omega t \\ \ddot{x}_2 + (0.02\dot{x}_2 - 0.01\dot{x}_1) + (2x_2 - x_1) &= 0 \end{aligned} \tag{4}$$

are depicted in Figures 8 and 9 for F varying between 0.002 N and 0.2 N.

The modal curves and frequencies of oscillation of NNMs also depend on the total energy in the system. In contrast to linear theory, this energy dependence prevents the direct separation of space and time in the governing equations of motion, which complicates the analytical calculation of the NNMs.

Returning to the undamped system (2), Figure 10 shows the time series, the configuration space, the power spectral density (PSD) and two-dimensional projections of the phase space of three in-phase NNM motions of increasing energies. The NNM motion at low energy resembles that of the in-phase LNM of the underlying linear system (3). The modal curve is a straight line, there is one main harmonic component in the system response, and the motion in phase space is a circle. For the motion at moderate energy, the NNM is now a curve, and the presence of two harmonic components can be detected. A clear departure from the LNM (harmonic) motion is observed. At high energy, this is even more enhanced. For instance, the motion in phase space is a strongly deformed ellipse. When moving from the low- to the high-energy NNM, the period of the motion decreases from 6.28 s to 4.755 s. This is due to the hardening characteristic of the cubic spring. Another noticeable characteristic of the NNMs is that the modes at higher energies are not the geometric continuation of those at lower energies.

To further illustrate the frequency-energy dependence of the NNMs, the harmonic balance method can be applied to system (2). This method expresses the periodic motion of a system by means of a finite Fourier series (46). For simplicity, a series with a single harmonic component is considered

$$x_1(t) = A\cos\omega t, \quad x_2(t) = B\cos\omega t \tag{5}$$

This *ansatz* is plugged into the equations of motion (2). Expanding $\cos^3\omega t$ in terms of $\cos\omega t$ and $\cos 3\omega t$, and balancing all the coefficients of the $\cos\omega t$ terms yields

$$\begin{aligned} -A\omega^2 + (2A - B) + 0.5\,\frac{3A^3}{4} &= 0 \\ -B\omega^2 + (2B - A) &= 0 \end{aligned} \tag{6}$$

Analytic expressions for coefficients A and B are then readily obtained

$$A = \pm\sqrt{\frac{8(\omega^2-3)(\omega^2-1)}{3(\omega^2-2)}} \tag{7}$$

$$B = \frac{A}{2-\omega^2} \tag{8}$$

The square root exists in the two frequency intervals

$$\omega_1 \in [1, \sqrt{2}[\quad \text{and} \quad \omega_2 \in [\sqrt{3}, +\infty[\tag{9}$$

noting that $\omega = 1$ rad/s and $\omega = \sqrt{3}$ rad/s are the two natural frequencies of the underlying linear system (3). In the first (second) frequency interval, B has the same (opposite) sign as A; an in-phase (out-of-phase) NNM motion is observed for initial conditions $[x_1(0)\ x_2(0)\ \dot{x}_1(0)\ \dot{x}_2(0)] = [A\ B\ 0\ 0]$).

The (conserved) total energy during the free response of system (2) is

$$\text{Total Energy} = \frac{A^2}{2} + \frac{(B-A)^2}{2} + \frac{B^2}{2} + 0.5\,\frac{A^4}{4} \tag{10}$$

which, according to equations (7) and (8), demonstrates the frequency-energy dependence of NNM motions.

An appropriate graphical depiction of the NNMs is key to their exploitation. The usual representation in the literature is to plot the motion amplitude at a particular DOF as a function of frequency. Due to the frequency-energy dependence, the representation of NNMs in a *frequency-energy plot* (FEP) is particularly convenient (26; 33). An NNM is represented by a point in the FEP, which is drawn at a frequency corresponding to the minimal period of the periodic motion and at an energy equal to the conserved total energy during the motion. A branch, represented by a solid line, is a family of NNM motions possessing the same qualitative features (e.g., the in-phase NNM motions of a 2DOF system).

As a point of comparison, the FEP of the linear system (3) is shown in Figure 11. Because the natural frequencies do not depend on energy, this FEP comprises two horizontal lines at the two resonant frequencies of the system. The LNM motions represented in the configuration space are inset and are obviously unaltered by the energy level.

The FEP of the nonlinear system (2) was computed using the method proposed in Chapter 5 and is shown in Figure 12. The backbone of the plot is formed by two branches, which represent in-phase ($S11+$) and out-of-phase ($S11-$) synchronous NNMs. The letter S refers to symmetric periodic solutions for which the displacements and velocities of the system at half period are equal but with an opposite sign to those at time $t = 0$. As shown in the next section, unsymmetric periodic solutions may also be encountered and are denoted by a letter U. The indices in the notations are used to mention that the two masses vibrate with the same dominant frequency. The FEP clearly shows that the nonlinear modal parameters have a strong dependence on the total energy in the system:

1. The frequency of both the in-phase and out-of-phase NNMs increases with the energy level, which reveals the hardening characteristic of the system.
2. The modal curves change for increasing energies. The in-phase NNM tends to localize to the second DOF (i.e., it resembles a vertical curve), whereas the out-of-phase NNM localizes to the first DOF (i.e., it resembles an horizontal curve). This localization property is a key feature of nonlinear systems and is discussed extensively in (10). It is also exploited for vibration mitigation in (26; 33; 47; 48).

The comparison between Figures 11 and 12 also reveals that NNMs have a clear and simple conceptual relation to the LNMs.

Modal Interactions — Internally Resonant Nonlinear Normal Modes

Another salient feature of nonlinear systems is that NNMs may interact during a general motion of the system. Nonlinear modal interactions have been studied extensively in the literature (see, e.g., the monograph (38)). A case of particular interest is when the linear natural frequencies are commensurate or nearly commensurate (44; 49; 50; 51). An energy exchange between the different modes involved may therefore be observed during the internal resonance. For instance, exciting a high-frequency mode may produce a large-amplitude response in a low-frequency mode. Vibration absorbers exploiting these energy transfers have been studied in (52).

Internally resonant NNMs have no counterpart in linear systems and are generated through bifurcations. Considering system (2) and according to the discussion in the previous section, the FEP in Figure 12 does not seem to feature internally resonant NNMs. However, when carrying out the NNM computation at higher energy levels, Figure 13 shows that another branch of periodic solutions, termed a tongue, emanates from the backbone branch $S11+$. On this tongue, denoted $S31$, there is a 3:1 internal resonance between the in-phase and out-of-phase NNMs.

Surprisingly, the ratio of the linear natural frequencies of system (2) is $\sqrt{3}$. Due to energy dependence, a 3:1 ratio between the two frequencies can still be realized, because the frequency of the in-phase NNM increases less rapidly than that of the out-of-phase NNM. This clearly shows that *NNMs can be internally resonant without necessarily having commensurate linear natural frequencies*, a feature that is rarely discussed in the literature. This also underlines that important nonlinear phenomena can be missed when resorting to perturbation techniques, which are limited to small-amplitude motions.

There exists a smooth transition from $S11+$ to $S11-$ that occurs on tongue $S31$. This transition is depicted in Figure 14 where the evolution of the configuration space and of the Fourier coefficients is shown for several points on $S31$ or in its vicinity. Starting from NNM (a), an in-phase motion characterized by two perceptible harmonic components is observed. From (a) to (d), the relative importance of the third harmonics grows, as clearly confirmed by the motion in the configuration space. Moving from (d) to (e) corresponds to a drastic qualitative change in the dynamics. Firstly, the first harmonics has completely disappeared for both oscillators. Secondly, the signs of the coefficients of the third harmonics are opposite. Overall, this means that an out-of phase motion with a three times as large frequency is realized. Eventually, through a 3:1 internal resonance, the motion ends up on $S33-$ or, equivalently, on $S11-$. From (f) to (h), the relative importance of the third harmonics diminishes, and a motion qualitatively similar to that at (a) is observed. However, the configuration space of NNM (h) reveals the presence of a fifth harmonics, which is a precursor to the gradual development of tongue $S51$.

This indicates that other resonance scenarios exist in this seemingly simple system. The frequency of the in-phase NNM motions on $S11-$ steadily increases for increasing energies, whereas the NNM motions on $S11+$ have their frequency asymptotically approaching a value close to $\sqrt{3}\,\mathrm{rad/s}$. Following this reasoning, we expect the existence of a countable infinity of internal resonance cases (e.g., 2:1, 4:1, 5:1, etc.). To confirm this conjecture, additional tongues have been computed numerically and are represented in Figure 15. These tongues emanate from $S11+$ and coalesce into $S11-$ following a mechanism similar to that described above (Figure 14). To illustrate the rich dynamics, a few representative NNMs of system (2) are depicted in Figure 16. Such a complex dynamics was first observed in (26) for a system with an essential nonlinearity. It is interesting that this can also be reproduced for more generic nonlinear systems.

Mode Bifurcations and Stability

A third fundamental property of NNMs is that their number may exceed the number of DOFs of the system. Due to mode bifurcations, not all NNMs can be regarded as nonlinear continuation of normal modes of linear systems

(10; 53; 54). Internally resonant NNMs are one example. Another possible example corresponds to the NNM bifurcations of the system

$$\begin{aligned}\ddot{x}_1 + x_1 + x_1^3 + K(x_1 - x_2)^3 &= 0\\ \ddot{x}_2 + x_2 + x_2^3 + K(x_2 - x_1)^3 &= 0\end{aligned} \tag{11}$$

for variations of the coupling spring K (7). This system possesses similar NNMs that obey to the relation $x_2(t) = cx_1(t)$. Eliminating x_2 from equations (11) yields

$$\begin{aligned}\ddot{x}_1 + x_1 + \left[1 + K(1-c)^3\right] x_1^3 &= 0\\ \ddot{x}_1 + x_1 - \frac{1}{c}\left[K(1-c)^3 + c^3\right] x_1^3 &= 0\end{aligned} \tag{12}$$

Because both equations must lead to the same solution, it follows

$$K(1+c)(c-1)^3 = c(1-c^2), \quad c \neq 0 \tag{13}$$

Equation (13) means that system (11) always possesses two modes characterized by $c = \pm 1$ that are direct extension of the LNMs. However, this system can possess two additional similar NNMs that cannot be captured using linearization procedures. At $K = 0.25$, these NNMs bifurcate from the out-of-phase mode, as shown in Figure 17.

Another important characteristic of NNMs is that they can be stable or unstable, which is in contrast to linear theory where all modes are neutrally stable. In this context, instability means that small perturbations of the initial conditions that generate the NNM motion lead to the elimination of the mode oscillation. Therefore, unstable NNMs are not physically realizable. The NNM stability analysis can be performed numerically or analytically. In Figure 18, stability is computed numerically through the eigenvalues of the monodromy matrix. In other studies, analytical results are obtained through Floquet theory after adding small perturbations to the periodic solutions. For a detailed stability analysis of the NNMs, the reader can refer to (7; 9; 10; 55; 56).

Bifurcations and stability are interrelated concepts, because a change in stability occurs through a bifurcation. For instance, the bifurcation in system (11) generates a pair of stable/unstable NNMs (Figure 17). Returning to system (2), another illustration of NNM stability is shown in the FEP of Figure 18. When the tongue $U21$ bifurcates from $S11+$, the NNMs on this latter branch lose stability. A detailed description of this tongue and the related dynamical mechanisms (e.g., symmetry-breaking bifurcation) is beyond the scope of this paper. This figure also shows that stability can be lost when a turning point is encountered.

3 Nonlinear Normal Modes: How to Compute Them ?

Different methods for computing NNMs of discrete and continuous systems are briefly described in this section. They are classified in two categories, according to whether the computation relies on analytical or numerical methods. This discussion is by no means a comprehensive overview of the past and current approaches and focuses on the free response of nonlinear vibrating systems. There is no attempt to summarize the methods dealing with the forced response case.

3.1 Analytical Techniques

Rosenberg was the first to develop constructive techniques for computing NNMs of discrete conservative oscillators. Rand obtained explicit approximate expressions for modal curves in 2DOF systems by applying a perturbation method to a modal equation (3). Manevitch and Mikhlin reduced the problem of computing nonsimilar NNMs to a set of singular boundary value problems, which were then solved using power series expansions (5). The book by Vakakis et al. (10) summarizes the developments until the 1990s.

The early 1990s witnessed a resurgence of interest in the NNMs with the works of Vakakis (6; 7; 8; 9) and Shaw and Pierre (12; 13; 14; 15). Simple discrete systems were first studied (8; 12; 14), but the generalization to continuous systems (9; 15) soon followed. For continuous systems, two main approaches exist in the literature. The first approach is to study directly the original partial differential equations (9; 15). An alternative method is to discretize the governing nonlinear partial differential equations into an infinite set of ordinary differential equations that is truncated to a finite number of equations (13). The two alternatives are compared in (57) using the invariant manifold approach.

An Energy-Based Formulation

This formulation relies on Rosenberg's work (1) and expresses an NNM as a modal curve in the configuration space. It was further developed by Manevitch and Mikhlin for discrete conservative oscillators (5) and exploited in a few other studies (8; 10). To illustrate the method, it is applied to system (2)

$$\begin{aligned} \ddot{x}_1 + (2x_1 - x_2) + 0.5\,x_1^3 &= 0 \\ \ddot{x}_2 + (2x_2 - x_1) &= 0 \end{aligned} \tag{14}$$

When the system vibrates along an NNM, the displacement x_2 is linked to x_1 through the expression of the modal curve $\hat{x}_2$

$$x_2 = \hat{x}_2(x_1) \tag{15}$$

The objective of the method is to eliminate the time derivatives from the equations of motion (14). To compute the second time derivative of x_2, relation (15) is differentiated twice using the chain rule

$$\ddot{x}_2 = \hat{x}_2''\dot{x}_1^2 + \hat{x}_2'\ddot{x}_1 \tag{16}$$

where prime denotes differentiation with respect to x_1. This expression involves the second time derivative of x_1, which is readily obtained from the equations of motion

$$\ddot{x}_1 = -2x_1 + \hat{x}_2 - 0.5x_1^3 \tag{17}$$

It then remains to compute the first time derivative of x_1 appearing in equation (16). To this end, a first integral of motion expressing explicitly the conservation of energy during the motion is written by multiplying equation (17) by $\dot{x}_1$ and integrating

$$\dot{x}_1^2 = 2\int_0^{\dot{x}_1} \dot{x}_1\,\mathrm{d}\dot{x}_1 = -2\int_{X_1}^{x_1} \left[2u - \hat{x}_2(u) + 0.5u^3\right]\,\mathrm{d}u \tag{18}$$

where X_1 is the maximum amplitude attained by x_1; i.e., when $\dot{x}_1 = 0$. The derivatives are substituted into the second of equations (14), which yields the equation governing the modal curve:

$$\begin{aligned}\hat{x}_2''\left\{-2\int_{X_1}^{x_1} \left[2u - \hat{x}_2(u) + 0.5u^3\right]\,\mathrm{d}u\right\} + \hat{x}_2'\left[-2x_1 + \hat{x}_2 - 0.5x_1^3\right]& \\ +(2\hat{x}_2 - x_1) = 0&\end{aligned} \tag{19}$$

Because the coefficient of the highest derivative vanishes when $x_1 = X_1$, this functional equation is singular at the maximum equipotential surface. It must therefore be supplemented by a boundary condition

$$\left\{\hat{x}_2'\left[-2x_1 + \hat{x}_2 - 0.5x_1^3\right] + (2\hat{x}_2 - x_1)\right\}_{x_1 = X_1} = 0 \tag{20}$$

which expresses that the nonlinear mode intersects orthogonally the maximum equipotential surface in the configuration space. Equation (19) does not depend on the time variable, and its solution is amenable to a power series expansion:

$$\hat{x}_2(x_1) = \hat{x}_2^{(0)}(x_1) + \epsilon\hat{x}_2^{(1)}(x_1) + \epsilon^2\hat{x}_2^{(2)}(x_1) + O(\epsilon^3) \tag{21}$$

This formulation was extended to undamped continuous systems in (9). The displacement of any point of the system is expressed in terms of a single reference displacement $x_0(t) = x(s_0, t)$ by the functional relation

$$x(s,t) = X\left[s, x_0(t)\right] \tag{22}$$

where s is the spatial coordinate, and X is a modal function characterizing the considered NNM. Then, an integral equation expressing the conservation of energy during the motion is used in conjunction with equation (22) to eliminate the time derivatives from the equations of motion. Eventually, the equation governing the modal function X is obtained and is solved using power series.

In the presence of internal resonances, the folding of the NNMs in the configuration space may result in multivalued relationship among the various coordinates (see Figure 6). This has been nicely addressed in (51) by considering NNMs in an appropriately defined modal space.

The Invariant Manifold Approach

The invariant manifold approach (12; 13; 14; 15) is similar in spirit to the energy-based formulation and is described in detail in the next chapter. The difference with the previous approach is that a pair of state variables (i.e., both the displacement and the velocity) are chosen as master coordinates, the remaining variables being functionally related to the chosen pair:

$$x(s,t) = X_1\left[s, x_0(t), \dot{x}_0(t)\right] \quad \text{and} \quad \dot{x}(s,t) = X_2\left[s, x_0(t), \dot{x}_0(t)\right] \tag{23}$$

These relations define a two-dimensional invariant manifold in phase space. By taking the time derivative of these constraint equations and using the chain rule differentiation, the explicit time dependence from the equations of motion can be eliminated. Eventually, this yields a set of partial differential equations governing the modal functions X_1 and X_2. These equations are as difficult to solve as the original problem, but the solution can be approximated using power series.

For systems with internal resonances, a multi-mode invariant manifold is considered in (49) to account for the influence of several modes. For instance, when two modes are resonant, the master coordinates comprise two pairs of state variables, and the resulting invariant manifold is four-dimensional. The invariant manifold approach was also reformulated using a complex framework (42), which was then extended to systems with internal resonances (43).

The Multiple Scales Method

One perturbation method that has received considerable attention for the NNM computation is the method of multiple scales (24; 76; 42; 44; 45; 58; 59). Governing partial differential equations can be attacked directly (i.e., without spatial discretization) with this method. The first step is to introduce a small nondimensional parameter ϵ to indicate the smallness of the nonlinear terms. The solution is then sought in the form of an asymptotic expansion (46).

The underlying idea of the multiple scales method is to consider expansions in terms of multiple independent time scales, instead of a single independent variable t

$$x(s,t) = \epsilon x_1(s, T_0, T_1, T_2, ...) + \epsilon^2 x_2(s, T_0, T_1, T_2, ...) + ... \quad \text{with } T_i = \epsilon^i t \tag{24}$$

where T_0 is a time scale characterizing the fast motion of the system (i.e., the motion occurring at the dominant frequency of the NNM). Because the generic motion of a nonlinear system is not harmonic, other time scales are necessary to describe the motion; these are the slow time scales $T_1, T_2, ...$. An increasingly accurate approximation is therefore obtained as additional time scales enter in the analysis. The approximating functions $x_i(s, T_0, T_1, T_2, ...)$ are then determined after integration of ordinary linear differential equations for each order of ϵ, and imposition of solvability conditions, which correspond to the elimination of secular terms.

Other Approaches

The method of normal forms was first employed by Lamarque and Jezequel (16) and Nayfeh (60) using a complex formulation. A real normal theory for the NNM computation was then proposed in (61; 62) for conservative systems and in (32) for nonconservative systems and is described in detail in Chapter 3 of this book. The philosophy of the method is to seek a nonlinear change of coordinates [1] that brings the equations of motion into the simplest possible form, termed the normal form. In other words, the objective of the method is to eliminate as many as possible of the nonlinear terms from the governing equations, which is similar in spirit to the decoupling of the equations of motion provided by the LNMs of a linear system. However, a complete decoupling of the equations is generally not possible for nonlinear systems, and only the invariance property can be enforced. Eventually, the

[1] To recover the linear results at small-amplitude motions, the applied coordinate transformations must be near-identity. As a result, the method can only be used in the neighborhood of an equilibrium point.

computed normal form dictates the dynamics of the system on the invariant manifold.

Another technique that assumes that the NNM motion is periodic in time and approximates it by means of a finite Fourier series is the harmonic balance method; e.g.,

$$x(s,t) = \sum_{n=0}^{N} \phi_1^n(s) \cos n\omega t + \sum_{n=0}^{N} \phi_2^n(s) \sin n\omega t \tag{25}$$

By substituting this relation into the governing equations of motion and 'balancing the harmonics', the nonlinear modes can be computed by solving nonlinear boundary value problems for the ϕ^n (63; 64; 65; 66; 67; 68; 69; 70). Because analytical solutions are available in a limited number of cases (mostly when a single harmonic component is considered; see Section 2.2), numerical methods are often used to solve the resulting equations. The harmonic balance method can therefore be viewed as a semi-analytical technique.

A method similar in spirit to the harmonic balance method and to the Galerkin-based approach was introduced in (34; 71). The most distinctive feature of this formulation is that the modal vector and the corresponding frequency depend on the amplitude but also on the total phase variable. The dynamics is defined by a one-dimensional differential equation, governing the total phase motion, from which the period of the oscillations is deduced.

3.2 Numerical Techniques

Most existing constructive techniques for computing NNMs are based on asymptotic approaches. Despite that robust algorithms for the computation of isolated periodic orbits and for the continuation of a family of orbits have been developed, it is somewhat surprising that there have been few attempts to compute NNMs using numerical methods (26; 72; 73; 74; 75; 76; 77).

One of the first approaches was proposed by Slater in (72). Based on Rosenberg's definition, the procedure integrates directly the governing equations of motion over one period using numerical algorithms (e.g, Runge-Kutta and Newmark). It comprises two steps:

1. An isolated periodic solution corresponding to a specific energy level is computed by modifying iteratively the initial conditions governing the free response of the system. This is carried out using optimization algorithms that minimize a periodicity condition (i.e., a cost function representing the lack of periodicity of the current iterate).
2. Low-energy modal curves and the corresponding periods of oscillation are first computed, taking the normal modes and natural frequencies

of the underlying linear system as initial guesses. The energy is then gradually increased with the previously computed NNM as an initial guess for the next NNM.

This step-wise type of continuation of periodic solutions is called sequential continuation (78). Similarly, shooting algorithms coupled with sequential continuation were considered in Lee et al. (26; 79) and Bajaj et al. (76; 80) to numerically solve the nonlinear boundary value problem that defines a family of NNM motions.

A more sophisticated continuation method is the so-called asymptotic-numerical method (81), described in Chapter 6. It is a semi-analytical technique that is based on a power series expansion of the unknowns parameterized by a control parameter. It is utilized to follow the NNM branches in conjunction with the harmonic balance method in (69) or with finite difference methods in (75; 82). Another well-established method implemented in the AUTO software is the pseudo-arclength continuation. It is the approach used for the NNM calculation in (83) and is also described in Chapter 5.

Based on the invariant manifold approach, Pesheck et al. (73; 84) developed a meaningful numerical extension of it. In the original formulation, the master variables are the position and velocity in Cartesian coordinates, and the solution is sought using a polynomial expansion. In the proposed Galerkin-based approach, an alternative set of coordinates is defined (i.e., the amplitude and phase of the considered mode), and the polynomial approach is replaced by a Galerkin method. Eventually, a set of nonlinear algebraic equations is obtained and solved using local optimization algorithms.

Finally, we note that computer implementation of both the multiple scales and the invariant manifold approach have been carried out in (58; 86) and applied to finite element models of planar frames and beams.

3.3 Assessment of the Different Methodologies

Analytical methodologies have the advantage that NNMs can be constructed symbolically, which is useful for gaining insight into the dynamics and for performing parametric studies. Among other things, they clearly highlight the frequency-energy dependence of the NNMs. The fundamental drawbacks of these techniques is that (i) they are quite analytically involved and require a careful treatment in the presence of internal resonances; (ii) the resultant dynamics are only accurate for small-amplitude motions; and (iii) the upper bound for these motions is not known a priori.

The energy-based formulation is an elegant approach, but, because it is based on symmetry arguments, it requires that the nonlinearities be of odd order. It is also a priori limited to undamped systems. The invariant manifold approach does not present these limitations. However, though its basic definition allows for large-amplitude motions, the constructive technique, which relies on power series expansions, is limited to small motions. Regarding perturbation analysis (e.g., the multiple scales method), it can now be performed using symbolic manipulation programs, but their application to nonlinear systems with more than a few DOFs remains involved. Different analytical methods have been compared in (60; 80), and the results obtained were consistent. In (60), the author reports that the method of multiple scales is the simplest and involves the least algebra. Finally, we note that the harmonic balance method yields solutions which are generally valid over a much larger domain compared to the aforementioned methodologies. However, because analytic expressions of the resulting equations are available only in a limited number of cases, it should be regarded more as a numerical technique.

Analytical approaches may become inaccurate in the moderate to strongly nonlinear range of motion and are limited to systems with low dimensionality. In this context, numerical methods have certainly the potential to make nonlinear modal analysis more accessible to the practicing structural engineer. The key advantage of these methods is that they lend themselves fairly easily to strongly nonlinear systems with large-amplitude motions, which is nicely evidenced in (84; 85). In addition, most of them provide an *exact* solution to the NNM calculation. Their fundamental drawback is that they rely on extensive numerical simulations and are still computationally intensive.

The Galerkin-based invariant manifold approach is one of the most effective techniques for building reduced-order models of nonlinear vibrating systems. It can be applied to a large variety of nonlinear dynamic systems, including nonconservative, gyroscopic and piecewise-linear systems, with an accuracy controlled over the chosen amplitude range. One possible limitation is that the interpretation of the NNMs is complicated when multi-mode invariant manifolds, which are higher-dimensional surfaces, are computed.

On the contrary, a particularly appealing feature of the continuation of periodic solutions is that the resulting NNMs have a clear conceptual relation to the LNMs, with which practicing structural engineers are familiar (see section 2.2). As discussed in Section 4.2, this makes it a promising technique for developing a practical nonlinear analog of experimental modal analysis, which is well-established for linear systems. In this framework, the implementation of sequential continuation techniques is truly straightforward, and the calculations can be performed with limited user interaction. They represent the ideal starting point for the dynamicist not necessarily acquainted with the numerical calculation of the NNMs. However, their computational efficiency is limited, and they are likely to fail when a turning point or a bifurcation is encountered. Effective alternatives are those based on more sophisticated continuation techniques (e.g., the asymptotic-numerical method and the pseudo-arclength continuation). For instance, the NNMs of a real-life aircraft are computed in (87). One limitation of the continuation of periodic solutions is that it is not clear how they can be extended to nonconservative systems. Nevertheless, as shown in Section 4.2, the damped dynamics can be interpreted based on the topological structure and bifurcations of the NNMs of the underlying undamped system.

4 Nonlinear Normal Modes: Why Are They Useful ?

The objective of this section is to illustrate the usefulness of NNMs for modal analysis. Nonlinear model reduction and the study of localization phenomena are also discussed later in this book.

4.1 'Linear' Modal Analysis

Modal analysis and testing of linear mechanical structures has been developed over the past 40-50 years, and the techniques available today are mature and advanced. Clearly, though, linearity is an idealization, an exception to the rule; nonlinearity is a frequent occurrence in real-life applications. In the presence of nonlinear phenomena, the structural dynamicist should therefore ask the question: *can I still use the linear modes ?* Obviously, the answer depends on the type of the nonlinearity and on the excitation level.

In this context, the computation of the NNMs and their representation in a FEP is a robust tool to decide whether or not the linear framework is still applicable. It can be used to determine which modes (and to what extent) are sensitive to the nonlinearity. Going back to Figure 12, it is clear that, until an energy of 10^{-1}, the mode shapes and natural frequencies are unaffected by the nonlinearity and can safely be used. Beyond this critical energy level, both the in-phase and out-of-phase modes show a significant departure from the LNMs and become dependent on the total energy in the system.

4.2 Nonlinear Modal Analysis

When it is certain that the system is excited in the nonlinear range, the linear framework should be abandoned in favor of a nonlinear modal analysis.

Considering again system (2) as a first example, its FEP in Figure 12 greatly helps to understand how the modal curves deform under the action of the nonlinearity. The in-phase NNM tends to localize to the second DOF, whereas the out-of-phase NNM localizes to the first DOF. Regarding the corresponding frequency of oscillation, both modes are characterized by a hardening behavior due to the presence of the cubic spring.

As a second example, a planar cantilever beam discretized by 20 finite elements and with a cubic spring at the free end is now considered. This models a real nonlinear beam that was used as a benchmark for nonlinear system identification during the European action COST F3 (88). The first two modes are plotted in the FEPs of Figures 19 and 20, respectively. Considering the same energy level, the first modal curve seems somewhat more affected by the nonlinearity compared to the second modal curve. Their frequencies of oscillation undergo a strong increase with increasing energy levels. The FEPs also highlight the presence of two tongues, revealing the existence of internal resonances. The tongue in Figure 19 corresponds to a 5:1 internal resonance between the first and second modes of the beam. When the energy gradually increases along the tongue, a smooth transition from the first mode to the second mode occurs following a dynamical mechanism similar to that described in Section 2.2. Similarly, a 5:1 internal resonance between the second and fourth modes is observed in Figure 20. These internal resonances occur despite that the linear natural frequencies are not commensurate, as also discussed in Section 2.2.

These two examples demonstrate that such a nonlinear modal analysis is an important tool for thoroughly understanding the system's vibratory response in the nonlinear regime. Clearly, this cannot be achieved using linearization procedures. However, because the general motion of a nonlinear system cannot be expressed as a superposition of individual NNM motions and because the modes in all these figures are computed based on the underlying undamped system, the practical utility of the nonlinear modal analysis might appear, at first, questionable.

A first motivation to compute and exploit the NNMs is that forced resonances in nonlinear systems occur in their neighborhoods. The knowledge of the NNMs can therefore provide valuable insight into the structure of the resonances, a feature of considerable engineering importance (10). For illustration, system (4) is considered. In Figures 21 and 22, the backbone of the FEP of Figure 12 is superposed to the nonlinear frequency response functions of Figure 8 and 9. It can be observed that the backbone of the FEP traces the locus of the frequency response peaks for both the in-phase and out-of-phase modes. Furthermore, Figure 23 compares the forced response of the system close to the first resonance (for $F = 0.1$, see the square in Figure 21) to the free response of the corresponding point of the backbone. An excellent agreement is obtained between the two types of motion.

A second motivation is that the damped dynamics closely follows the NNMs of the underlying undamped system. To demonstrate this, a time-frequency analysis method, the continuous wavelet transform (CWT) (89), is used. In contrast to the Fourier transform, which assumes signal stationarity, the CWT involves a windowing technique with variable-sized regions. Small time intervals are considered for high-frequency components, whereas the size of the interval is increased for lower-frequency components. The CWT can therefore track the temporal evolution of the instantaneous frequencies, which makes it an effective tool for analyzing nonlinear signals. The usual representation of the transform is to plot its modulus as a function of time and frequency in a three-dimensional or contour plot. To use the CWT in conjunction with the FEP, a different representation is proposed herein. The CWT is represented in a frequency-energy plot by substituting the instantaneous energy in the system for time.

The free response of system

$$\begin{aligned} \ddot{x}_1 + 0.03\dot{x}_1 + (2x_1 - x_2) + 0.5\,x_1^3 &= 0 \\ \ddot{x}_2 + 0.01\dot{x}_2 + (2x_2 - x_1) &= 0 \end{aligned} \qquad (26)$$

is depicted in Figures 24 and 25 for an excitation of an in-phase and out-of-phase NNM, respectively. The top plot is the theoretical FEP, that is the FEP computed from the equations of motion. The bottom plot is the 'experimental' FEP, calculated directly from the time series: (i) the backbone is provided by the CWT, and (ii) the modal curves are obtained by representing the time series in the configuration space for one oscillation around a specific energy level. For comparison, the theoretical backbone is represented by a solid line in the experimental FEP. A perfect agreement is obtained between the two FEPs, which shows that the undamped NNMs are attractors for the damped trajectories. In the present case, the modal damping ratios are 1% and 0.6%, but we note that this result holds for higher damping ratios.

Figure 26 displays the free response of the planar cantilever beam excited at its first mode (with a damping matrix equal to the mass matrix, $\mathbf{C} = \mathbf{M}$). It shows that similar conclusions can also be reached for more complex systems.

Even if a possible criticism of the proposed approach is that it defines an NNM as a periodic solution of the underlying undamped system, these two examples support that they still give a very accurate picture of the damped dynamics. These results also show that the CWT is the ideal companion to the NNMs. We believe that the combined use of the FEP and the CWT represents a suitable framework for developing a new nonlinear system identification method, which could be viewed as a practical nonlinear analog of experimental modal analysis.

For instance, one specific application that could ultimately benefit from the proposed advancements is aircraft ground vibration testing (GVT) (90). GVTs are performed on aircraft prototypes before their first flight and provide critical data for flutter analysis. Identification of an accurate model during GVTs could allow the effects of nonlinearity on flutter to be explored by calculation prior to the flight test programme. Such an improvement would increase the aeroelastic prediction capabilities.

5 Conclusion

To robustly and accurately model nonlinearity in realistic vibrating structures is one of the greatest challenges in structural engineering. In this context, NNMs certainly represent a useful framework for the dynamicist. They have a clear conceptual relation to the linear normal modes, yet they can highlight nonlinear phenomena that are unexpected (and unexplainable) from a linear viewpoint.

The two main definitions, the fundamental properties and different analytical and numerical methods for computing NNMs were reviewed and illustrated with numerical examples. We have also highlighted that even seemingly simple nonlinear systems can exhibit very complicated dynamics. The 2DOF system investigated herein is characterized by an intricate NNM structure with (presumably) a countable infinity of internal resonances and strong motion localization in either oscillators. One interesting finding is that the internal resonances occur without necessarily having commensurate linear natural frequencies.

Bibliography

[1] R.M. Rosenberg, On nonlinear vibrations of systems with many degrees of freedom, *Advances in Applied Mechanics* 9 (1966), 155-242.

[2] R. Rand, Nonlinear normal modes in two-degree-of-freedom systems, *Journal of Applied Mechanics* 38 (1971), 561.

[3] R. Rand, A higher-order approximation for nonlinear normal modes in two-degree-of-freedom systems, *International Journal of Non-Linear Mechanics* 6 (1971), 545-547.

[4] R. Rand, A direct method for nonlinear normal modes, *International Journal of Non-Linear Mechanics* 9 (1974), 363-368.

[5] L.I. Manevitch, Y.V. Mikhlin, On periodic solutions close to rectilinear normal vibration modes, *PMM* 36 (1972), 1051-1058.

[6] A.F. Vakakis, *Analysis and Identification of Linear and Nonlinear Normal Modes in Vibrating Systems*, Ph.D. Dissertation, California Institute of Technology, 1990.

[7] T.K. Caughey, A.F. Vakakis, J.M. Sivo, Analytical study of similar normal modes and their bifurcations in a class of strongly nonlinear systems, *International Journal of Non-Linear Mechanics* 25 (1990), 521-533.

[8] A.F. Vakakis, Non-similar normal oscillations in a strongly non-linear discrete system, *Journal of Sound and Vibration* 159 (1992), 341-361.

[9] M.E. King, A.F. Vakakis, An energy-based formulation for computing nonlinear normal-modes in undamped continuous systems, *Journal of Vibration and Acoustics* 116 (1994), 332-340.

[10] A.F. Vakakis, L.I. Manevitch, Y.V. Mikhlin, V.N. Pilipchuk, A.A. Zevin, *Normal Modes and Localization in Nonlinear Systems*, John Wiley & Sons, New York, 1996.

[11] A.F. Vakakis, Non-linear normal modes and their applications in vibration theory: an overview, *Mechanical Systems and Signal Processing* 11 (1997), 3-22.

[12] S.W. Shaw, C. Pierre, Non-linear normal modes and invariant manifolds, *Journal of Sound and Vibration* 150 (1991), 170-173.

[13] S.W. Shaw, C. Pierre, On nonlinear normal modes, *ASME Winter Annual Meeting* 1992.

[14] S.W. Shaw, C. Pierre, Normal modes for non-linear vibratory systems, *Journal of Sound and Vibration* 164 (1993), 85-124.

[15] S.W. Shaw, C. Pierre, Normal modes of vibration for non-linear continuous systems, *Journal of Sound and Vibration* 169 (1994), 319-347.

[16] L. Jezequel, C.H. Lamarque, Analysis of nonlinear dynamic systems by the normal form theory, *Journal of Sound and Vibration*, 149 (1991), 429-459.

[17] A.H. Nayfeh, S.A. Nayfeh, Nonlinear normal modes of a continuous system with quadratic nonlinearities, *Journal of Vibration and Acoustics* 117 (1995), 199-205.

[18] Y.V. Mikhlin, Normal vibrations of a general class of conservative oscillators, *Nonlinear Dynamics* 11 (1996), 1-15.

[19] M.I. Qaisi, Non-linear normal modes of a continuous system, *Journal of Sound and Vibration* 209 (1998), 561-569.

[20] G.S. Agnes, D.J. Inman, Performance of nonlinear vibration absorbers for multi-degrees-of-freedom systems using nonlinear normal modes, *Nonlinear Dynamics* 25 (2001), 275-292.

[21] H. Yabuno, A.H. Nayfeh, Nonlinear normal modes of a parametrically excited cantilever beam, *Nonlinear Dynamics* 25 (2001), 6577.

[22] W.C. Xie, H.P. Lee, S.P. Lim, Nonlinear dynamic analysis of MEMS switches by nonlinear modal analysis, *Nonlinear Dynamics* 31 (2003), 243256.

[23] C.E.N. Mazzilli, M.E.S. Soares, G.P. Baracho Neto, Non-linear normal modes of a simply supported beam: continuous system and finite-element models, *Computer and Structures* 82 (2004), 26832691.

[24] O.V. Gendelman, Bifurcations of nonlinear normal modes of linear oscillator with strongly nonlinear damped attachment, *Nonlinear Dynamics* 37 (2004), 115-128.

[25] W. Lacarbonara, R. Camillacci, Nonlinear normal modes of structural systems via asymptotic approach, *International Journal of Solids and Structures* 41 (2004), 5565-5594.

[26] Y.S. Lee, G. Kerschen, A.F. Vakakis, P.N. Panagopoulos, L.A. Bergman, D.M. McFarland, Complicated dynamics of a linear oscillator with a light, essentially nonlinear attachment, *Physica D* 204 (2005), 41-69.

[27] S.C. Sinha, S. Redkar, E.A. Butcher, Order reduction of nonlinear systems with time periodic coefficients using invariant manifolds, *Journal of Sound and Vibration* 284 (2005), 985-1002.

[28] C.V. Serra Villa, J.J. Sinou, F. Thouverez, The invariant manifold approach applied to nonlinear dynamics of a rotor-bearing system, *European Journal of Mechanics A/Solids* 24 (2005), 676-689.

[29] A.I. Musienko, L.I. Manevitch, Short wavelength dynamics of the system of nonlinear oscillators coupled by stretched weightless beam, *Chaos, Solitons and Fractals* 26 (2005), 107-116.

[30] A.J. Dick, B. Balachandran, C.D. Mote, Nonlinear vibration modes in micro-resonator arrays, *Smart Structures and Materials 2006: Modeling, Signal Processing, and Control, Proceedings of the SPIE* Volume 6166 (2006), 206-217.

[31] C.H. Pak, On the coupling of non-linear normal modes, *International Journal of Non-Linear Mechanics* 41 (2006), 716-725.

[32] C. Touzé, M. Amabili, Nonlinear normal modes for damped geometrically nonlinear systems: Application to reduced-order modelling of harmonically forced structures, *Journal of Sound and Vibration* 298 (2006), 958-981.

[33] G. Kerschen, Y.S. Lee, A.F. Vakakis, D.M. McFarland, L.A. Bergman, Irreversible passive energy transfer in coupled oscillators with essential nonlinearity, *SIAM Journal on Applied Mathematics* 66 (2006), 648-679.

[34] S. Bellizzi, R. Bouc, An amplitude-phase formulation for nonlinear modes and limit cycles through invariant manifolds, *Journal of Sound and Vibration* 300 (2007), 896-915.

[35] N. Srinil, G. Rega, Two-to-one resonant multi-modal dynamics of horizontal/inclined cables. Part II: Internal resonance activation, reduced-order models and nonlinear normal modes, *Nonlinear Dynamics* 48 (2007), 253-274.

[36] S. Lenci, G. Rega, Dimension reduction of homoclinic orbits of buckled beams via the non-linear normal modes technique, *International Journal of Non-Linear Mechanics* 42 (2007), 515-528.

[37] W. Lacarbonara, A. Paolone, F. Vestroni, Non-linear modal properties of non-shallow cables, *International Journal of Non-Linear Mechanics* 42 (2007), 542-554.

[38] A.H. Nayfeh, *Nonlinear Interactions: Analytical, Computational and Experimental Methods*, Wiley-Interscience, New York, 2000.

[39] C. Pierre, D. Jiang, S.W. Shaw, Nonlinear normal modes and their application in structural dynamics, *Mathematical Problems in Engineering* 10847 (2006), 1-15.

[40] K.V. Avramov, Y.V. Mikhlin, Review of applications of nonlinear normal modes for vibrating mechanical systems, *Applied Mechanics Reviews* 65 (2013).

[41] G. Kerschen, M. Peeters, J.C. Golinval, A.F. Vakakis, Nonlinear normal modes, Part I: A useful framework for the structural dynamicist, *Mechanical Systems and Signal Processing* 23 (2009), 170-194.

[42] A.H. Nayfeh, S.A. Nayfeh, On nonlinear modes of continuous systems, *Journal of Vibration and Acoustics* 116 (1994), 129-136.

[43] A.H. Nayfeh, C.M. Chin, S.A. Nayfeh, On nonlinear normal modes of systems with internal resonance, *Journal of Vibration and Acoustics* 118 (1996), 340345.

[44] W. Lacarbonara, G. Rega, A.H. Nayfeh, Resonant non-linear normal modes, Part I: analytical treatment for structural one-dimensional systems, *International Journal of Non-Linear Mechanics* 38 (2003), 851-872.

[45] W. Lacarbonara, G. Rega, Resonant non-linear normal modes, Part II: Part II: activation/orthogonality conditions for shallow structural systems, *International Journal of Non-Linear Mechanics* 38 (2003), 873-887.

[46] A.H. Nayfeh, D.T. Mook, *Nonlinear Oscillations*, Wiley-Interscience, New York, 1979.

[47] A.F. Vakakis, Inducing passive nonlinear energy sinks in vibrating systems, *Journal of Vibration and Acoustics* 123 (2001), 324-332.

[48] Y.S. Lee, A.F. Vakakis, L.A. Bergman, D.M. McFarland, G. Kerschen, Suppression of aeroelastic instability by means of broadband passive TET: Part I, theory, *AIAA Journal* 45 (2007), 693-711.

[49] N. Boivin, C. Pierre, S.W. Shaw, Non-linear modal analysis of structural systems featuring internal resonances, *Journal of Sound and Vibration* 182 (1995), 336-341.

[50] D. Jiang, C. Pierre, S.W. Shaw, The construction of non-linear normal modes for systems with internal resonance, *International Journal of Non-linear Mechanics* 40 (2005), 729-746.

[51] M.E. King, A.F. Vakakis, An energy-based approach to computing resonant nonlinear normal modes, *Journal of Applied Mechanics* 63 (1995), 810-819.

[52] S.S. Oueini, C.M. Chin, A.H. Nayfeh, Dynamics of a cubic nonlinear vibration absorber, *Nonlinear Dynamics* 20 (1999), 283295.

[53] A.F. Vakakis, R.H. Rand, Normal modes and global dynamics of a 2-degree-of-freedom nonlinear-system; Part I: low energies, *International Journal of Non-Linear Mechanics* 27 (1992), 861-874.

[54] A.F. Vakakis, R.H. Rand, Normal modes and global dynamics of a 2-degree-of-freedom nonlinear-system; Part I: high energies, *International Journal of Non-Linear Mechanics* 27 (1992), 875-888.

[55] C.H. Pak, Synge's concept of stability applied to non-linear normal modes, *International Journal of Non-Linear Mechanics* 41 (2006), 657-664.

[56] G. Recktenwald, R. Rand, Stability of strongly nonlinear normal modes, *Communications in Nonlinear Science and Numerical Simulation* 12 (2007), 1128-1132.

[57] N. Boivin, C. Pierre, S.W. Shaw, Non-linear normal modes, invariance, and modal dynamics approximations of non-linear systems, *Nonlinear Dynamics* 8 (1995), 315-346.

[58] C.E.N. Mazzilli, O.G.P. Baracho Neto, Evaluation of non-linear normal modes for finite-element models, *Computer and Structures* 80 (2002), 957-965.

[59] X. Li, J.C. Ji, C.H. Hansen, Non-linear normal modes and their bifurcation of a two DOF system with quadratic and cubic non-linearity, *International Journal of Non-linear Mechanics* 41 (2006), 1028-1038.

[60] A.H. Nayfeh, On direct methods for constructing nonlinear normal modes of continuous systems, *Journal of Vibration and Control* 1 (1995), 389-430.

[61] C. Touzé, O. Thomas, A. Huberdeau, Asymptotic non-linear normal modes for large-amplitude vibrations of continuous structures, *Computers & Structures* 82 (2004), 2671-2682.

[62] C. Touzé, O. Thomas, A. Chaigne, Hardening/softening behaviour in non-linear oscillations of structural systems using non-linear normal modes, *Journal of Sound and Vibration* 273 (2004), 77-101.

[63] R. Lewandowski, Application of the Ritz method to the analysis of non-linear free vibrations of beams, *Journal of Sound and Vibration* 114 (1987), 91-101.

[64] W. Szemplinska-Stupnicka, Non-linear normal modes and the generalized ritz method in the problems of vibrations of non-linear elastic continuous systems, *International Journal of Non-linear Mechanics* 18 (1983), 149-165.

[65] W. Szemplinska-Stupnicka, *The Behavior of Nonlinear Vibrating Systems — Volume II: Advanced Concepts and Application to Multi-degree-of-freedom Systems*, Kluwer Academic Publishers, Dordrecht, 1990.

[66] R. Benamar, M.M.K. Bennouna, R.G. White, The effects of large vibration amplitudes on the mode shapes and natural frequencies of thin elastic structures, Part I: Simply supported and clamped-clamped beams, *Journal of Sound and Vibration* 149 (1991), 179-195.

[67] T.D. Burton, M.N. Hamdan, On the calculation of non-linear normal modes in continuous systems, *Journal of Sound and Vibration* 197 (1996), 117-130.

[68] C.H. Pak, *Nonlinear Normal Mode Dynamics for Two-degree-of-freedom Systems*, Inha University Press, Seoul, 1999.

[69] F. Pérignon, Vibration forcées de structures minces, élastiques, non-linéaires, *PhD Thesis*, Université de la Méditerrané, Marseille, 2004.

[70] M. Haterbouch, R. Benamar, Geometrically nonlinear free vibrations of simply supported isotropic thin circular plates, *Journal of Sound and Vibration* 280 (2005), 903-924.

[71] S. Bellizzi, R. Bouc, A new formulation for the existence and calculation of nonlinear normal modes, *Journal of Sound and Vibration* 287 (2005), 545-569.

[72] J.C. Slater, A numerical method for determining nonlinear normal modes, *Nonlinear Dynamics* 10 (1996), 19-30.

[73] E. Pesheck, Reduced-order modeling of nonlinear structural systems using nonlinear normal modes and invariant manifolds, *PhD Thesis*, University of Michigan, Ann Arbor, 2000.

[74] D. Jiang, C. Pierre, S.W. Shaw, Large-amplitude non-linear normal modes of piecewise linear systems, *Journal of Sound and Vibration* 272 (2004), 869-891.

[75] R. Arquier, S. Bellizzi, R. Bouc, B. Cochelin, Two methods for the computation of nonlinear modes of vibrating systems at large amplitudes, *Computers & Structures* 84 (2006), 1565-1576.

[76] F.X. Wang, A.K. Bajaj, Nonlinear normal modes in multi-mode models of an inertially coupled elastic structure, *Nonlinear Dynamics* 47 (2007), 25-47.

[77] T.D. Burton, Numerical calculation of nonlinear normal modes in structural systems, *Nonlinear Dynamics* 49 (2007), 425-441.
[78] A.H. Nayfeh, B. Balachandran, *Applied Nonlinear Dynamics: Analytical, Computational and Experimental Method*, John Wiley and Sons, New York, 1995.
[79] G. Kerschen, J.J. Kowtko, D.M. McFarland, L.A. Bergman, A.F. Vakakis, Theoretical and experimental study of multimodal targeted energy transfer in a system of coupled oscillators, *Nonlinear Dynamics* 47 (2007), 285-309.
[80] F.X. Wang, A.K. Bajaj, K. Kamiya, Nonlinear normal modes and their bifurcations for an inertially coupled nonlinear conservative system, *Nonlinear Dynamics* 42 (2005), 233-265.
[81] B. Cochelin, N. Damil, M. Potier-Ferry, Asymptotic numerical methods and Padé approximants for nonlinear elastic structures, International Journal of Numerical Methods in Engineering 37 (1994), 11871213.
[82] R. Arquier, Une méthode de calcul des modes de vibrations nonlinéaires de structures, *PhD Thesis*, Université de la Méditerrané, Marseille, 2007.
[83] M. Peeters, R. Viguié, G. Sérandour, G. Kerschen, J.C. Golinval, Nonlinear normal modes, Part II: Toward a practical computation using numerical continuation, *Mechanical Systems and Signal Processing* 23 (2009), 195-216.
[84] E. Pesheck, C. Pierre, S.W. Shaw, A new Galerkin-based approach for accurate non-linear normal modes through invariant manifolds, *Journal of Sound and Vibration* 249 (2002), 971-993.
[85] E. Pesheck, C. Pierre, S.W. Shaw, Accurate reduced-order models for a simple rotor blade model using nonlinear normal modes , *Mathematical and Computer Modelling* 33 (2001), 1085-1097.
[86] M.E.S. Soares, C.E.N. Mazzilli, Nonlinear normal modes of planar frames discretised by the finite element method, *Computers and Structures* 77 (2000), 485-493.
[87] G. Kerschen, M. Peeters, J.C. Golinval, C. Stephan, Nonlinear modal analysis of a full-scale aircraft, *AIAA Journal of Aircraft* 50 (2013), 1409-1419.
[88] F. Thouverez, Presentation of the ECL benchmark, *Mechanical Systems and Signal Processing* 17 (2003), 195-202.
[89] S. Mallat, *A Wavelet Tour of Signal Processing*, Academic Press, 1999, 2nd edition.
[90] D. Göge, M. Böswald, U. Füllekrug, P. Lubrina, Ground vibration testing of large aircraft — State-of-the-art and future perspectives, *Proceedings of the International Modal Analysis Conference*, Orlando, 2007.

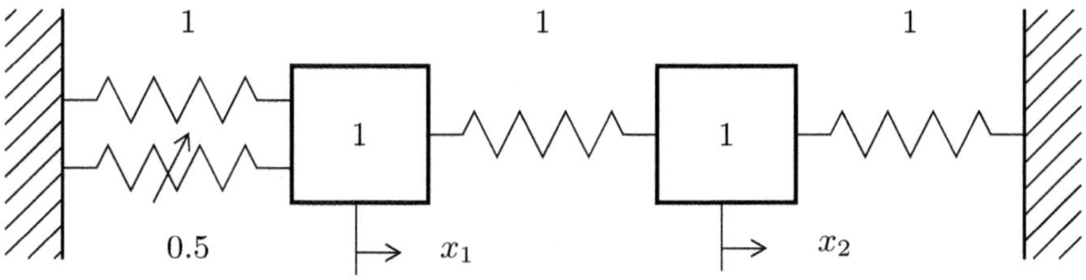

Figure 1. Schematic representation of the 2DOF system example.

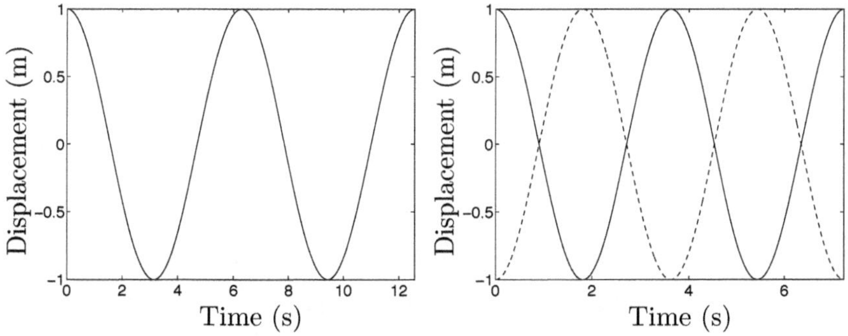

Figure 2. Time series of LNM motions of system (3) (——: $x_1(t)$; – – –: $x_2(t)$). Left plot: in-phase LNM ($[x_1(0)\ x_2(0)\ \dot{x}_1(0)\ \dot{x}_2(0)] = [1\ 1\ 0\ 0]$); right plot: out-of-phase LNM ($[x_1(0)\ x_2(0)\ \dot{x}_1(0)\ \dot{x}_2(0)] = [1\ -1\ 0\ 0]$).

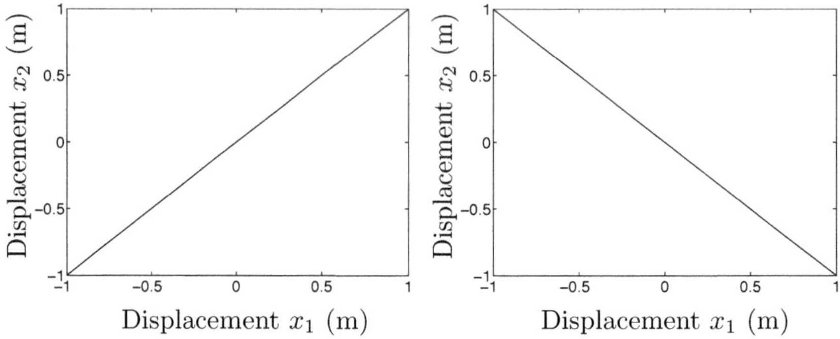

Figure 3. LNM motions of system (3) in the configuration space. Left plot: in-phase LNM; right plot: out-of-phase LNM.

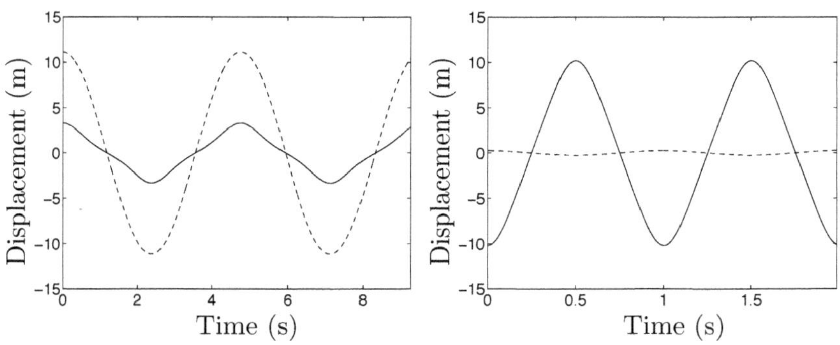

Figure 4. Time series of NNM motions of system (2) (——: $x_1(t)$; – – –: $x_2(t)$). Left plot: in-phase NNM ($[x_1(0)\ x_2(0)\ \dot{x}_1(0)\ \dot{x}_2(0)] = [3.319\ 11.134\ 0\ 0]$); right plot: out-of-phase NNM ($[x_1(0)\ x_2(0)\ \dot{x}_1(0)\ \dot{x}_2(0)] = [-10.188\ 0.262\ 0\ 0]$).

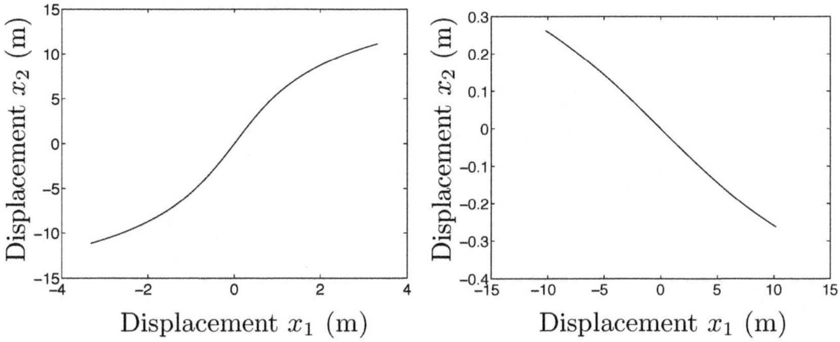

Figure 5. NNM motions of system (2) in the configuration space. Left plot: in-phase NNM; right plot: out-of-phase NNM.

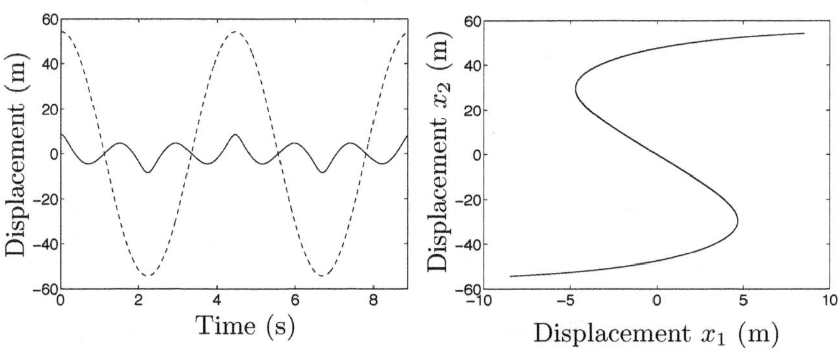

Figure 6. Internally resonant NNM (3:1 internal resonance; $[x_1(0)\ x_2(0)\ \dot{x}_1(0)\ \dot{x}_2(0)] = [8.476\ 54.263\ 0\ 0]$). Left plot: time series (——: $x_1(t)$; – – –: $x_2(t)$); right plot: configuration space.

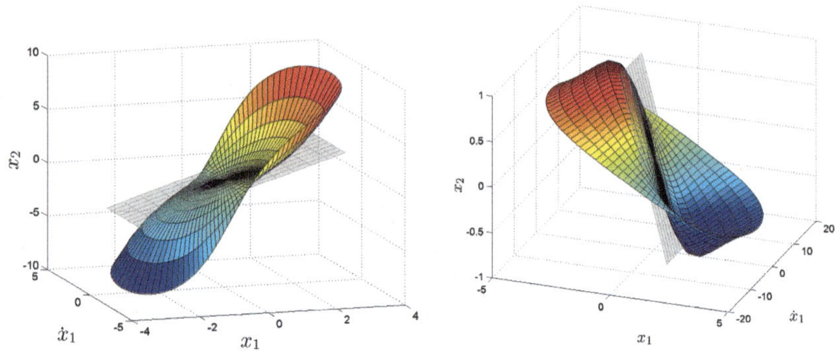

Figure 7. Two-dimensional invariant manifolds of system (2) with the corresponding LNMs. Left plot: in-phase LNM and NNM; right plot: out-of-phase LNM and NNM.

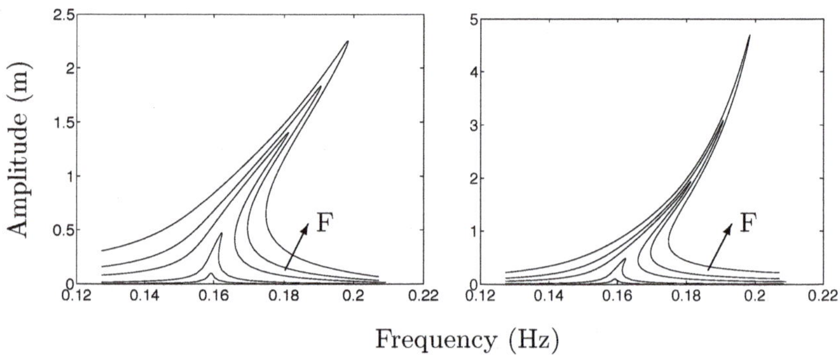

Figure 8. Nonlinear frequency response functions close to the first resonant frequency (5 different forcing amplitudes: 0.002N, 0.01N, 0.05N, 0.1N, 0.2N). Left plot: x_1; right plot: x_2.

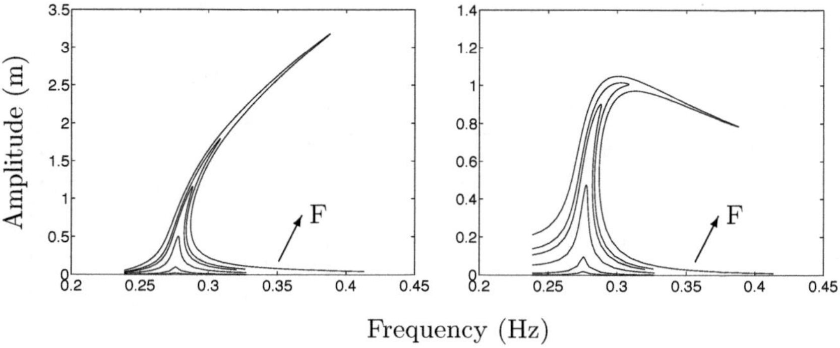

Figure 9. Nonlinear frequency response functions close to the second resonant frequency (5 different forcing amplitudes: 0.002N, 0.01N, 0.05N, 0.1N, 0.2N). Left plot: x_1; right plot: x_2.

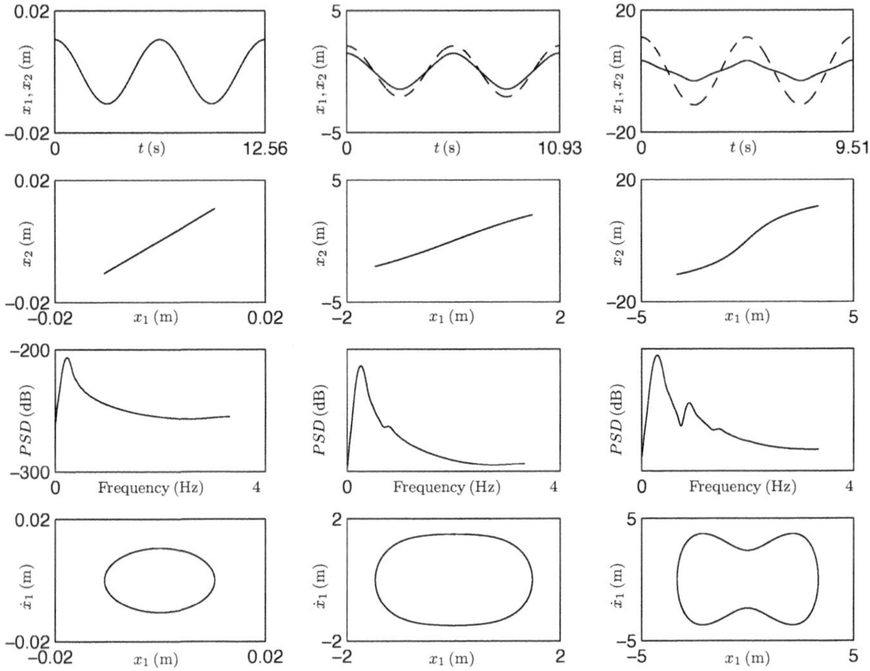

Figure 10. Frequency-energy dependence of NNMs. From left to right: low-($[x_1(0)\ x_2(0)\ \dot{x}_1(0)\ \dot{x}_2(0)] = [0.105\ 0.105\ 0\ 0]$), moderate-($[x_1(0)\ x_2(0)\ \dot{x}_1(0)\ \dot{x}_2(0)] = [1.476\ 2.100\ 0\ 0]$) and high-($[x_1(0)\ x_2(0)\ \dot{x}_1(0)\ \dot{x}_2(0)] = [3.319\ 11.134\ 0\ 0]$) energy in-phase NNMs. From top to bottom: time series (——: $x_1(t)$; – – –: $x_2(t)$); configuration space; power spectral density of $x_1(t)$; two-dimensional projections of the phase space.

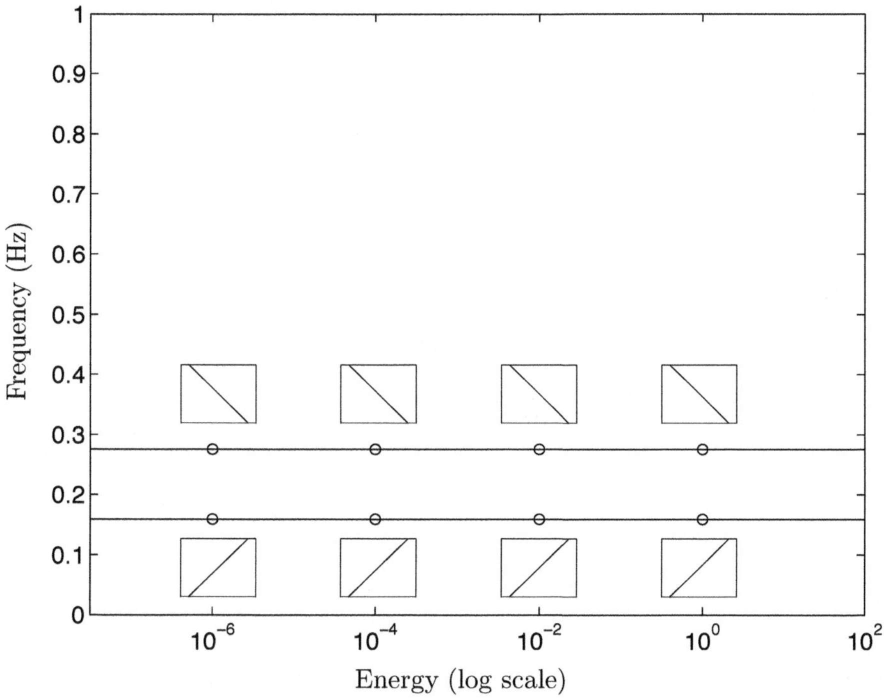

Figure 11. Frequency-energy plot of system (3). LNM motions depicted in the configuration space are inset.

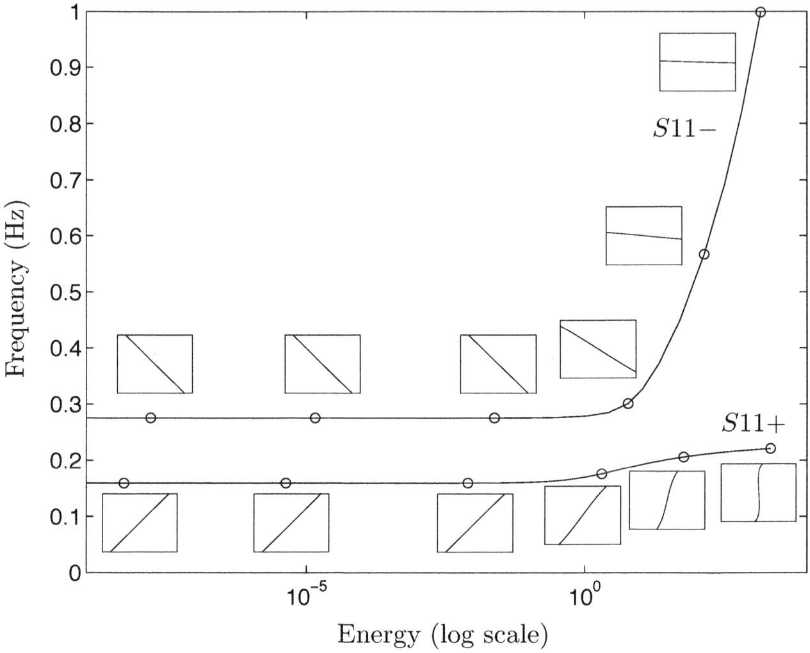

Figure 12. Frequency-energy plot of system (2). NNM motions depicted in the configuration space are inset. The horizontal and vertical axes in these plots are the displacements of the first and second DOFs, respectively; the aspect ratio is set so that increments on the horizontal and vertical axes are equal in size to indicate whether or not the motion is localized to a particular DOF.

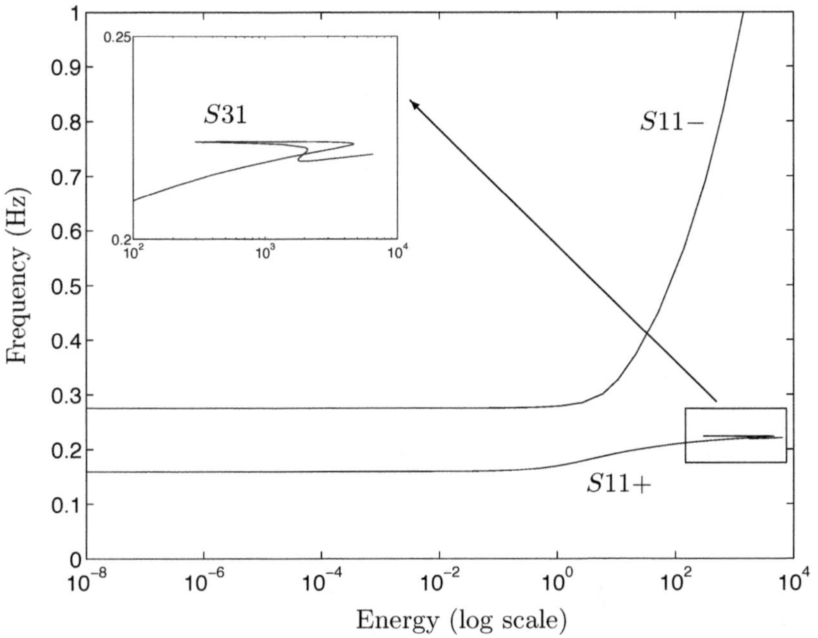

Figure 13. Frequency-energy plot of system (2) featuring a 3:1 internal resonance between the in-phase and out-of-phase NNMs.

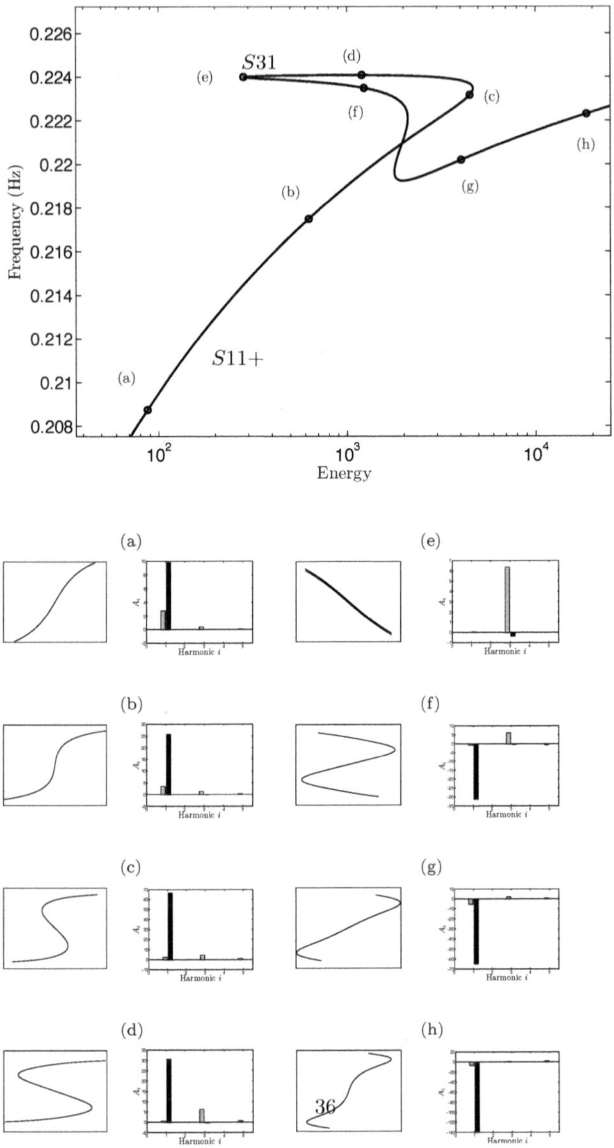

Figure 14. Internally resonant NNMs (3:1 resonance). Top plot: close-up of the tongue $S31$ in the frequency-energy plot. Bottom plots: configuration space (horizontal axis: x_1; vertical axis: x_2) and Fourier coefficients of a series containing cosine terms only (grey: x_1; black: x_2).

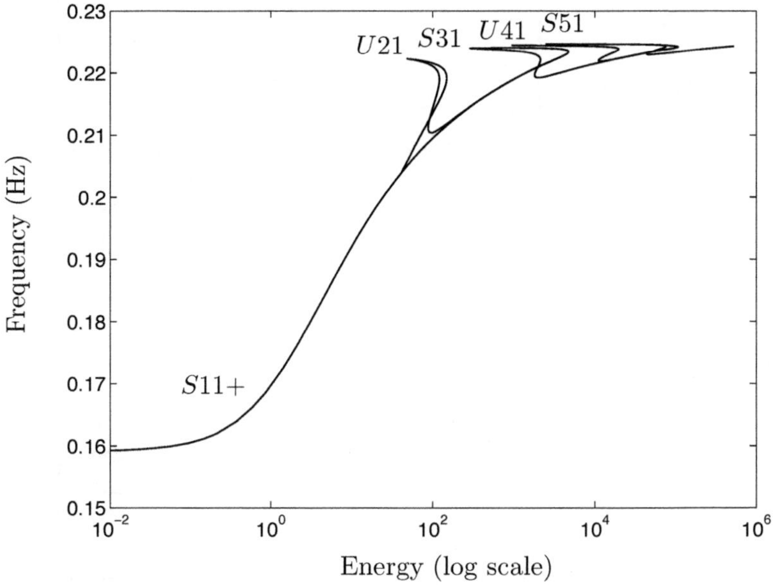

Figure 15. Close-up of $S11+$ at higher energy levels.

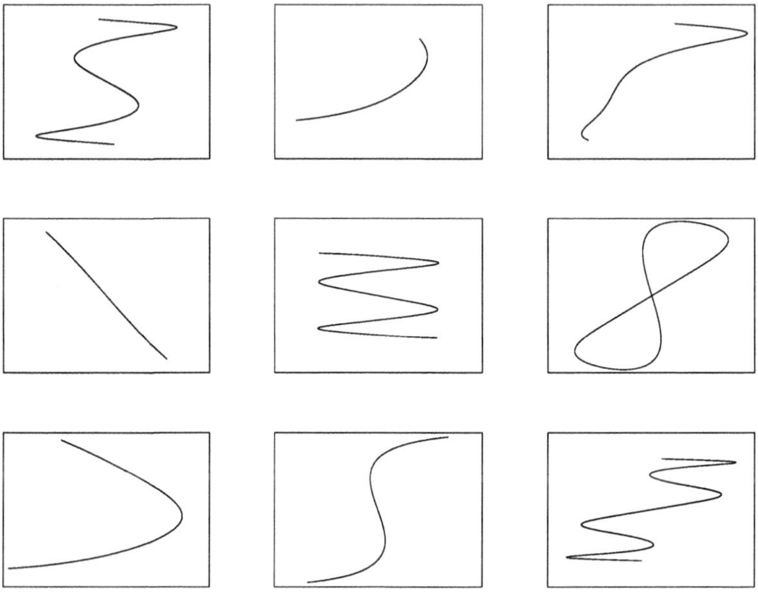

Figure 16. A few representative NNMs of system (2) in the configuration space.

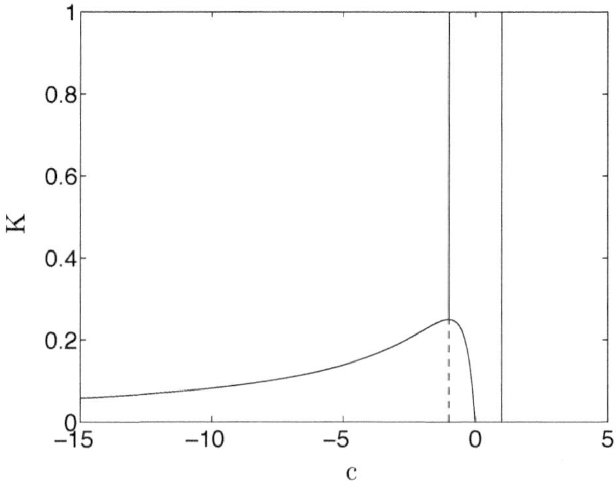

Figure 17. NNM bifurcations of system (11) (7) (——: stable NNMs; – – –: unstable NNMs).

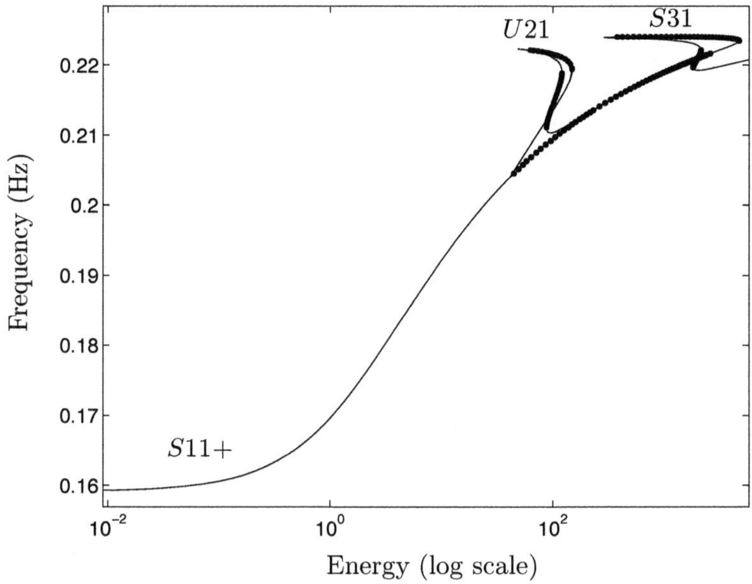

Figure 18. Close-up of $S11+$ with stability results (——: stable NNMs; • • •: unstable NNMs).

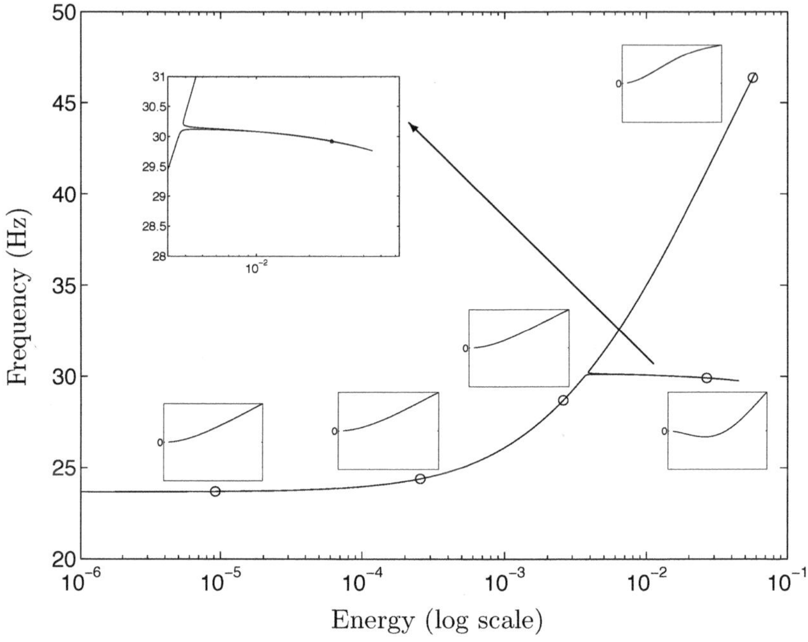

Figure 19. Frequency-energy plot of the cantilever beam; close-up of the first mode.

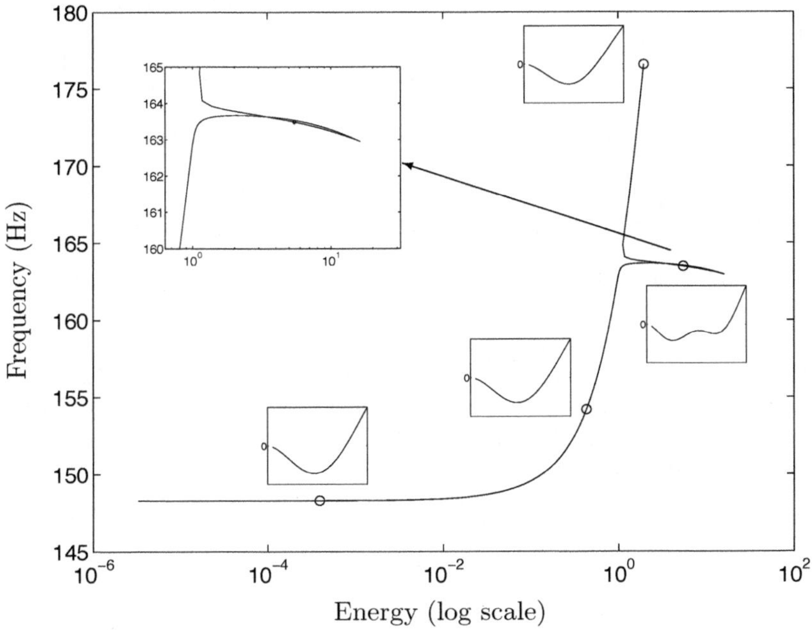

Figure 20. Frequency-energy plot of the cantilever beam; close-up of the second mode.

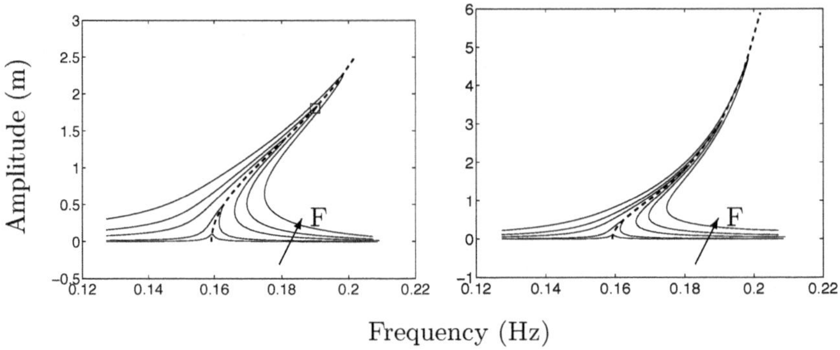

Figure 21. Nonlinear frequency response functions close to the first resonant frequency (5 different forcing amplitudes: 0.002N, 0.01N, 0.05N, 0.1N, 0.2N). The dashed line is the backbone $S11+$ of the frequency-energy plot. Left plot: x_1; right plot: x_2.

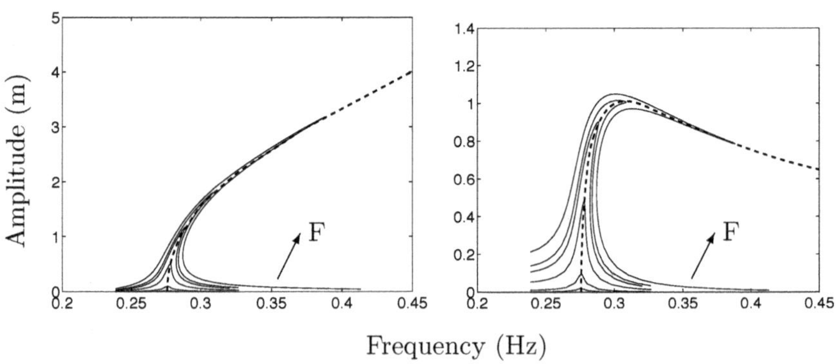

Figure 22. Nonlinear frequency response functions close to the second resonant frequency (5 different forcing amplitudes: 0.002N, 0.01N, 0.05N, 0.1N, 0.2N). The dashed line is the backbone $S11-$ of the frequency-energy plot. Left plot: x_1; right plot: x_2.

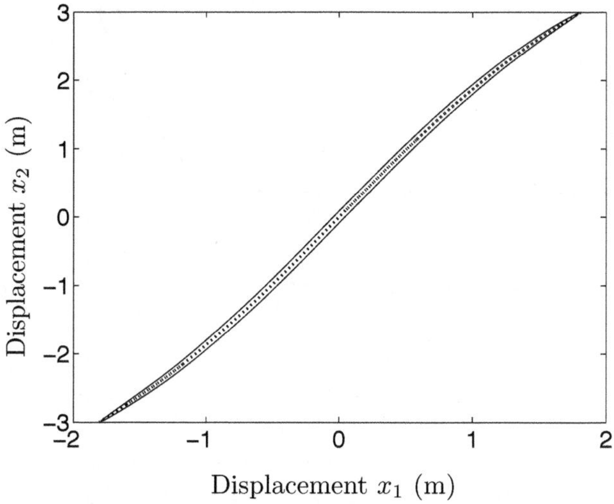

Figure 23. Free ($F = 0$) and forced responses ($F = 0.1$) of system (4) in the configuration space. ——: forced response; - - -: free response

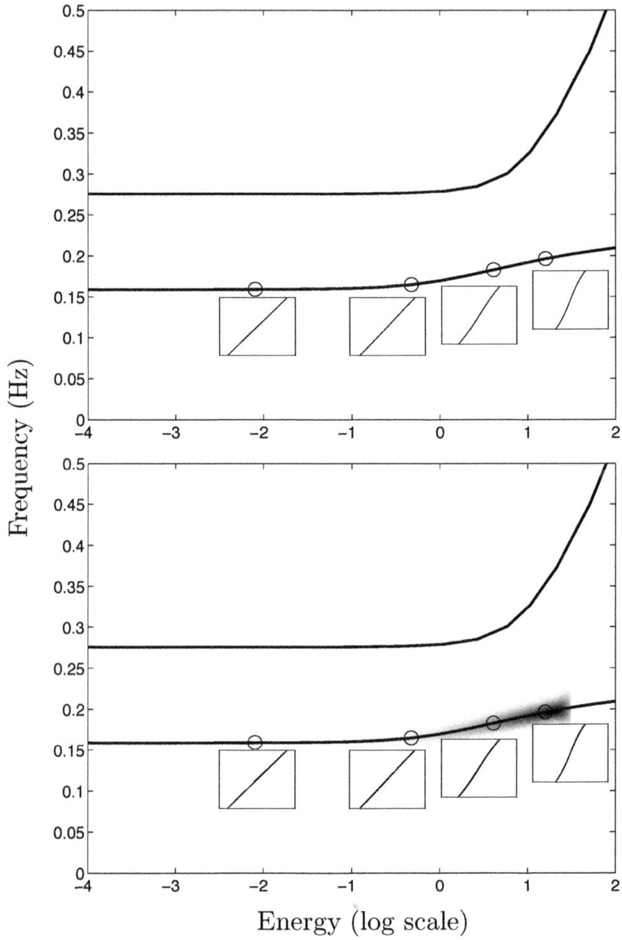

Figure 24. Frequency-energy plot of system (2). Top plot: theoretical FEP; bottom plot: experimental FEP for an excitation of an in-phase NNM ($[x_1(0)\ x_2(0)\ \dot{x}_1(0)\ \dot{x}_2(0)] = [2.500\ 5.895\ 0\ 0]$).

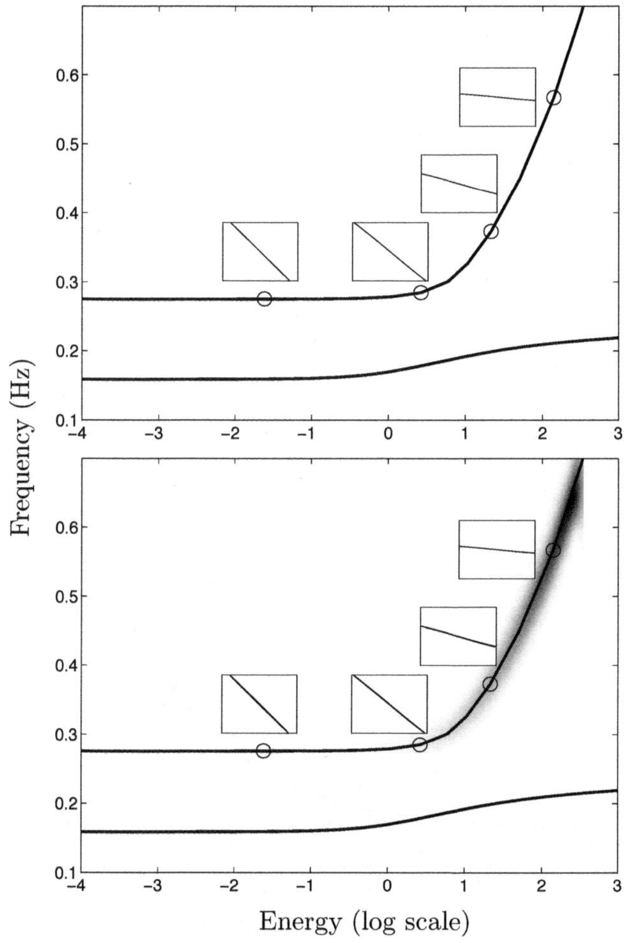

Figure 25. Frequency-energy plot of system (2). Left plot: theoretical plot; right plot: experimental plot for an excitation of an out-of-phase NNM ($[x_1(0)\ x_2(0)\ \dot{x}_1(0)\ \dot{x}_2(0)] = [-6.842\ 0.389\ 0\ 0]$).

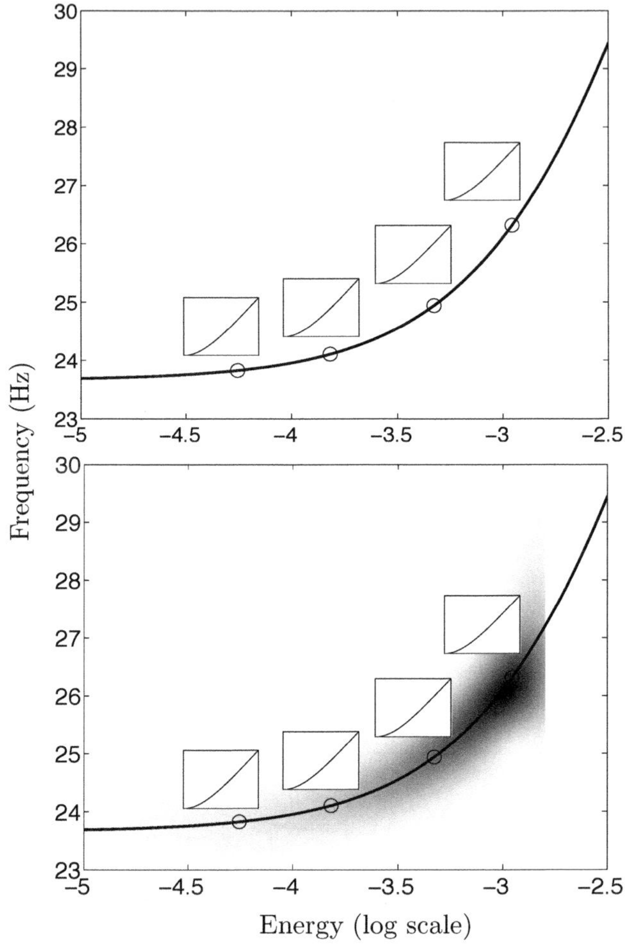

Figure 26. Frequency-energy plot of the planar cantilever beam. Left plot: theoretical plot; right plot: experimental plot for an excitation of the first mode.

Invariant Manifold Representations of Nonlinear Modes of Vibration

Steven W. Shaw
Department of Mechanical Engineering
Michigan State University
East Lansing, MI, 48824, USA

Abstract This chapter describes the definition and use of modes of vibration in terms of invariant manifolds. The goal of this chapter is to provide some basic background in terms of the terminology, ideas, and constructive methods for nonlinear modes using invariant manifolds. A range of formulations and applications are summarized and an example is presented that demonstrates the main ideas for a system with nonsychronous modes.

1 Introduction

For linear models of vibrating systems the concept of normal modes provide a means of describing complex systems in terms of *uncoupled* lower order systems. The power of the mode concept is that it: (i) provides a way to understand vibratory resonances, and (ii) it provides methods for systematic reduction of model size, based on frequency ranges of interest. The underlying basis for these tools is superposition, which essentially defines linear systems. Thus, a generalization of these notions to nonlinear systems must, necessarily, abandon any hope of applying or using superposition. It is natural to ask, then, if is there anything useful that can be done for nonlinear systems. In fact, the basic concept of modes can be generalized; these modes help define nonlinear resonances, and can be used for model reduction, although superposition does not hold. The concept of "nonlinear normal modes," or NNMs, of vibration is over a half-century old now, having been pioneered by Rosenberg (20; 21; 22). A review of the basic concepts of NNMs from Rosenberg's work and its extensions up to the mid-1990's, can be found in (26; 27). Those sources, and more recent reviews, describe a broad range of approaches and applications of NNMs (19; 9; 16; 11; 12; 1). In this chapter we emphasize the use of invariant manifolds, which provides a generalization of Rosenberg's definition that

G. Kerschen (Ed.), *Modal Analysis of Nonlinear Mechanical Systems*, CISM International Centre for Mechanical Sciences DOI 10.1007/ 978-3-7091-1791-0_2 © CISM Udine 2014

allows one to treat non-conservative and non-self-adjoint systems. Other chapters in this volume offer other useful methods for constructing and utilizing NNMs.

The chapter is organized as follows. In the next section, we give some basic definitions, supported by simple examples, both of which provide the required conceptual framework for the remainder of the chapter. We avoid mathematical formalism as much as possible, in order to focus on concepts and constructive tools. Section 3 develops some of the ideas in the context of linear vibration problems, in order to set the stage for nonlinear systems. Section 4 provides the development of individual nonlinear modes, discusses various formulations, and provides an example that demonstrates their construction. The paper closes in Section 5 with a discussion of this approach and some personal reflections about the utility of various approaches to NNMs, including those presented here and elsewhere.

It should be noted that the references cited in this chapter are far from complete. We point the reader to some general reviews and a few papers to specific topics; but, for a broader perspective on the NNM literature the reader should consult those papers and the other chapters of this volume.

2 Basic Concepts

2.1 Some Definitions

Here we provide some definitions and concepts useful for our development. We use minimal mathematical formalism in order to present the concepts in a manner that is accessible to readers with a wide range of backgrounds. More detailed presentations of this material can be found in many standard books such as (5; 6).

A *fixed point* of an autonomous dynamical system $\dot{z} = F(z)$, $z \in R^m$, is a point $\bar{z}$ that satisfies $F(\bar{z}) = 0$. That is, if the system is started at $\bar{z}$ it remains there for all time.

A fixed point $\bar{z}$ is said to be *stable* if, roughly speaking, all solutions started sufficiently near $\bar{z}$ remain near $\bar{z}$ for all time.

A fixed point $\bar{z}$ is said to be *asymptotically stable* if it is stable and if all solutions started sufficiently near $\bar{z}$ tend towards $\bar{z}$ as $t \to \infty$.

The *linearization* of $\dot{z} = F(z)$ near $\bar{z}$ is the linear time-invariant system $\dot{y} = Ay$ where $y = z - \bar{z}$ and $A = D_z F(\bar{z})$, that is, A is the $m \times m$ Jacobian of $F(z)$ evaluated at $\bar{z}$.

An *invariant manifold* of an autonomous dynamical system $\dot{z} = F(z)$, $z \in R^m$, is a set of points S in the phase space defined by the property that if an initial condition is taken in S, that is, $z(0) \in S$, the resulting solution, denoted $z(t; z(0))$, remains in S for all time, $-\infty < t < \infty$. Mathematically this can be expressed as

$$S = \{z(0) \mid z(t) \in S \; \forall \; t\}.$$

Locally, an invariant manifold of dimension p "looks like" (formally, is topologically equivalent to) R^p. The concept of asymptotic stability can be generalized to invariant manifolds. Examples of invariant manifolds of practical interest include fixed points (zero dimensional) and limit cycles (one dimensional), which, when asymptotically stable, represent steady-state responses for models of physical systems. Another group of useful invariant manifolds are the stable, unstable, and center manifolds of fixed points, developed next.

The *stable/unstable/center eigenspace*, denoted $E^{s/u/c}$, of a fixed point $\bar{z}$ is the linear space spanned by the eigenvectors of A corresponding to eigenvalues of A with negative/positive/zero real parts. Note that if the eigenvalues are complex, the real and imaginary parts of the corresponding eigenvector are used to span the corresponding part of the eigenspace. These spaces are invariant for the linearized system $\dot{y} = Ay$, and are of dimension $n_{s/u/c}$ where $n_s + n_u + n_c = m$.

A *stable/unstable/center manifold*, denoted $W^{s/u/c}$, of a fixed point $\bar{z}$ is the nonlinear generalization of the eigenspaces defined above, based on extending the eigenspaces away from the fixed point. Specifically, $W^{s/u/c}$ are invariant manifolds of $\dot{z} = F(z)$ that are tangent to $E^{s/u/c}$ at $\bar{z}$, and have the same dimensions as $E^{s/u/c}$. Note that points on W^s (W^u) are asymptotic to $\bar{z}$ as $t \to +\infty$ ($-\infty$). The properties of W^c are more subtle, since solutions on it are dominated by nonlinear behavior that can be stable or unstable, and in fact center manifolds need not be unique (5; 3). Readers familiar with the phase plane for nonlinear second order systems can recognize that the stable manifold of a saddle point is also known as a *separatrix*, since it separates initial conditions that branch away from the saddle point in opposite directions. Stable (unstable) manifolds for saddle points are also referred to as the "inset" ("outset") of the saddle point.

For vibrations of mechanical systems we focus on stable fixed points and, in the case of harmonically force vibration, stable periodic motions. It is key to note that since our focus is on vibrations, the motions are oscillatory and thus must take place on invariant manifolds that are (at least) two-dimensional. Before turning to that case, we present a simple example that illustrates the utility and construction of invariant manifolds.

2.2 Example Illustrating Invariant Manifolds

Here we present results for a simple planar system, selected for instructive purposes, that demonstrates many of the ideas described above. In particular, this example demonstrates the utility of center manifolds. The nonlinear system of interest is

$$\dot{z}_1 = bz_1z_2 + cz_1^3 \tag{1}$$
$$\dot{z}_2 = -z_2 + az_1^2 \tag{2}$$

where a, b, c are constants. This system can be thought of as arising from a series expansion near a fixed point. Here the origin $(z_1, z_2) = (0, 0)$ is the equilibrium of interest, and in fact, it is the only equilibrium of this system. The origin has eigenvalues $\lambda_{1,2} = (0, -1)$, which implies that linearization cannot determine its stability (the origin is said to be *non-hyperbolic*). Here one refers to z_2 as a fast mode and z_1 as a slow mode, and in fact it is nonlinearly slow and its dynamics dominate the system behavior. It is tempting to proceed as follows: Note that the second equation indicates that z_2 is asymptotically (in fact, exponentially) stable, so one might simply take $z_2 \to 0$, so that the first equation yields $\dot{z}_1 = cz_1^3$, which implies that the z_1 component is stable (unstable) for $c < 0$ $(c > 0)$. This approach is referred to as the *tangent space approximation*, and it is valid in some cases (specifically, if either a or b is zero, as we will prove below). This approach is, however, incorrect for this system if $ab \neq 0$. We show next that coefficients a and b must be systematically included in the analysis, which employs a center manifold, to determine the stability of the origin.

The correct approach is to assume there exists an invariant (in this case, center) manifold of the form,

$$z_2 = h(z_1) = \alpha z_1^2 + \ldots \tag{3}$$

where the form of the power series approximations accounts for the fact that the center manifold must pass through the origin and be tangent to the center eigenspace, in this case the z_1 axis. The constant α is to be determined. In this case only the quadratic term is required; but, one can take more terms in the expansion if needed. Taking d/dt of this equation yields

$$\dot{z}_2 = \frac{dh}{dz_1}\dot{z}_1 = 2\alpha z_1 \dot{z}_1 + \ldots\,. \tag{4}$$

Substituting the $\dot{z}$'s from the original Equations (1,2) and z_2 from the center manifold expansion in Equation (3) into expression (4), and expanding in z_1 results in an equation that involves only series terms in z_1 and does not include any time-dependence. Note that solutions that satisfy this equation must satisfy both the dynamical system and the center manifold constraint. The resulting equation has the form,

$$-(\alpha z_1^2 + \dots) + a z_1^2 = 2\alpha z_1 \left(b z_1 (\alpha z_1^2 + \dots) + c z_1^3 \right) + \dots \tag{5}$$

One can now gather like powers of z_1, where in this case only the quadratic term is of interest, which yields $\alpha = a$ (only terms from the left hand side of the equation are needed here). If one repeats this process and includes cubic and quadratic terms, it is found that the local expansion of the center manifold near the origin is given by $h(z_1) = a z_1^2 - 2a(ab + c) z_1^4 + \dots$. Note that to leading (quadratic) order the center manifold is the curve along which $\dot{z}_2 = 0$, according to the Equation (2). However, inclusion of higher order terms shows that this approach does not provide the correct higher order terms for the center manifold.

Theorems state that the dynamics near the origin initially have behavior such that z_2 collapses exponentially fast towards the center manifold, and then slow dynamics occur near the center manifold. These are governed approximately by the z_1 dynamics as projected onto the center manifold, that is, by Equation (1), with the substitution $(z_1, z_2) \to (w, h(w))$ (3). This process, including only the leading order nonlinear terms, yields the slow dynamics in the form,

$$\dot{w} = (ab + c) w^3 + \dots \tag{6}$$

where w is a coordinate tangent to h (3). It is now seen that the stability of the z_1 dynamics is dictated by the coefficient $(ab + c)$, that is, the modal coupling terms come into play in an essential manner. Specifically, for $(ab + c) > 0$ the origin will be unstable, and for $(ab + c) < 0$ the origin will be asymptotically stable, albeit very slow, converging like $t^{-1/2}$. Note that if $(ab + c) = 0$, higher order terms are critical and one must redo the analysis keeping higher order terms in all expressions.

Figure 1 shows sample phase planes, including the center manifold approximated by $z_2 = h(z_1)$ to quartic order, for this system for three sets of parameter values. These cases depict various combinations of stability conditions and center manifold curvatures. Note the fast convergence of trajectories onto the center manifold and the slow flow on it, which indicates the stability of the origin.

Center manifold analysis is a very powerful tool for the investigation of bifurcations, since it allows for the systematic reduction of models to their essential modes near critical points (5). It also allows one to determine the behavior of systems when parameters are varied around the critical point,

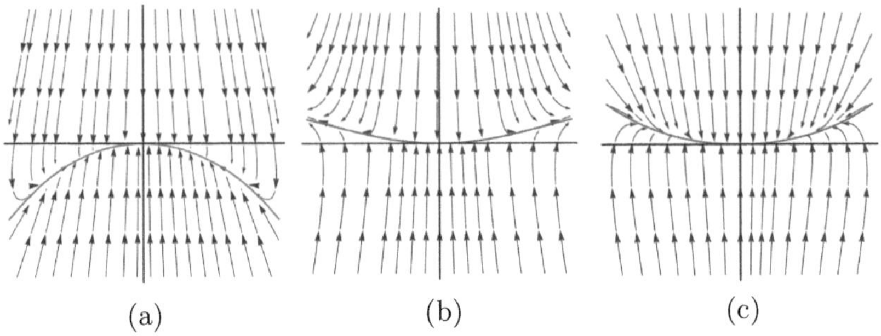

Figure 1. Sample (z_1, z_2) phase plane trajectories for Equations (1,2), and the center manifold $z_2 = h(z_1)$ shown as the green curve. Scale is $-0.25 < (z_1, z_2) < 0.25$. (a) $(a, b, c) = (-2, 1, 1)$, a stable case with $(ab + c) = -1$. (b) $(a, b, c) = (1, 1, 1)$, an unstable case with $(ab + c) = 2$. (c) $(a, b, c) = (1, -1, -1)$, a stable case with $(ab + c) = -2$.

using the so-called "suspension trick" (3). We do not pursue these issues here, but rather focus on the general framework outlined in the example for constructing invariant manifolds and generating the dynamics on them.

2.3 Key Points About Invariant Manifolds

The key points regarding invariant manifolds are that (i) in a dynamical system with m states there exist invariant sets of lower dimension k ($< m$) on which the system is governed by a k-dimensional system; (ii) one can develop differential equations for the form of invariant manifolds expressed in terms of the state variables (without time); (iii) these equations can be solved locally by expansion methods, and possibly by other methods; and (iv) one must account for modal coupling when doing model reduction in nonlinear systems, even weakly nonlinear systems. We now turn to equations and similar analysis relevant for vibratory systems.

2.4 Nonlinear Vibration Equations of Motion

In the field of mechanical vibrations, the equations of motion are generated by Lagrange's method or Hamilton's principle, resulting in second order equations typically of the form,

$$M\ddot{x} + C\dot{x} + Kx + g(x, \dot{x}) = f(x, \dot{x}, t) \tag{7}$$

where $x \in R^n$ is a displacement of a lumped mass degree of freedom (DOF) or, more generally, a linear combination of the degrees of freedom, M, C, K are the usual mass, damping, and stiffness matrices, $g(x, \dot{x})$ represents nonlinear stiffness and damping terms, and $f(x, \dot{x}, t)$ represents time-dependent excitation. This model is developed such that $x = 0$ is the unforced ($f = 0$) equilibrium point, so that the nonlinear terms satisfy $g(0, 0) = 0$. This can be achieved for a general case by a translation of coordinates. Note that for the first order (state space) form, $\dot{z} = F(z)$, one takes the state vector $z = (x, \dot{x})$, $z \in R^{2n}$, and uses standard methods for putting the equations of motion into first order form.

We begin our development for vibratory systems by considering linear systems, that is, with $f = 0$, $g = 0$. This sets the stage for the subsequent application of invariant manifolds to nonlinear systems.

3 Normal Modes as Invariant Manifolds for Linear Systems

For convenience, we will assume that M and K are symmetric, but allow C to have both symmetric (dissipative) and anti-symmetric (gyroscopic, non-dissipative) components. The unforced ($f = 0$) linear ($g = 0$) system is known to have solutions of the form $x(t) = u\, e^{\lambda t}$, $u \in R^n$, resulting in the eigenproblem,

$$\det\left(\lambda^2 M + \lambda C + K\right) = 0 \quad (8)$$
$$\left(\lambda^2 M + \lambda C + K\right) u = 0. \quad (9)$$

Since we are interested in the case of vibratory motions about the equilibrium, we assume the system has underdamped linear modes, that is, eigenvalues λ are complex conjugate pairs of the form $\omega_k(-\zeta_k \pm i\sqrt{1-\zeta_k^2})$ ($k = 1, \ldots, n$) where ω_k are the undamped modal frequencies, ζ_k are the modal damping ratios, and $i^2 = -1$. The results presented can be generalized to cases where some eigenvalues are real, but we do not consider that case here. We will often assume that the damping matrix C is Caughey (4), that is, C is made diagonal by the same similarity transformation that diagonalizes M and K, the most common example of which is the case of proportional damping, where C is a linear combination of M and K. When C is Caughey, the system modes u_k are real and the system has a modal matrix P formed by taking the u_k's as columns, and P is the matrix for the transformation that renders $P^T M P$, $P^T C P$, and $P^T K P$ diagonal. In this case the vibration modes are standing waves, with nodal points of the u_k's fixed in time. When C is not Caughey the system has complex modes u_k

that represent traveling waves with moving nodes. In all cases we assume that the modes have distinct eigenvalues in order to avoid issues related to degenerate modes. The topic of degenerate modes, which is important in terms of modal interactions and stability, is considered in other chapters in this volume.

When the system is Caughey and the eigenvalues are distinct, the modes are orthogonal with respect to M, C, and K. If we normalize the modes to satisfy $u_j^T M u_k = \delta_{jk}$, where δ_{jk} is the Kronecker delta, then $u_k^T K u_k = \omega_k^2$. Employing the standard transformation to modal coordinates via $x = P\eta = \sum_{k=1}^{n} u_k \eta_k$ and projection onto the individual modes using the orthogonality conditions renders uncoupled equations of motion of the form,

$$\ddot{\eta}_j + 2\zeta_j \omega_j \dot{\eta}_j + \omega_j^2 \eta_j = 0 \qquad j = 1, \ldots, n \tag{10}$$

A fundamental property of linear modes is that if one starts with a set of initial conditions in a single mode, that is, $(x(0), \dot{x}(0)) = (\alpha u_k, \beta u_k)$, where α and β are scalar constants, then the response remains in that mode for all time, that is, $(x(t), \dot{x}(t)) = (\eta_k(t) u_k,\ \dot{\eta}_k(t) u_k)$. In other words, the individual modes are invariant. In this modal response $\eta_k = A_k \exp(-\zeta_k \omega_k t) \cos(\omega_k \sqrt{1 - \zeta_k^2}\ t + \phi_k)$, where A_k and ϕ_k depend on the initial conditions, and the displacements and velocities of the system DOF are mutually synchronous, so that peaks and zeros of displacements and velocities are reached simultaneously by all degrees of freedom, resulting in a standing wave response of the system (a damped wave if $\zeta_k > 0$). In this case one can represent a mode by a linear relationship between only the displacements of the DOF (the ratios of amplitudes are fixed), and the relationships between velocities of the DOF necessarily follow suit. It is precisely these types of modes that Rosenberg extended to the case of nonlinear systems (22). In this case, he envisioned a nonlinear relationship between the amplitudes of the DOF that must be satisfied, one that ensures synchronous motion. That formalism results in a relationship that depends on the total energy of the response, that is, there is not a single nonlinear relationship that works for all amplitudes of vibration of a given mode (except in special cases).

A key insight into these linear modes is that since all DOF have the same time behavior, one is sufficient to determine the response of the entire system. In this way one can express a modal response in terms of a single DOF, that is, $x_\ell(t) = a_{\ell j} x_j(t)$ where $x_j(t)$ is considered the master DOF and all others are subordinate to it. In this notation the constants $a_{\ell j}$ represent the elements of the associated modal vector u normalized so that $a_{\ell\ell} = 1$.

If C is not Caughey, this synchronous property is lost, even for the linear system, and one must consider a state space formulation to develop the notion of invariance. Specifically, in this case one must allow the displacements and velocities of all DOF to be subordinate to a pair of states, typically taken to be the displacement and velocity of a given DOF. Introducing $y_\ell = \dot{x}_\ell$ and taking (x_j, y_j) for a particular DOF j as the master states, these conditions are given by,

$$\begin{aligned} x_\ell &= a_{\ell j} x_j + b_{\ell j} y_j \\ y_\ell &= c_{\ell j} x_j + d_{\ell j} y_j \end{aligned} \tag{11}$$

where constants $(a, b, c, d)_{\ell j}$ must obviously satisfy $a_{\ell\ell} = d_{\ell\ell} = 1$ and $b_{\ell\ell} = c_{\ell\ell} = 0$. Note that there will be one set of coefficients for each mode, so that another index is required in the eventual formulation. For the case of Caughey damping, the displacements and velocities are independent, so that $b_{\ell j} = c_{\ell j} = 0$, and each have identical relative ratios, so that $d_{\ell j} = a_{\ell j}$. This returns to the case described previously where the modes are real and only one set of coefficients, the $a_{\ell j}$'s, are required.

The constraints expressed in Equation (11) represent a two dimensional plane in the full state space parameterized by the master states. For a system with n DOF and distinct eigenvalues there will be n such planes. The system response restricted to such an invariant plane is equivalent to a single DOF linear system, and its time behavior is the attendant modal response. Since the plane is parameterized by two variables the responses on this plane represent the family of modal responses resulting from all possible initial conditions. It is precisely this picture that is extended to define modes for nonlinear systems.

Example: A Linear Gyroscopic System Before turning to the nonlinear case, we demonstrate the main ideas as applied to linear systems using a two DOF gyroscopic system. It is shown below that this process can be generalized to nonlinear systems with several DOF. The example linear system consists of a mass m connected via springs with homogeneous stiffness k to a disk that rotates at a constant rate Ω, as shown in Figure 2. We will restrict $\Omega < \sqrt{k/m}$ since the system will become unstable above this speed. The equations of motion for this undamped system expressed in coordinates fixed to the disk are given by

$$\begin{aligned} \ddot{x}_1 + (\omega_0^2 - \Omega^2)x_1 - 2\Omega\dot{x}_2 &= 0 \\ \ddot{x}_2 + (\omega_0^2 - \Omega^2)x_2 + 2\Omega\dot{x}_1 &= 0, \end{aligned} \tag{12}$$

or, in state space form with $y_j = \dot{x}_j$,

$$\begin{aligned}
\dot{x}_1 &= y_1 \\
\dot{y}_1 &= -(\omega_0^2 - \Omega^2)x_1 + 2\Omega y_2 \\
\dot{x}_2 &= y_2 \\
\dot{y}_2 &= -(\omega_0^2 - \Omega^2)x_2 - 2\Omega y_1,
\end{aligned} \tag{13}$$

where $\omega_0^2 = k/m$ is the (repeated) natural frequency associated with the responses of the stationary ($\Omega = 0$) disk. Here the system is normalized so that the mass matrix M is the 2×2 identity matrix I_2, the stiffness matrix is $K = (\omega_0^2 - \Omega^2)I_2$, and the skew-symmetric matrix $C = -C^T$ arises from the gyroscopic nature of the system. Thus the system is not Caughey and we expect complex modes if the system is formulated in the traditional second order form.

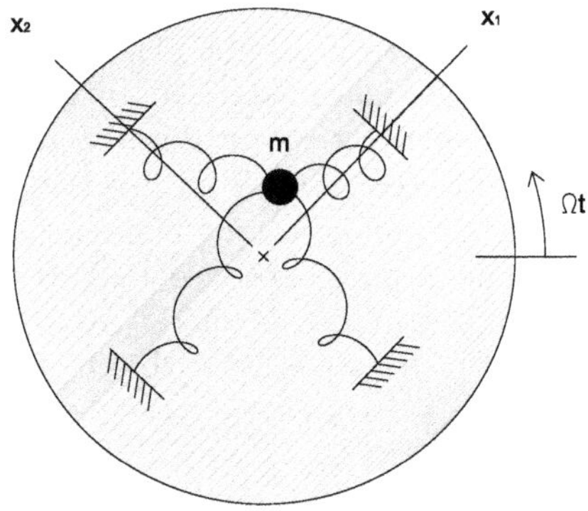

Figure 2. A mass suspended by springs with homogeneous stiffness that are fixed to a disk rotating about a fixed point at a constant rate Ω.

The choice of master coordinate is arbitrary (so long as it is not a node of any mode), so here we use (x_1, y_1) as masters and (x_2, y_2) as subordinates. The process of finding the normal modes using invariance is one that removes time dependence using the constraint Equations (11) and the equations of motion, yielding equations for the unknown coefficients. We begin with Equation (11) and its time derivative, specialized to this two DOF case, for which we can drop the indices on the coefficients, which are given by,

$$\begin{aligned} x_2 &= ax_1 + by_1 \\ y_2 &= cx_1 + dy_1 \end{aligned} \tag{14}$$

and

$$\begin{aligned} \dot{x}_2 &= a\dot{x}_1 + b\dot{y}_1 \\ \dot{y}_2 &= c\dot{x}_1 + d\dot{y}_1. \end{aligned} \tag{15}$$

Coefficients (a, b, c, d) are to be determined, and there will be one set for each mode. We substitute the expressions for the time derivatives given in the equations of motion (13) into Equations (15), which results in the follow equations relating the states without time behavior,

$$\begin{aligned} y_2 = &= ay_1 + b\left(-(\omega_0^2 - \Omega^2)x_1 + 2\Omega y_2\right) \\ -(\omega_0^2 - \Omega^2)x_2 - 2\Omega y_1 &= cy_1 + d\left(-(\omega_0^2 - \Omega^2)x_1 + 2\Omega y_2\right). \end{aligned} \tag{16}$$

The final step for obtaining the equations for the coefficients is to eliminate x_2 and y_2 using the constraints given in Equations (14). Doing so and gathering the terms for x_1 and y_1 in the results, one obtains the following coupled quadratic equations for (a, b, c, d),

$$\begin{aligned} c - 2bc\Omega - b\Omega^2 + b\omega_0^2 &= 0 \\ -a + d - 2bd\Omega &= 0 \\ -2cd\Omega + a\Omega^2 - d\Omega^2 - a\omega_0^2 + d\omega_0^2 &= 0 \\ -c - 2\Omega - 2d^2\Omega + b\Omega^2 - b\omega_0^2 &= 0 \end{aligned} \tag{17}$$

It is interesting to note that the solution to these equations will yield the linear modes of vibration, but the equations themselves are nonlinear, as they must be since there are multiple modes (two in the present case). It is also interesting to note that by this procedure one will find the mode shapes (eigenvectors) before determining the eigenvalues (the coefficients for the exponential time behavior).

There are multiple solutions of Equations (17), but only two have real coefficients, and these are given by

$$\begin{aligned} a_{1,2} &= 0 \\ b_{1,2} &= \frac{1}{\Omega \mp \omega_0} \\ c_{1,2} &= \pm\omega_0 - \Omega \\ d_{1,2} &= 0 \end{aligned} \tag{18}$$

An invariant modal for this linear system can be represented by two planes $X_2(x_1, y_1)$ and $Y_2(x_1, y_1)$. Examples of these for the first mode are shown in Figure 3.

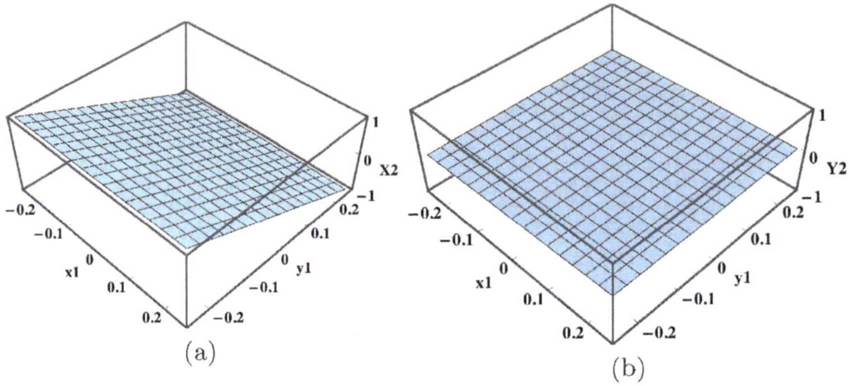

Figure 3. Constraints for the first mode of the gyroscopic system for $\omega_0 = 1$ and $\Omega = 0.75$: (a) $X_2^{(1)}(x_1, y_1)$; (b) $Y_2^{(1)}(x_1, y_1)$.

For this system the normal modes are not synchronous, since the displacements do not simultaneously reach zero and extrema, nor do the velocities. In fact, in these modal responses the subordinate displacement x_2 is *independent* of master coordinate x_1 (since $a_j = 0$) and, in fact, they are simply proportional to master coordinate y_1 via the b_j. Thus, in these modal responses the displacements are phase shifted from one another by $\pi/2$, as an analysis of the linear system response would also reveal. Similarly, the velocity y_2 in a given modal response is independent of y_1 (since $d_j = 0$) and is proportional to the displacement x_1 via the c_j.

To obtain the time behavior of these modal responses one simply enforces the constraints on either of the equations of motion, which must yield the same time response for (x_1, y_1) for each mode. Using the first equation of motion and the constraint on $\dot{x}_2 = y_2$ yields the second order equation,

$$\ddot{x}_1 + (\omega_0^2 - \Omega^2 - 2\Omega c)x_1 - 2\Omega d\dot{x}_1 = 0.$$

Letting u_j be the displacement of mode j and using the values for the mode coefficients in Equation (18), gives the equations of motion governing the dynamics of the modes,

$$\ddot{u}_j + \left(\omega_0^2 - \Omega^2 - 2\Omega(\pm\omega_0 - \Omega)\right) u_j = 0 \qquad j = 1, 2 \tag{19}$$

from which one can compute the natural frequencies of the modes,

$$\omega_{1,2} = \omega_0 \mp \Omega \tag{20}$$

which are valid for $\Omega < \omega_0$. Note that the rotation splits the repeated frequencies and that ω_1 reaches zero when $\Omega = \omega_0$, at which point an instability occurs.

An interesting feature of these responses is that the amplitudes of the two DOF are equal in the modal responses, that is, if $x_1(t) = A\cos(\omega_j t)$ where A is the amplitude for mode j, then by this formulation $x_2(t) = b_j y_2(t) = b_j \dot{x}_1(t) = -b_j \omega_j A \sin(\omega_1 t)$. For the first mode one finds that the amplitude of x_2 is also A since $b_1\omega_1 = -1$, and thus this mode has $(x_1, x_2)^{(1)} = (A\cos(\omega_1 t), A\sin(\omega_1 t))$, which is a counter-clockwise circular motion at frequency ω_1 in the (x_1, x_2) plane. Similarly, for the second mode one finds that $b_2\omega_2 = +1$, resulting in $(x_1, x_2)^{(2)} = (A\cos(\omega_2 t), -A\sin(\omega_2 t))$, which is a clockwise circular motion at frequency ω_2.

The modes for this system are in contrast with modes of a Caughey system which are represented by lines (the eigenvectors) in the (x_1, x_2) plane. For non-Caughey systems, including this gyroscopic example, the modes are complex in the usual linear formulation. The present approach accounts for the phase relationships between the DOF in a modal response using the displacements and velocities, without the need for complex coordinates.

Of course, this system has radial symmetry and one can use polar coordinate defined by $(x_1, x_2) = (r\cos\theta, r\sin\theta)$ to analyze the response. In this case the equations of motion are given by,

$$\begin{aligned} \ddot{r} - (\Omega + \dot{\theta})^2 r + \omega_0^2 r &= 0 \\ r\ddot{\theta} + 2\dot{r}(\Omega + \dot{\theta}) &= 0 \end{aligned} \tag{21}$$

where the second equation is equivalent to conservation of angular momentum, that is, $h = r^2(\Omega + \dot{\theta})$ is constant. For a modal response we have constant amplitude $r = A$ and constant angular speed $\dot{\theta}$. Using these conditions it is seen that the θ equation of motion becomes trivially satisfied, and one finds that for $r \neq 0$ only two values of $\dot{\theta}$ satisfy the r equation of motion, namely, $\dot{\theta}_{1,2} = \pm\omega_0 - \Omega$. For $\Omega < \omega_0$ these represent a counter-clockwise ($\dot{\theta} > 0$) rotation at $\omega_0 - \Omega$, that is, at ω_1, and a clockwise ($\dot{\theta} < 0$) rotation at a rate $\omega_0 + \Omega$, that is, at ω_2. While the analysis using these coordinates is considerably cleaner, its simplicity does not generalize to cases where the stiffness is not homogeneous, whereas the (x_1, x_2) formulation does.

A general procedure of this type for linear vibration systems can be shown to be identical to the usual steps for determining modes and natural frequencies (24). We next develop the theory of modes for nonlinear systems based on invariant manifolds using a generalization of the approach described above.

4 Normal Modes for Nonlinear Systems

Based on the formulation for a linear system described above, we provide a definition for a NNM based on invariant manifolds and describe some of its properties and methods of determining these modes. We formulate the general problem first and then demonstrate its use on an example.

A *nonlinear normal mode* is defined by a parameterized family of responses that lie on a two-dimensional invariant manifold that is tangent to a linear mode eigenspace. It can be visualized as a two dimensional, generally curved, surface that constrains all DOF to a single pair of master coordinates, typically the displacement and velocity of a DOF, or of a modal coordinate. In this way it provides a natural extension to linear normal modes that is a direct result of the extension of linear eigenspaces into invariant manifolds, such as those shown in the linear system example presented above. In fact, this concept goes back to Lyapunov, who proved the existence of these invariant manifolds for conservative systems under certain non-resonance conditions (10). Cases with internal resonance were considered by Weinstein (28) and Moser (13). The line of work described here was carried out to extend these notions to systems with damping and other nonconservative effects and to provide constructive methods for determining NNMs in more general cases, with a view towards their application to the nonlinear vibration of mechanical systems.

The key to our formulation is to impose invariance of an individual mode. That is, we assume that the entire system behaves like a single DOF system for special sets of initial conditions. These conditions are determined by assuming there exists a response in which all system DOF are subordinate to a single DOF. As shown above, in linear systems these conditions can be expressed as linear combinations of the amplitudes of displacements and velocities of the system DOF as they relate to a pair of master coordinates, and these give rise to a purely modal response. These relations are equivalent, of course, to the eigenvectors of the linearized system, and therefore the amplitude is arbitrary. In nonlinear systems the relationships are generally nonlinear and the time dependent behavior depends on the amplitude of motion. These features are universal for NNMs. We now describe the formulation for invariant manifold NNMs using different types of coordinates for the system dynamics, and also discuss some extensions of the method.

4.1 State Space Formulation

The formulation proceeds as follows for a system with n DOF and states (x_ℓ, y_ℓ), $\ell = 1, 2, \ldots, n$, with equations of motion in the form

$$\begin{aligned} \dot{x}_\ell &= y_\ell \\ \dot{y}_\ell &= f_\ell(x, y) \qquad \ell = 1, 2, \ldots, n \end{aligned} \tag{22}$$

where $x = (x_1, x_2, \ldots, x_n)$ and $y = (y_1, y_2, \ldots, y_n)$. Note that if the inertia (mass) matrix of the system is not diagonal, and/or depends on the states in some way, it must be inverted in order to isolate the accelerations to reach this form of the equations of motion.

A pair of master coordinates (x_j, y_j) are selected, and in a modal response all coordinates are subordinate to the masters, expressed as

$$\begin{aligned} x_\ell &= X_\ell(x_j, y_j) \\ y_\ell &= Y_\ell(x_j, y_j) \qquad \ell = 1, 2, \ldots, n \end{aligned} \tag{23}$$

where clearly $X_j = x_j$ and $Y_j = y_j$. Taking time derivatives of these conditions yields

$$\begin{aligned} \dot{x}_\ell &= \frac{\partial X_\ell}{\partial x_j}\dot{x}_j + \frac{\partial X_\ell}{\partial y_j}\dot{y}_j \\ \dot{y}_\ell &= \frac{\partial Y_\ell}{\partial x_j}\dot{x}_j + \frac{\partial Y_\ell}{\partial y_j}\dot{y}_j \qquad \ell = 1, 2, \ldots, n. \end{aligned} \tag{24}$$

Now, the equations of motion (22) are employed in Equation (24) in order to remove time dependence, obtaining,

$$\begin{aligned} y_\ell &= \frac{\partial X_\ell}{\partial x_j} y_j + \frac{\partial X_\ell}{\partial y_j} f_j(x, y) \\ f_\ell(x, y) &= \frac{\partial Y_\ell}{\partial x_j} y_j + \frac{\partial Y_\ell}{\partial y_j} f_j(x, y) \qquad \ell = 1, 2, \ldots, n. \end{aligned} \tag{25}$$

The modal invariance constraints (23) are now imposed, to make Equation (25) as follows,

$$\begin{aligned} Y_\ell &= \frac{\partial X_\ell}{\partial x_j} y_j + \frac{\partial X_\ell}{\partial y_j} f_j(X, Y) \\ f_\ell(X, Y) &= \frac{\partial Y_\ell}{\partial x_j} y_j + \frac{\partial Y_\ell}{\partial y_j} f_j(X, Y) \qquad \ell = 1, 2, \ldots, n. \end{aligned} \tag{26}$$

One pair of these Equations ($\ell = j$) are trivial, and the remaining $2n-2$ are partial differential equations for the $2n-2$ unknown functions (X_ℓ, Y_ℓ), $\ell \neq$

j, with independent variables (x_j, y_j). Solving these equations is equivalent to solving the original equations of motion, but approximate methods can be employed, as described below.

In nondegenerate cases these modes are local extensions of the linear modes. The more complicated cases of internal resonances and strongly nonlinear modes are not considered here, but are presented in other chapters of this volume. To construct the local nonlinear modes that are extensions of the linear modes one can use numerical methods (18; 16) or series expansions in (x_j, y_j) to approximate the solutions for the unknown (X_ℓ, Y_ℓ). The series expansions are assumed to have the form

$$\begin{aligned} X_\ell &= a_{\ell 1} x_j + a_{\ell 2} y_j + a_{\ell 3} x_j^2 + a_{\ell 4} x_j y_j + a_{\ell 5} y_j^2 + \cdots \\ Y_\ell &= b_{\ell 1} x_j + b_{\ell 2} y_j + b_{\ell 3} x_j^2 + b_{\ell 4} x_j y_j + b_{\ell 5} y_j^2 + \cdots \\ & \ell = 1, 2, \ldots, n; \ \ell \neq j, \end{aligned} \tag{27}$$

with the a and b coefficients to be determined. This is a nonlinear extension of the approach described previously for linear systems. Of course, one can include nonlinear terms up to the desired order.

These expansions are substituted into Equation (26) and expanded in terms of (x_j, y_j). Terms of like powers of $x_j^{n_x} y_j^{n_y}$ are gathered to provide a set of equations for the a's and b's. At each order (linear, quadratic, etc), these equations contain coefficients from that order and all lower orders, but none from higher orders. As demonstrated in the linear gyroscopic example, solution of the coefficients of the linear terms, for which $n_x + n_y = 1$, involves solving a set of nonlinear equations for the $(a_{1\ell}, a_{2\ell}, b_{1\ell}, b_{2\ell})$, as it must since there exist n linear modes and thus the equations must have multiple solutions. The equations for the coefficients of the nonlinear terms are solved sequentially at each order, that is at order $m = n_x + n_y$ for $m = 2, 3, \ldots$. These equations are linear, but contain lower order coefficients in their formulation. Thus, solving at quadratic order, $m = 2$, the equations will had have non homogeneous terms that depend on $(a_{1\ell}, a_{2\ell}, b_{1\ell}, b_{2\ell})$, so that each nonlinear mode has its own set of nonlinear coefficients. These features are demonstrated in the example that follows.

A visualization of these modes is that they represent two-dimensional surfaces whose tangent plane at the origin is the eigenspace corresponding to the linear mode. In the terminology of Rosenberg, if these surfaces are flat and identical to the linear modes, the nonlinear modes are referred to as "similar," which occurs in systems with certain types of symmetries.

Note that if one converts the equations of motion using the usual transformation to linear modal coordinates, then in Equation (22) the function f_ℓ will contain linear terms that depend on (x_ℓ, y_ℓ), and the nonlinear terms will depend generally on all states. In this formulation the form of the invariant manifolds is simplified since the eigenspaces are already known and the NNM manifolds are tangent to them. The constraints in Equation (23) are now purely nonlinear, except for $\ell = j$, since the dynamics are uncoupled at linear order. This implies that the linear coefficients in Equation (27) are all zero. It is quite natural to investigate systems in this manner since the transformation to linear coordinates is quite standard and it simplifies the analysis of the NNM manifolds.

4.2 Amplitude/Phase Coordinates

For numerical solutions of the invariant manifolds it is often convenient to use polar coordinates, namely $(x_\ell, \dot{x}_\ell) = (a_\ell \cos(\phi_\ell), -a_\ell \omega_\ell \sin(\phi_\ell))$ where ω_ℓ is the linear frequency of mode ℓ. Here one can use Fourier series as basis functions for the angles ϕ_ℓ, and there are several possibilities for basis functions for expansion in the amplitudes a_ℓ. In this case for a particular j the masters are taken to be (a_j, ϕ_j), and the modal invariant manifold constraints take the form,

$$\begin{aligned} \dot{a}_\ell &= A_\ell(a_j, \phi_j) \\ \dot{\phi}_\ell &= \Phi_\ell(a_j, \phi_j) \qquad \ell = 1, 2, \ldots, n \end{aligned} \tag{28}$$

where $A_j = a_j$ and $\Phi_j = \phi_j$. These coordinates are similar to those used in the van der Pol transformation for the method of averaging (5). This formulation uses expansions and projections to obtain a set of nonlinear equations for unknown coefficients for the NNM invariant manifolds. These are solved using standard techniques, from which the manifolds, and the equations of motion on them, can be constructed. See (18) for examples of these calculations.

4.3 Systems with Harmonic Excitation

All of the formulations described to this point are for free vibrations. These systems do not have explicit time dependence, that is, they are autonomous. In many applications it is useful to know the system frequency response, that is, to understand its response to harmonic excitation over a range of frequencies. The formulation described above can be extended to that case by the addition of an auxiliary state that generates the harmonic excitation. Specifically, for a system with n DOF and harmonic excitation, we supplement the equations of motion with an additional state, ψ such that $\dot{\psi} = \omega$ is the excitation frequency. The equations of motion in this formulation become

$$\begin{aligned}\dot{x}_\ell &= y_\ell \\ \dot{y}_\ell &= f_\ell(x, y, \psi) \qquad \ell = 1, 2, \ldots, n \\ \dot{\psi} &= \omega\end{aligned} \tag{29}$$

which allows for any type of harmonic forcing, direct, parametric, etc. With this addition, the invariant manifolds of Equation (23) now become

$$\begin{aligned}x_\ell &= X_\ell(x_j, y_j, \psi) \\ y_\ell &= Y_\ell(x_j, y_j, \psi) \qquad \ell = 1, 2, \ldots, n\end{aligned} \tag{30}$$

which can be thought of as two-dimensional manifolds that vary periodically in time (they are really three dimensional in the full extended state space). If one prefers to not mix polar and state coordinates, a state model for the harmonic excitation can be used where the auxiliary state dynamics are of the form $\dot{x}_{n+1} = y_{n+1}$, $\dot{y}_{n+1} = -\omega^2 x_{n+1}$, in which case x_{n+1} can play the role of the harmonic excitation of frequency ω. See (8) for details of this development and examples.

While this formulation allows one to address these problems in the invariant manifold framework, it must be admitted that usual perturbation techniques are preferable in terms of computations and interpretation.

4.4 Multi-Nonlinear Mode Models

It must be stated that the concept of modes in nonlinear systems is not nearly as powerful as it is for linear systems, primarily due to the lack of superposition in nonlinear systems. The utility of NNMs is that system with many DOF do respond with only a few active states under many circumstances. The NNM formulation described above allows one to systematically reduce a nonlinear system to a single mode of vibration. In linear systems all system responses can be constructed using combinations of these individual modes. While this is not the case for nonlinear systems, there exists a formulation that allows one to construct invariant manifolds for multi-mode dynamics.

This formulation is particularly interesting in the case where internal resonances occur. In fact, when formulating individual NNMs, as described above, the equations for the nonlinear terms of the invariant manifold become singular under conditions when the mode of interest cannot be dynamically uncoupled from one or more other modes. This is precisely what occurs in conditions of internal resonance, where nonlinear coupling between modes allows energy exchange between those modes. In these cases, one must generalize the concept of NNMs to account for multi-mode interactions.

We briefly outline the development here; the reader can find more details in (2; 17). The system is expressed in terms of linear modal coordinates as,

$$\ddot{\eta}_i + \omega_i^2 \eta_i = f_i(\eta), \qquad i = 1, 2, \ldots, n \tag{31}$$

where the η_i's are the linear modal coordinates, η is the vector with elements η_i, the ω_i's are the linear natural frequencies, and the $f_i(\eta)$'s are the nonlinear forces expressed in terms of the η's. (The method can easily be extended to the case where the forces also depend on the $\dot{\eta}$'s, although the computational effort for developing NNMs is much higher for those systems.) A multi-NNM invariant manifold that accounts for M modes is taken to be a $2M$-dimensional manifold in the full state space, such that any motion started on the manifold will remain on it for all times.

Generalizing the method for individual NNMs, a multi-NNM manifold is developed as follows. We define a set of indices, S_M, which described a subset of the modes in the full dynamic model. For example, if one has interactions between the first and third modes, then $M = 2$ and $S_M = \{1, 3\}$. In the general case, we require $2M$ master coordinates. Again using the notation $\mu_j = \dot{\eta}_j$, the masters are denoted by (η_k, μ_k), $k \in S_M$, and all other DOF as subordinate to these, according to

$$\left\{ \begin{array}{l} \eta_i = X_i(\eta_k, \mu_k) \\ \mu_i = Y_i(\eta_k, \mu_k) \end{array} \right. , \qquad \text{for } i \notin S_M, \quad k \in S_M \tag{32}$$

Pesheck *et al.* (17) developed this method, using expansions similar to those of Equation (27) to develop local approximations of these manifolds. Of course, in this case there are many more coefficients to obtain. Jiang *et al.* (7) developed these equations using the polar coordinates formulation and used it to solve some example problems. The equations of motion are obtained by selecting M pairs of state equations and restricting them to the invariant manifolds. This leads to a system with M DOF in which the modes are coupled via nonlinear terms. If the modes are nonresonant, then a nonlinear (near identity) coordinate transformation can eliminate the coupling terms, providing the normal form. If the modes are internally resonant, transformations can eliminate some coupling terms, but the essential ones that describe the mode interactions cannot be removed, and this is precisely when the multi-mode models are most useful.

It must be admitted that the invariant manifold approach for multi-mode systems is quite cumbersome, and its utility is mostly conceptual. Standard methods that involve projection onto the linear modes and perturbation techniques, which result in the normal form for the particular internal resonance at hand are generally more practical for investigating specific models.

4.5 Continuous Systems

The approach can be extended to the case when the model is a partial differential equation for a distributed parameter, or continuous, system. In this case, the method assumes that the response in a NNM is slaved to that of a single point on the system, so long as that point is not a node of the mode of interest. For this case we focus on one-dimensional structures of length ℓ with displacement field $u(s,t)$ where s is the spatial variable. The equations of motion are taken to be of the form,

$$\begin{aligned} u_{tt} + L[u, u_t] + N[u, u_t] &= 0 \ , \quad 0 \leq s \leq \ell \\ B_j[u] &= 0 \quad j = 1, \ldots, m \ , \quad s = 0 \ \& \ s = \ell \end{aligned} \tag{33}$$

where subscripts t refer to time derivatives, $B_j[u]$ are the boundary conditions, L is a linear operator, and N is a nonlinear operator. In this case we assume that the master states are the displacement and velocity at a given point s_o on the structure, namely $(u(s_o,t), u_t(s_o,t))$, which we denote as $(u_o(t), v_o(t))$. In a modal response the entire structure follows these states according to constraints

$$\begin{aligned} u(s,t) &= U(u_o(t), v_o(t), s, s_o) \\ u_t(s,t) &= V(u_o(t), v_o(t), s, s_o). \end{aligned} \tag{34}$$

When these conditions are forced onto the equations of motion, one ends up with equations that must be solved for (U, V). One can use expansions in powers of $(u_o(t), v_o(t))$, of the form $U(u_o(t), v_o(t), s, s_o) = a(s, s_o)u_o(t) + \cdots$, which includes linear and nonlinear terms in $(u_o(t), v_o(t))$. The process then leads to differential equations in s for the coefficients in the expansions, such as $a(s, s_o)$. For linear systems this leads to an alternative formulation for the eigenvalues and eigenfunctions, and its generalization allows one to compute NNMs for this class of problems. Once the (U, V) are determined, the modal dynamics are obtained by imposing them on the equations of motion, which leads to ordinary differential equations for $(u_o(t), v_o(t))$ or, equivalently, a second order ordinary differential equation for $u_o(t)$. Details for this approach can be found in (25).

Another approach for continuous systems is to carry out an expansion and projection of the equations of motion (33) onto some assumed modes, which then convert the system into a finite dimensional form that can be handled by the methods described above (23). NNMs for continuous systems can also be handled using perturbation methods (15; 14).

4.6 Example: A Nonlinear Gyroscopic System

To illustrate the ideas and calculation for a relatively simple system, we extend the linear gyroscopic model shown in Figure 2 to include nonlinear stiffness effects. The restoring force is assumed to depend on the radial displacement $r = \sqrt{x_1^2 + x_2^2}$ from the equilibrium, and have linear and cubic terms, that is, it has magnitude $F_r = kr + \hat{\gamma}r^3$. Again, one can use polar coordinates to analyze this system, but we show results for (x_1, x_2) to demonstrate the methodology.

The projections of the restoring force onto the x_j coordinates are given by $F_{rj} = -x_j\left(k + \hat{\gamma}(x_1^2 + x_2^2)\right)$, and so the equations of motion are given by

$$\begin{aligned} \ddot{x}_1 + (\omega_0^2 - \Omega^2)x_1 + \gamma x_1(x_1^2 + x_2^2) - 2\Omega\dot{x}_2 &= 0 \\ \ddot{x}_2 + (\omega_0^2 - \Omega^2)x_2 + \gamma x_2(x_1^2 + x_2^2) + 2\Omega\dot{x}_1 &= 0, \end{aligned} \tag{35}$$

where $\gamma = \hat{\gamma}/m$. In state space form with $y_j = \dot{x}_j$ these become,

$$\begin{aligned} \dot{x}_1 &= y_1 \\ \dot{y}_1 &= -(\omega_0^2 - \Omega^2)x_1 - \gamma x_1(x_1^2 + x_2^2) + 2\Omega y_2 \\ \dot{x}_2 &= y_2 \\ \dot{y}_2 &= -(\omega_0^2 - \Omega^2)x_2 - \gamma x_1(x_1^2 + x_2^2) - 2\Omega y_1. \end{aligned} \tag{36}$$

The process for determining the NNMs for this system follow the steps of the linear system, using (x_1, y_1) as masters and (x_2, y_2) as subordinates. Here the constraints must be nonlinear, so the form given in Equation (23) is used, and expansions similar to those described in Equation (27) are used to implement them. In the present case we will use some knowledge about the system to reduce the number of coefficients required in the NNM formulation. First, we have already solved the linear coefficients of the manifold and found that two of the four are zero. Also, since the nonlinearities are cubic in nature, no quadratic terms will appear in the expansion. Finally, from symmetry considerations one finds that four of the eight cubic coefficients are zero. Based on these facts we can assume a simplified expansion of the form

$$\begin{aligned} X_2 &= by_1 + ex_1^2y_1 + fy_1^3 + \cdots \\ Y_2 &= cx_1 + gx_1^3 + hx_1y_1^2 + \cdots \end{aligned} \tag{37}$$

These expansions are substituted into Equation (26) and expanded in powers of $x_1^{n_x}y_1^{n_y}$. For $n_x + n_y = 1$ the linear coefficients are obtained and, of course, match those in the analysis of the linearized system as presented above in Equation (18). Gathering the cubic ($n_x + n_y = 3$) terms and matching coefficients leads to a set of four equations that are linear in (e, f, g, h) and have nonhomogeneous terms that depend on b and c. Thus, there are two sets of solutions for these equations, one for each mode. These coefficients are proportional to the nonlinear stiffness coefficient γ, and are given by

$$\begin{aligned} e_{1,2} &= \frac{\pm\gamma}{2\omega_0(\Omega \mp \omega_0)^2} \\ f_{1,2} &= \frac{\pm\gamma}{2\omega_0(\Omega \mp \omega_0)^4} \\ g_{1,2} &= \frac{\pm\gamma}{2\omega_0} \\ h_{1,2} &= e_{1,2} \end{aligned} \tag{38}$$

The corresponding modal surfaces for the two modes, represented by the constraints (X_2, Y_2) as functions of (x_1, y_1) for each mode, are two dimensional surfaces. These are plotted in Figures 4 and 5 for the two modes. Note that the tangent planes of these curved surfaces at the origin represent the linear normal modal eigenspaces, and that only one of these surfaces has significant nonlinear effects for these parameter conditions and amplitudes. The curved nature of the surfaces are precisely the conditions required to maintain invariance, so that if one starts with initial conditions on a pair modal surfaces, the response will move on trajectories that remain on these surfaces for all time. Since this system is conservative, these responses will be periodic, described by the equation derived below. The invariant manifolds are precisely the surfaces formed by this family of periodic solutions in the state space, parameterized by the amplitude of vibration, which can be quantified in several ways, for example, the energy of the response.

As noted for the linear version of this system, the normal modes are not synchronous, since the displacements do not simultaneously reach zero and extrema, nor do the velocities. For the NNMs, from the form in Equation (37) it is seen that the phase shift of $\pi/2$ between the displacements in a NNM response persists, namely that when the displacement x_1 is zero, the velocity y_2 is zero, and similarly with x_2 and y_1. However, in contrast to the linear modes, the displacement of x_2 is not independent of master coordinate x_1 due to nonlinear coupling effects, and similarly y_2 is not independent of y_1. While the modal responses will be close to harmonic, they will have small higher order harmonics (these will be odd due to the symmetric nature of the nonlinearity), and their frequencies of oscillation will depend on amplitude, as we now show.

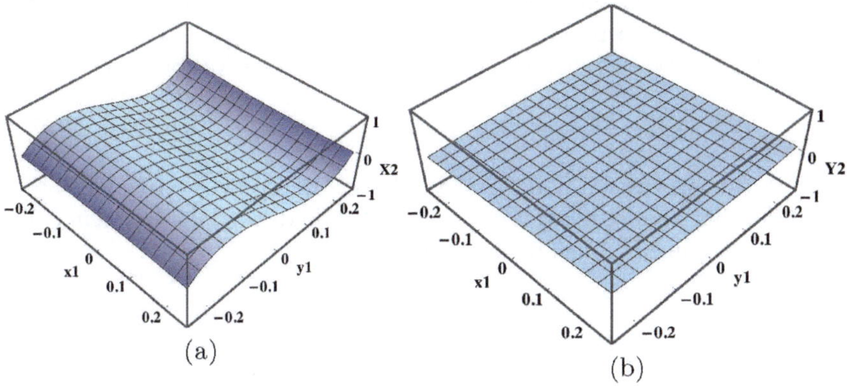

Figure 4. Constraints for the first NNM of the gyroscopic system for $\omega_0 = 1$, $\gamma = 0.5$, and $\Omega = 0.75$: (a) $X_2^{(1)}(x_1, y_1)$; (b) $Y_2^{(1)}(x_1, y_1)$.

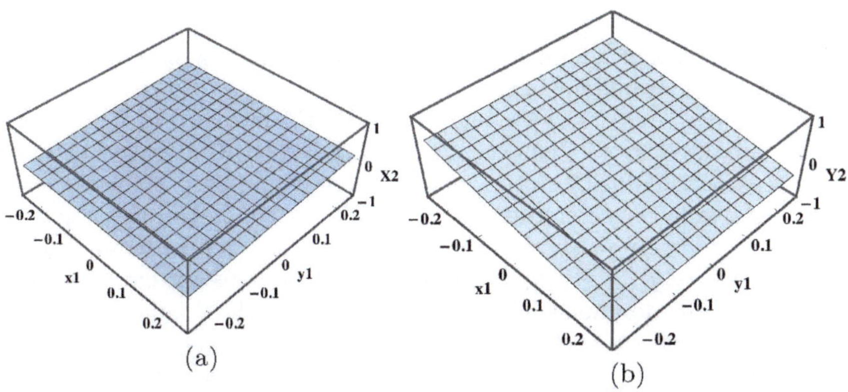

Figure 5. Constraints for the second NNM of the gyroscopic system for $\omega_0 = 1$, $\gamma = 0.5$, and $\Omega = 0.75$: (a) $X_2^{(2)}(x_1, y_1)$; (b) $Y_2^{(2)}(x_1, y_1)$.

As in the linear case, the time behavior of these modal responses are obtained by enforcing the constraints on either of the equations of motion. Here, the equations of motion and the constraints are nonlinear. Using the first equation of motion and imposing the modal constraints yields,

$$\ddot{x}_1 + (\omega_0^2 - \Omega^2)x_1 + \gamma x_1(x_1^2 + X_2^2(x_1, y_1)) - 2\Omega Y_2(x_1, y_1) = 0.$$

Using the series expansions for X_2 and Y_2, keeping only cubic terms, noting that $y_1 = \dot{x}_1$, rearranging terms, and letting $(x_1, y_1) \to (u_j, \dot{u}_j)$ where $j = 1, 2$ for the two modes, the modal response equation becomes,

$$\ddot{u}_j + (\omega_0^2 - \Omega^2 - 2\Omega c_j)u_j + (\gamma - 2\Omega g_j)\, u_j^3 + \left(\gamma b_j^2 - 2\Omega h_j\right) u_j \dot{u}_j^2 = 0,$$

which is a conservative nonlinear oscillator. Using standard methods one can show that this oscillator has an amplitude-dependent frequency given to leading nonlinear order in the amplitude A by

$$\omega_j(A) = \omega_j + A^2 \frac{3\,(\gamma - 2\Omega g_j) + \omega_j^2\left(\gamma b_j^2 - 2\Omega h_j\right)}{8\omega_j} + \cdots$$

where $\omega_j = \sqrt{\omega_0^2 - \Omega^2 - 2\Omega c_j}$ is the linear natural frequency of mode j. Substitution of the coefficients for the two modes and simplification yields,

$$\omega_{1,2}(A) \quad = \quad \omega_0 \mp \Omega + A^2 \frac{\gamma}{2\omega_0} + \cdots \tag{39}$$

as the frequencies of nonlinear oscillation. Note that the nonlinear adjustment term is the same for both modes.

To consider the time behavior of these NNM motions we consider the relative amplitudes of x_1 and x_2 in a NNM response. Recall that in the linear case the amplitudes of x_1 and x_2 were equal, even though the displacements were phase shifted by $\pi/2$. Due to symmetry, this must still be the case, as we now show. For a NNM response the constraint $x_2(t) = b_j y_1(t) + e_j x_1^2(t) y_1(t) + f_j y_1^3(t)$ must hold. From this we can obtain the amplitude of the dominant harmonic of $x_2(t)$ using the dominant harmonic approximation $(x_1(t), y_1(t)) = (A\cos(\omega_j t), -A\omega_j \sin(\omega_j t))$, along with the known expressions for the coefficients (b_j, e_j, f_j), and some trig identities. This process shows that the main harmonic of x_2 is $A\sin(\omega_1 t)$ for the first NNM and $-A\sin(\omega_1 t)$ for the second NNM. Thus the first NNM is a counterclockwise circular motion at $\omega_1(A)$ in the (x_1, x_2) plane and the second mode is a clockwise circular motion at $\omega_2(A)$. This is valid for a range of amplitudes A for which the system is weakly nonlinear, that is, $|\omega_0^2 A| >> |\gamma A^3|$, or, equivalently, $A < \omega_0/\sqrt{\gamma}$.

Again, using polar coordinates for this symmetric system is cleaner and provides a check on the preceding results. Extending the analysis used for the linear model, we obtain for the equations of motion in (r, θ), as follows,

$$\begin{aligned} \ddot{r} - (\Omega + \dot{\theta})^2 r + \omega_0^2 r + \gamma r^3 &= 0 \\ r\ddot{\theta} + 2\dot{r}(\Omega + \dot{\theta}) &= 0 \end{aligned} \tag{40}$$

For a NNM response we have constant amplitude $r = A$ and constant angular speed $\dot{\theta}$. Using these conditions, one finds that for $r \neq 0$ only two values of $\dot{\theta}$ satisfy the r equation of motion, namely, $\dot{\theta}_1 = \omega_1(A)$, a counterclockwise rotation for $\Omega < \omega_0$ and small A, and $\dot{\theta}_2 = -\omega_2(A)$ a clockwise rotation, verifying the NNM analysis.

5 Discussion

The topic of NNMs has an interesting history. Originally developed by Rosenberg to extend the notion of modes to nonlinear systems, it stayed in the realm of nonlinear dynamics for several decades, where many interesting results were obtained about the stability and bifurcations of NNMs. The area was relatively dormant for some time but has recently expanded into tools for dealing with nonlinear vibrations in mechanical systems. There are many opinions about the practical utility of NNMs as a tool for analysis. It is true that for weakly nonlinear systems perturbation methods yield the same results and are preferred by many researchers (including the author) for most calculations. The NNM concept can be extended to strongly nonlinear systems, and typically internal resonances with other modes cause NNM responses to become complicated by coupling to other modes, as described in other chapters in this volume. Furthermore, there is a looming question about the conditions under which any given NNM may or may not be observed in a system response with general initial conditions. All of these criticisms are valid and relevant to the utility of NNMs for experimental work. However, the fact remains that NNMs provide a very powerful and useful framework for thinking about nonlinear vibrations of mechanical systems. They can be used for developing useful low order models for systems that systematically account for nonlinear effects, including coupling between linear modes, and they are closely related to the system responses one sees when they are excited near resonances. From this point of view, the concept of NNMs is something that should be in the mindset of researchers working in nonlinear vibrations.

Acknowledgements

The author is grateful to the organizers of this CISM course, in particular, Gaetan Kerschen, for inviting him to contribute. The author's work on this topic was originally instigated by discussions with Alex Vakakis while on sabbatical leave at Caltech in 1989, and also based on his previous exposure to these problems by Richard Rand while a graduate student at Cornell. The research described in the author's papers was carried out in close collaboration with Christophe Pierre and several joint graduate students, including Polarit Apiwattanalunggarn, Dongying Jiang, Nicholas Boivin, and Eric Pesheck. Assistance for the preparation of this chapter by Scott Strachan and Ming Mu is gratefully acknowledged. The work was funded during those years by grants from the US Army Research Office and the US National Science Foundation (NSF). The author's current related work on nonlinear resonances in MEMS and torsional vibration absorbers is supported by the US NSF under grants CMMI-1234067 and CMMI-1100260.

Bibliography

[1] K. V. Avramov and Y. V. Mikhlin. Review of applications of nonlinear normal modes for vibrating mechanical systems. *Applied Mechanics Reviews*, 65, 2013.

[2] N. Boivin, C. Pierre, and S. W. Shaw. Non-linear modal analysis of structural systems featuring internal resonances. *Journal of Sound and Vibration*, 182(2):336–341, 1995.

[3] Jack Carr. *Applications of centre manifold theory*, volume 35. Springer-Verlag New York, 1981.

[4] TK Caughey. Classical normal modes in damped linear dynamic systems. *Journal of Applied Mechanics*, 27:269, 1960.

[5] John Guckenheimer and Philip Holmes. Nonlinear oscillations, dynamical systems, and bifurcations of vector fields. 1983.

[6] Morris W Hirsch, Robert L Devaney, and Stephen Smale. *Differential equations, dynamical systems, and linear algebra*, volume 60. Access Online via Elsevier, 1974.

[7] D Jiang, C Pierre, and SW Shaw. The construction of non-linear normal modes for systems with internal resonance. *International Journal of Non-linear mechanics*, 40(5):729–746, 2005.

[8] D Jiang, C Pierre, and SW Shaw. Nonlinear normal modes for vibratory systems under harmonic excitation. *Journal of sound and vibration*, 288(4):791–812, 2005.

[9] Gaëtan Kerschen, Maxime Peeters, Jean-Claude Golinval, and Alexander F Vakakis. Nonlinear normal modes, part i: A useful framework for the structural dynamicist. *Mechanical Systems and Signal Processing*, 23(1):170–194, 2009.

[10] Aleksandr Mikhailovich Lyapunov. The general problem of the stability of motion. *International Journal of Control*, 55(3):531–534, 1992.

[11] Yuri V Mikhlin and Konstantin V Avramov. Nonlinears normal modes for vibrating mechanical systems. review of theoretical developments. *Applied Mechanics Reviews*, 63(6):0802, 2010.

[12] YV Mikhlin and NV Perepelkin. Non-linear normal modes and their applications in mechanical systems. *Proceedings of the Institution of Mechanical Engineers, Part C: Journal of Mechanical Engineering Science*, 225(10):2369–2384, 2011.

[13] Jürgen Moser. Periodic orbits near an equilibrium and a theorem by Alan Weinstein. *Communications on Pure and Applied Mathematics*, 29(6):727–747, 1976.

[14] A. H. Nayfeh. On direct methods for constructing nonlinear normal modes of continuous systems. *Journal of Vibration and Control*, 1:389–430, 1995.

[15] A. H. Nayfeh and S. A. Nayfeh. On nonlinear modes of continuous systems. *Journal of Vibration and Acoustics*, 116:129–136, 1994.

[16] Maxime Peeters, Régis Viguié, Guillaume Sérandour, Gaëtan Kerschen, and J-C Golinval. Nonlinear normal modes, part ii: Toward a practical computation using numerical continuation techniques. *Mechanical systems and signal processing*, 23(1):195–216, 2009.

[17] E. Pesheck, N. Boivin, C. Pierre, and S. W. Shaw. Nonlinear modal analysis of structural systems using multi-mode invariant manifolds. *Nonlinear Dynamics*, 25:183–205, 2001.

[18] E. Pesheck, C. Pierre, and S. W. Shaw. A new galerkin-based approach for accurate nonlinear normal modes through invariant manifolds. *Journal of Sound and Vibration*, 249(5):971–993, 2002b.

[19] Christophe Pierre, Dongying Jiang, and Steven Shaw. Nonlinear normal modes and their application in structural dynamics. *Mathematical Problems in Engineering*, 2006, 2006.

[20] R. M. Rosenberg. On normal vibrations of a general class of nonlinear dual-mode systems. *Journal of Applied Mechanics*, 28:275–283, 1961.

[21] R. M. Rosenberg. The normal modes of nonlinear n-degree-of-freedom systems. *Journal of Applied Mechanics*, 29:7–14, 1962.

[22] R. M. Rosenberg. On nonlinear vibrations of systems with many degrees of freedom. *Advances in Applied Mechanics*, 9:155–242, 1966.

[23] S. W. Shaw. An invariant manifold approach to nonlinear normal modes of oscillation. *Journal of Nonlinear Science*, 4:419–448, 1994.

[24] S. W. Shaw and C. Pierre. Normal modes for non-linear vibratory systems. *Journal of Sound and Vibration*, 164(1):85–124, 1993.

[25] S. W. Shaw and C. Pierre. Normal modes of vibration for non-linear continuous systems. *Journal of Sound and Vibration*, 169(3):319–347, 1994.

[26] Alexander F Vakakis. Non-linear normal modes (nnms) and their applications in vibration theory: an overview. *Mechanical systems and signal processing*, 11(1):3–22, 1997.

[27] Alexander F Vakakis, Leonid I Manevitch, Yuri V Mikhlin, Valery N Pilipchuk, and Alexandr A Zevin. *Normal modes and localization in nonlinear systems*. Wiley. com, 2008.
[28] Alan Weinstein. Normal modes for nonlinear hamiltonian systems. *Inventiones mathematicae*, 20(1):47–57, 1973.

Normal form theory and nonlinear normal modes: Theoretical settings and applications

Cyril Touzé
Unité de Mécanique (UME)
ENSTA-ParisTech
828 Boulevard des Maréchaux
91762 Palaiseau Cedex, France
cyril.touze@ensta-paristech.fr
http://www.ensta.fr/~touze

1 Introduction and notations

These lecture notes are related to the CISM[1] course on "Modal Analysis of nonlinear Mechanical systems", held at Udine, Italy, from June 25 to 29, 2012. The key concept at the core of all the lessons given during this week is the notion of Nonlinear Normal Mode (NNM), a theoretical tool allowing one to extend, through some well-chosen assumptions and limitations, the linear modes of vibratory systems, to nonlinear regimes. More precisely concerning these notes, they are intended to show the explicit link between Normal Form theory and NNMs, for the specific case of vibratory systems displaying polynomial type nonlinearities. After a brief introduction reviewing the main concepts for deriving the normal form for a given dynamical system, the relationship between normal form theory and nonlinear normal modes (NNMs) will be the core of the developments. Once the main results presented, application of NNMs to vibration problem where geometric nonlinearity is present, will be highlighted. In particular, the developments of reduced-order models based on NNMs expressed asymptotically with the formalism of real normal form, will be deeply presented. Applications are devoted to thin structures vibrating at large amplitudes, with a special emphasis on thin shells of different geometry (from plates to closed circular cylindrical shells). Effective reduced-order models for the prediction of the type of nonlinearity (hardening/softening behaviour), or the computation of complete bifurcation diagrams for the case of forced vibrations, will be

[1]CISM stands for "Centre International des Sciences Mécaniques" (in French), or "International Centre for Mechanical Sciences", see www.cism.it

G. Kerschen (Ed.), *Modal Analysis of Nonlinear Mechanical Systems*, CISM International Centre for Mechanical Sciences DOI 10.1007/ 978-3-7091-1791-0_3 © CISM Udine 2014

shown, for toy models including a small number of degrees-of-freedom (typicaly two to three dofs), as well as for continuous models such as beams, plates and shells.

The following notations will be used throughout the lecture notes. For generic (nonlinear) dynamical systems, $\mathbf{X}$ will denote the state vector, $\mathbf{X} \in \mathcal{E}$, where $\mathcal{E}$ is the phase space, of dimension n. Generally, the simple choice $\mathcal{E} \equiv \mathbb{R}^n$ is retained. The dynamical system is denoted as:

$$\dot{\mathbf{X}} = \mathbf{F}(\mathbf{X}), \tag{1}$$

with $\mathbf{F}$ the vector field, which could depend on one or more parameters.

For the particular case of vibratory system, geometric nonlinearity is considered, so that only quadratic and cubic type nonlinearities are present in the equations of motions. Hence for nonlinear vibration problems, the generic equations of motion consist in a set of N oscillator equations, denoted under the general formulation:

$$\forall p = 1, ..., N \; : \quad \ddot{X}_p + \omega_p^2 X_p + \sum_{i=1}^{N} \sum_{j \geq i}^{N} g_{ij}^p X_i X_j + \sum_{i=1}^{N} \sum_{j \geq i}^{N} \sum_{k \geq j}^{N} h_{ijk}^p X_i X_j X_k = 0, \tag{2}$$

where g_{ij}^p denotes the generic quadratic nonlinear coupling coefficient, and h_{ijk}^p the cubic one. In this notation, the upperscript p refers to the oscillator-equation considered, while the subscripts (i, j, k) denotes the coupling monom $X_i X_j X_k$. Note that damping is not considered in Eq. (2). Inclusion of dissipative mechanism will be included in some parts of the notes. Note also that the linear part is diagonal, which means that the variable X_p is the modal amplitude of the pth linear normal mode. In cases where the linear part is not given as diagonal, a linear change of coordinate can be performed to fit the framework presented here. Finally, to recover the first-order dynamical system formalism, the velocity $Y_p = \dot{X}_p$ is used as complementary variable, so that, for a collection of N oscillators, the state variable writes: $\mathbf{X} = [X_1 \; Y_1 \; X_2 \; Y_2 \; ... \; X_N \; Y_N \;]$, so that $\dim(\mathcal{E}) = 2N$.

The first section is devoted to normal form theory. The main idea of introducing a nonlinear transform in order to simplify as much as possible the equations of motion, is first introduced in an illustrative manner where the reverse problematic is considered. Very simple examples in dimension 1 and 2 allows introducing the key concept of resonance. The core of the theory, the theorems of Poincaré and Poincaré-Dulac, are then given in a general context, and the result is then specialized to the case of vibratory problems. The main results are given for a conservative problem, and the link to NNMs, is illustrated, allowing derivation of important ideas such as reduced-order modeling and classification of nonlinear terms for physical interpretation. Finally, the case of damped mechanical systems is tackled.

The second section focuses on the prediction of the type of nonlinearity (hardening/softening behaviour) for a system of oscillator of the form (2). It is shown that NNMs and normal form gives the correct framework for an efficient and accurate prediction. Examples on two-dofs system, and spherical-cap shells with varying radius of curvature (from the flat plate case to spherical shells), illustrates the method. Finally the influence of the damping on the type of nonlinearity is discussed.

The third section tackles the problem of deriving accurate reduced-order models (ROMs) for thins structures harmonically excited at resonance, in the vicinity of one of its eigenfrequency, and vibrating at large amplitudes. NNMs and normal form are used to derive the ROMs, and examples on different shells are shown: a doubly-curved (hyperbolic paraboloid) panel illustrates a case without internal resonance, while a closed circular cylindrical shell allows illustrating a more complicated case with a 1:1 internal resonance. the complete bifurcation diagram with different kind of bifurcation points are clearly recovered by the ROM, and a comparison with the most routinely used Proper Orthogonal Decomposition method (POD) is shown to conclude the notes.

2 Normal form theory

Normal form theory is a classical tool in the analysis of dynamical systems, and general introductions can be found in many textbooks, see *e.g.* (Guckenheimer and Holmes, 1983; Iooss and Adelmeyer, 1998; Wiggins, 2003; Manneville, 1990). It is generally used in bifurcation theory in order to define the simplest form of dynamical systems generating classical bifurcations of increasing co-dimension. Here however, the general theory will be used for another purpose: defining a nonlinear change of coordinates allowing one to express the dynamics in an invariant-based span of the phase space, where the (curved) generating axis are the invariant manifolds arising from the linear eigenspaces, *i.e.* the NNMs of the system.

Normal form theory is based on two major theorems, due to Poincaré and Poincaré-Dulac, which have been demonstrated in the beginning of the XXth century (Poincaré, 1892; Dulac, 1912). The main idea is to simplify, as far as possible, the equations of motion of a nonlinear dynamical systems, by means of nonlinear change of coordinates. The presentation will begin with a reverse problematic, in order to understand better the main issue: starting from a simple, linear problem, we will show how a nonlinear change of coordinate can make it appear as complicated. Then the normal form will be introduced gradually with two examples study, starting from the simplest cases with dimension 1 and 2. The "reverse" illustrative problematic is borrowed from Manneville (2004), and the presentation of normal form used follows closely that shown in Manneville (2004, 1990).

2.1 Problematic

Before entering the complicated calculations for tackling the general case, let us first introduce an illustrative example, allowing us to properly define the goal pursued, which is defined as trying to *simplify as far as possible*, in the vicinity of a particular solution (*e.g.* fixed point or periodic orbit), a given dynamical system.

To begin with, let us consider the simple initial-value problem:

$$\frac{\mathrm{d}^2 Y}{\mathrm{d}t^2} + Y = 0, \tag{3a}$$

$$Y(t=0) = Y_0, \quad \frac{\mathrm{d}Y}{\mathrm{d}t}(t=0) = 0, \tag{3b}$$

where Y is a real coordinate depending on time t. This equation is that of a single oscillator, with eigenfrequency equal to one. Its solution in time is known and reads:

$$Y(t) = Y_0 \cos t. \tag{4}$$

Now let us introduce the transformed variable:

$$X = \exp(Y) - 1. \tag{5}$$

Differentiating Eq. (5) two times with respect to time t, and inserting in the original dynamical equation (3a), one can show that the transformed variable X satisfy the following evolution equation:

$$(1+X)\ddot{X} - \dot{X}^2 + (1+X)^2 \ln(1+X) = 0, \tag{6}$$

where ln is the natural logarithm. In the case where one would have to face a physical problem expressed by Eq. (6), with appropriate initial conditions, then the solution would have been more difficult to find ! Maybe that with a great intuition and a bit of luck, one could have find the solution which, in our case, is known by construction, and is simply given by:

$$X(t) = \exp(Y_0 \cos t) - 1. \tag{7}$$

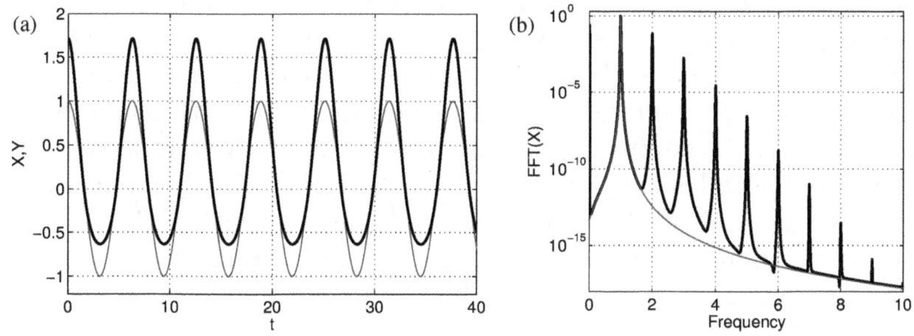

Figure 1. (a) Time series of the transformed variable X (black solid line), compared to the original Y (thin blue line), for the problem defined through Eqs (3)-(7). (b): Fourier Transform of X (black solid line) and Y (thin blue line), showing their harmonic content.

The solution for X is represented in Fig. 1. As awaited, it is 2π-periodic, but contrary to the initial solution for Y displaying only one harmonic component, the X solution shows an infinity of harmonics, with exponential decay in amplitude, as revealed by the Fourier transform of X, Fig. 1(b). This example shows that the apparent complexity of the problem for X only results from a nonlinear change of coordinates. Looking at the things in a reverse manner, the question arising is naturally: for a given problem, is it possible to find such a nonlinear transformation that could simplify, at best linearise, the initial system ? The main idea of normal form theory is to give an answer to this question. Without inspired intuition, one could at least try an asymptotic, power series expansion, in order to get an idea of the sought nonlinear change of coordinates. This functional relationship has to be defined in the vicinity of $Y = 0$ (or equivalently, $X = 0$), which is the fixed point of the original dynamical system (equivalently, of the transformed system). Once again the result is known and reads, in our case:

$$X = \sum_{n=1}^{+\infty} \frac{1}{n!} Y^n. \tag{8}$$

Note that $X \sim Y$ for X, Y small, which means that the transformation is identity-tangent, and aims at conserving the linear characteristic of the original system. In a general case where the solution is not known, one would like to find iteratively, order by order, the nonlinear transformation.

Let us close these introductory remarks by precising what is meant by "as far as possible", in the definition of the nonlinear transform. Consider the general dynamical system $\dot{X} = F(X)$, where $F(X) = X + X^p$, in the vicintiy of the fixed point $X = 0$. The common sense would state that this problem with p= 6 is "more nonlinear" than the same with p=2. However, as shown in Fig. 2, the linear approximation $F(X) \sim X$ is valid on a larger range of X-amplitudes for p=6 than for p=2. Hence the dynamical system with $F(X) = X+X^6$ is "less nonlinear" than the one with $F(X) = X+X^2$, because the nonlinearities are activated for larger values of the amplitude X, and one could give confidence to a linear approximation on a larger range of amplitudes.

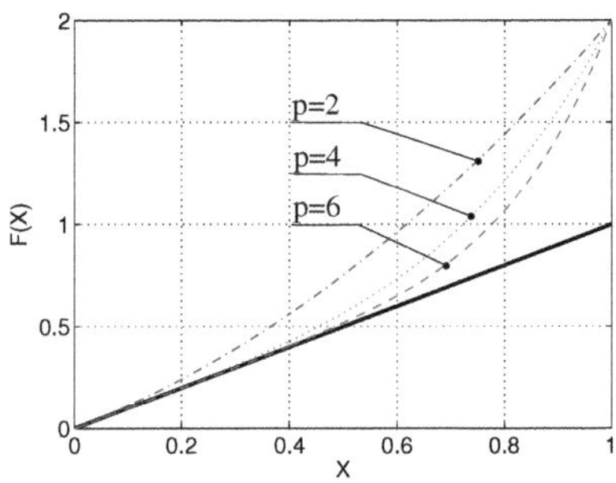

Figure 2. Representation of $F(X) = X + X^p$ for p=2, 4, 6, as compared to the linear case.

Finally, the goal can now be stated properly: for a given dynamical system, in the vicinity of a particular solution (typically a fixed point), we would like to find out nonlinear transformations that aims at extending the validity range of the linear approximation, by cancelling a maximum number of monoms of increasing orders, in the power series expansion of the vector field. The next subsection introduces the normal form transform for simple problems where the phase space have dimension 1 and 2, before generalizing the results for any dimension and any dynamical system.

2.2 Example study

The problem in dimension $n = 1$

Let us consider a dynamical system:

$$\dot{X} = sX + a_2X^2 + a_3X^3 + ... = sX + \sum_{p>1} a_pX^p, \tag{9}$$

where $X \in \mathbb{R}$ (phase space of dimension n=1). X=0 is a hyperbolic fixed point as long as $s \neq 0$, otherwise a marginal case is at hand (bifurcation point).

Let us introduce a nonlinear transform:

$$X = Y + \alpha_2 Y^2, \tag{10}$$

where α_2 is introduced in order to cancel the quadratic monom of the original system, *i.e.* $a_2 X^2$ in Eq. (9). Here Y is the new variable, and the goal of the transformation is to obtain a dynamical system for the new unknown Y that is simpler than the original one. Differentiating (10) with respect to time and substituting in (9) gives:

$$(1 + 2\alpha_2 Y)\dot{Y} = sY + (s\alpha_2 + a_2)Y^2 + \mathcal{O}(Y^3) \tag{11}$$

All the calculation are realized in the vicinity of the fixed point. Hence $Y \ll 1$, so that one can mutliply both sides by:

$$(1 + 2\alpha_2 Y)^{-1} = \sum_{p=0}^{+\infty} (-2)^p \alpha_2^p Y^p \tag{12}$$

Rearranging the terms by increasing orders finally leads to:

$$\dot{Y} = sY + (a_2 - s\alpha_2)Y^2 + \mathcal{O}(Y^3) \tag{13}$$

From that equation, it appears clearly that the quadratic term can be cancelled by selecting:

$$\alpha_2 = \frac{a_2}{s}, \tag{14}$$

which is possible as long as $s \neq 0$. This condition has been assumed at the beginning (hyperbolicity of the fixed point). Hence with that choice, the objective is fulfilled, the nonlinearity has been repelled to order three. The only condition appearing is that of hyperbolicity, which means that we are not at a (singular) bifurcation point. The process can be continued further, let us examine what happens for the third order. First, one has to write the resulting dynamical system for Y, by expliciting the third-order term, which has been modified after the first transform that has cancelled the quadratic monom. Substituting for (10) in (9), with the retained choice $\alpha_2 = a_2/s$, leads to:

$$\dot{Y} = sY + \bar{a}_3 Y^3 + \mathcal{O}(Y^4), \quad \text{where} \quad \bar{a}_3 = a_3 + 2a_2^2/s \tag{15}$$

A cubic transform is now introduced in order to simplify (15):

$$Y = Z + \alpha_3 Z^3, \tag{16}$$

Repeating the same procedure leads to:

$$\dot{Z} = sZ + (\bar{a}_3 - 2s\alpha_3)Y^3 + \mathcal{O}(Y^4), \tag{17}$$

which shows that the cubic term can be cancelled by selecting:

$$\alpha_3 = \frac{\bar{a}_3}{2s}. \tag{18}$$

Once again, the only condition appearing in the calculation is the one already found, $s \neq 0$. The complete change of coordinate can be written up to third order by gathering together Eqs. (10) and (16):

$$X = Z + \frac{a_2}{s}Z^2 + \frac{\bar{a}_3}{2s}Z^3 + \mathcal{O}(Z^4), \tag{19}$$

and the resulting dynamical system for the knew unknown Z now reads:

$$\dot{Z} = sZ + \bar{a}_4 Z^4 + \mathcal{O}(Z^5), \tag{20}$$

where $\bar{a}_4$ has to be computed properly by replacing (10) in (9) and carefully selecting fourth-order terms. This process is more and more difficult with increasing orders, and rapidly suggest for helping oneself with symbolic computation softwares. In any case we can conclude that:

- The objective is fulfilled: nonlinearities have been repelled up to fourth order. The process can be continued until complete linearisation of the problem, which is possible as long as $s \neq 0$ (hyperbolicity condition).
- The computed nonlinear transform becomes singular when one approaches the marginality for $s \longrightarrow 0$, because the α_i's scale as $1/s$. This is a reflection of the fact that the nonlinearities dominates the dynamical behaviour in the vicinity of the bifurcation point; however the linearisation is still possible as long as one is away from that point.

The problem in dimension $n = 2$

Before generalizing the result, the problem in a phase space of dimension 2 is now tackled. In dimensions $n \geq 2$, *resonance conditions* appear, hence making the problem a little bit more complicated than what could be expected from the precedent subsection, where the hyperbolicity condition has been found to be sufficient for simplifying the original system.

We consider $\mathbf{X} = (X_1 \; X_2)^t \in \mathbb{R}^2$, and for simplicity the dynamical system $\dot{\mathbf{X}} = F(\mathbf{X})$ is considered quadratic in $\mathbf{X}$. We also assume that the linear part is diagonal and that the two eigenvalues reads (s_1, s_2). The system can be written explicitely as:

$$\dot{X}_1 = s_1 X_1 + a^1_{11} X_1^2 + a^1_{12} X_1 X_2 + a^1_{22} X_2^2, \quad (21a)$$

$$\dot{X}_2 = s_2 X_2 + a^2_{11} X_1^2 + a^2_{12} X_1 X_2 + a^2_{22} X_2^2. \quad (21b)$$

The nonlinear transform is introduced:

$$X_1 = Y_1 + \alpha^1_{11} Y_1^2 + \alpha^1_{12} Y_1 Y_2 + \alpha^1_{22} Y_2^2, \quad (22a)$$

$$X_2 = Y_2 + \alpha^2_{11} Y_1^2 + \alpha^2_{12} Y_1 Y_2 + \alpha^2_{22} Y_2^2. \quad (22b)$$

Differentiating (22) with respect to time and replacing in (21) leads to:

$$\begin{aligned}(1 + 2\alpha^1_{11} Y_1 + \alpha^1_{12} Y_2)\dot{Y}_1 + (\alpha^1_{12} Y_1 + 2\alpha^1_{22} Y_2)\dot{Y}_2 = s_1 Y_1 + \\ (a^1_{11} + s_1 \alpha^1_{11}) Y_1^2 + (a^1_{12} + s_1 \alpha^1_{12}) Y_1 Y_2 + \\ (a^1_{22} + s_1 \alpha^1_{22}) Y_2^2 + \mathcal{O}(Y^3_{1,2}) \quad (23a)\end{aligned}$$

$$\begin{aligned}(1 + 2\alpha^2_{22} Y_2 + \alpha^2_{12} Y_1)\dot{Y}_2 + (\alpha^2_{12} Y_2 + 2\alpha^2_{11} Y_1)\dot{Y}_1 = s_2 Y_2 + \\ (a^2_{11} + s_2 \alpha^2_{11}) Y_1^2 + (a^2_{12} + s_2 \alpha^2_{12}) Y_1 Y_2 + \\ (a^2_{22} + s_2 \alpha^2_{22}) Y_2^2 + \mathcal{O}(Y^3_{1,2}) \quad (23b)\end{aligned}$$

The calculation is pursued by noting that, at the lowest order, one has:

$$\dot{Y}_1 = s_1 Y_1 + \mathcal{O}(Y^2_{1,2}), \quad (24)$$

$$\dot{Y}_2 = s_2 Y_2 + \mathcal{O}(Y^2_{1,2}), \quad (25)$$

Hence the derivatives with respect to time involving a product in (23) can be replaced thanks to $Y_i \dot{Y}_j = s_j Y_i Y_j + \mathcal{O}(Y^3_{1,2})$, so that finally one obtains:

$$\begin{aligned}\dot{Y}_1 = s_1 Y_1 + &(a^1_{11} - s_1 \alpha^1_{11}) Y_1^2 + (a^1_{12} - s_2 \alpha^1_{12}) Y_1 Y_2 \\ &+ (a^1_{22} + (s_1 - 2s_2)\alpha^1_{22}) Y_2^2 \quad (26a)\end{aligned}$$

$$\begin{aligned}\dot{Y}_2 = s_2 Y_2 + &(a^2_{11} + (s_2 - 2s_1)\alpha^2_{11}) Y_1^2 + (a^2_{12} - s_1 \alpha^2_{12}) Y_1 Y_2 \\ &+ (a^2_{22} - s_2 \alpha^2_{22}) Y_2^2 \quad (26b)\end{aligned}$$

In the previous equations, the unkonwns $\{\alpha^p_{ij}\}$ can be found by setting:

$$\alpha^1_{11} = \frac{a^1_{11}}{s_1}, \qquad \alpha^1_{12} = \frac{a^1_{12}}{s_2}, \quad \alpha^1_{22} = \frac{a^1_{22}}{2s_2 - s_1}, \quad (27a)$$

$$\alpha^2_{11} = \frac{a^2_{11}}{2s_1 - s_2}, \quad \alpha^2_{12} = \frac{a^2_{12}}{s_1}, \quad \alpha^2_{22} = \frac{a^2_{22}}{s_2}, \quad (27b)$$

so that the complete cancellation of all the nonlinear terms present in the original equations (21) is possible if and only if:

$$s_1 \neq 0, \quad s_2 \neq 0, \quad s_2 \neq 2s_1, \quad s_1 \neq 2s_2. \tag{28}$$

The first two conditions have already been encoutered in the precedent subsection in dimension $n = 1$: once again, they are the consequence of the assumption of *hyperbolicity* of the fixed point, stating that we are not at a (marginal) bifurcation point. The last two conditions are new and result from extending the dimension to $n = 2$. They are called in the remainder *resonance condition*, and they reflect the fact that the eigenvalues may share a commensurability relationship. Hence they are completely different from the hyperbolicity condition. When such a relationship exists between the eigenvalues, *e.g.* when $s_2 = 2s_1$, the analysis shows that the system can not be linearized. More precisely, a monom, present in the original equations, can not be cancelled because it is resonant through the eigenvalue relationship, and thus strongly couple the two equations. In the analysis, the resulting normal form keeps the monom and the dynamics is different from a linearizable case. For example, in the case $s_2 = 2s_1$ (assuming $s_1,\ s_2 \neq 0$), the system can be simplified at best to:

$$\dot{Y}_1 = s_1 Y_1, \tag{29a}$$

$$\dot{Y}_2 = s_2 Y_2 + a^2_{11} Y_1^2. \tag{29b}$$

The resonance condition can be understood in the following manner: from the first equation we have $Y_1(t) \propto \exp s_1 t$. Reporting in the second equation, the nonlinear term makes appear a term proportional to $\exp 2s_1 t = \exp s_2 t$. Consequently the nonlinear term in (29b) may be interpreted as a forcing term, acting precisely at the resonance eigenvalue s_2 of the second equation. The solutions of this nonhomogeneous differential equation lead to secular terms, which explains why the monom can not be cancelled. These resonance conditions are further analyzed in subsection 2.4, devoted to vibratory systems.

This subsection reveals the core of Poincaré and Poincaré-Dulac theorems. When no resonance condition exist between the eigenvalues, then the system is equivalent to a linear one (Poincaré). In case of resonance relationships, the system can be simplified thanks to successive nonlinear transforms. Only the resonant monoms stay in the resulting so-called *normal form* of the problem (Poincaré-Dulac).

2.3 General case: Poincaré and Poincaré-Dulac theorems

We are now in position to extend the result to a general case in dimension n. Let us denote the dynamical system as:

$$\dot{\mathbf{X}} = \mathbf{F}(\mathbf{X}) = \mathbf{L}\mathbf{X} + \mathbf{N}_2(\mathbf{X}) + ... + \mathbf{N}_p(\mathbf{X}) \tag{30}$$

where $\mathbf{X} \in \mathcal{E}$ ($\mathcal{E} \equiv \mathbb{R}^n$ otherwise stated), $\mathbf{L}$, the linear part, is assumed to be diagonal[2], with $\mathbf{L} \equiv \text{diag}(s_p)$, $\mathbf{N}_p$ denotes the terms at order p.

The generalization of the resonance condition reads, for a given order of nonlinearity $p \geq 2$:

$$s_j = \sum_i n_i s_i, \quad n_i \geq 0, \quad \sum_i n_i = p. \tag{31}$$

Note that the cancellation of non-resonant monoms via nonlinear transformations, is realized sequentially, order by order, so that the resonance condition appears by increasing orders p. For p=2, Eq. (31) make appear the conditions $s_j = 2s_i$ encountered on the precedent example, as well as a case involving three eigenvalues: $s_i = s_j + s_k$, that can be found only when the dimension is such that $n \geq 3$.

When no resonance condition of the form (31) exist between the eigenvalues of the system, Poincaré theorem states that a nonlinear transform, $\mathbf{X} = \mathbf{Y} + \mathbf{g}(\mathbf{Y})$ exist, such that the dynamics for the new coordinate $\mathbf{Y}$ is linear and simply writes: $\dot{\mathbf{Y}} = \mathbf{L}\mathbf{Y}$. This theorem has been extended by Dulac in 1917 to the case of the existence of resonance condition, stating (Poincaré-Dulac theorem) that a nonlinear transform $\mathbf{X} = \mathbf{Y} + \mathbf{g}(\mathbf{Y})$ exist such that the resulting system for the new variable $\mathbf{Y}$ is still nonlinear but drastically simplified as compared to the original, as it contains only the resonant monoms that can not be cancelled. In both cases, the simplified system $\dot{\mathbf{Y}} = \mathbf{f}(\mathbf{Y})$ is called the normal form of the original system. In the simplest case the normal form is a linear system, in the other cases it contains only the resonant monoms, resulting from the resonance conditions between the eigenvalues.

These theorems are very strong in the sense that they underline the fact that, in the vicinity of a particular solution, the eigenspectrum completely determines the nonlinear part. This means in particular that, from the

[2]The particular cases where the linear operator is not diagonalisable are not treated here for the sake of brevity. The normal form theory applies in these cases as well, one has just to use the *Jordan* representation of the linear operator to obtain generic cases, some examples can be found in classical textbooks, see *e.g.* Iooss and Adelmeyer (1998); Manneville (2004); Wiggins (2003).

knowledge of the eigenvalues, one is able to construct the normal form of the system by deriving the resonant monoms. One must keep however in mind that the theory is *local* only, and is valid only in the vicinity of a particular solution. Finally, the method for demonstrating the theorems is a sequential calculations that treats by successive orders of nonlinearity the monoms $\mathbf{N}_p$ appearing in the dynamical system (30). This sequential calculation can be automatized by using symbolic toolbox softwares, nonetheless it becomes more and mode difficult with increasing orders. And it shows that the method do not lend itself to a numerical treatment if one would like to find numerically the nonlinear change of coordinates.

2.4 Application to vibratory systems, undamped case

The case of vibratory systems displaying quadratic and cubic nonlinearities, is now examined, following the general notation stated in the introduction, Eq. (2). At present damping is not considered, in order to introduce properly the concepts and the Nonlinear normal modes. The results are extended to dissipative vibratory systems in subsection 2.8. In this subsection, the presentation follows closely the theoretical results published in Touzé et al. (2004).

The eigenvalues of a vibratory system are complex conjugate, $\pm i\omega_p$, p=1, ...N. To compute the normal form of Eq. (2), a first idea would be to follow strictly the general framework sketched in the precedent subsection. This would mean to express Eq. (2) as:

$$\dot{\mathbf{X}} = \mathbf{F}(\mathbf{X}) = \mathbf{L}\mathbf{X} + \mathbf{N}_2(\mathbf{X}) + \mathbf{N}_3(\mathbf{X}), \tag{32}$$

where the linear part $\mathbf{L}$ is diagonal, and thus reads: $\mathbf{L} = \text{diag}\{\pm i\omega_p\}$. For example, this framework has been first used by Jézéquel and Lamarque (1991), to express how the nonlinear normal modes can be introduced from the framework provided by normal form theory. However a different choice is retained here. The linear part will be retained as it is when the system is written at first-order by using the velocity $Y_p = \dot{X}_p$ as auxiliary variable, *i.e.* a block-diagonal matrix, each block being written as:

$$\begin{pmatrix} 0 & 1 \\ -\omega_p^2 & 0 \end{pmatrix} \tag{33}$$

This choice is done in order to avoid introducing complex quantities in the calculations. It allows also to have equations that will be of oscillator form at each stage of the calculation, so that the auxiliary variable can always be cancelled in order to recover oscillator-like equations beginning with $\ddot{X} + X + ...$ This so-called real formulation allows simpler physical interpretation of the different terms, as it will be shown next.

A second point of interest is a consequence of the purely imaginary eigenspectrum. With regard to the resonance relationships highlighted in the previous subsection, it appears that a complete linearization is not possible in any case for a conservative vibratory system. Indeed, if only one oscillator equation is considered, then one can always fulfill the following relationship between the two complex conjugate eigenvalues $\{+i\omega, \ -i\omega\}$:

$$+i\omega = +i\omega + i\omega - i\omega. \tag{34}$$

i.e. a relationship of the form (31) with order p=3. Coming back to the oscillator equation, this means that the Duffing equation:

$$\ddot{X} + \omega^2 X + \alpha X^3 = 0, \tag{35}$$

is under its normal form, the cubic term (monom associated with the resonance relation (34)) can not be cancelled through a nonlinear transform. On a physical viewpoint, this stands as a good news. Indeed, one of the most important observed feature in nonlinear oscillations is the frequency dependence upon vibration amplitude. If the system could be linearized, this would mean that the underlying dynamics is linear, hence the frequency should not change with the amplitude. Note that this frequency dependence on amplitude is the main topic of section 3, and thus is deeply investigated there.

In the general case where N oscillator-equations are considered, numerous resonance relationships of the form:

$$+i\omega_p = +i\omega_p + i\omega_k - i\omega_k, \tag{36}$$

are possible, for arbitrary $p, k \in [1, N]^2$. This means that the original system (2) can be simplified, but numerous terms will remain at the end of the process, in the normal form, following Poincaré-Dulac theorem. However, as it will be shown next, the game is worth the candle, as numerous important terms will be cancelled, and also because the remaining terms can be easily interpreted.

Before stating the main result, a last point must be underlined. The resonance relationships put forward through Eqs (34)-(36) are denoted as *trivial*: they exist whatever the values of the frequencies of the studied structure are. A second family of resonance relationships may arise from *internal resonances* between the eigenfrequencies of the system. For example, order-two internal resonances reads, for arbitrary (p, i, j):

$$\omega_p = \omega_i + \omega_j, \quad \omega_p = 2\omega_i, \tag{37}$$

while third-order (linked to cubic nonlinear coupling terms) writes, for arbitrary (p, i, j, k):

$$\omega_p = \omega_i + \omega_j \pm \omega_k, \quad \omega_p = 2\omega_i \pm \omega_j, \quad \omega_p = 3\omega_i. \tag{38}$$

These *internal resonances* may exist, or not, depending on the eigenfrequencies of the structure. In the remainder of the lecture, a clear distinction is thus made between trivial resonance relationships and internal resonance. The main result is given for the case where *no internal resonance* exist in the system. The case of their presence is not problematic and is handled in subsection 2.5.

Let us give the general result. Considering a vibrating system of N oscillator-modes, written as a first-order dynamical system by keeping oscillator-blocks in the linear part (real formulation), and assuming no internal resonance between the eigenfrequencies $\{\omega_p\}_{p=1...N}$, a nonlinear transform can be found in order to cancel the maximum number of quadratic and cubic coupling terms present in the original system. The nonlinear transform reads, up to order three:

$$X_p = R_p + \sum_{i=1}^{N}\sum_{j\geq i}^{N}(a_{ij}^p R_i R_j + b_{ij}^p S_i S_j) + \sum_{i=1}^{N}\sum_{j\geq i}^{N}\sum_{k\geq j}^{N} r_{ijk}^p R_i R_j R_k + \sum_{i=1}^{N}\sum_{j=1}^{N}\sum_{k\geq j}^{N} u_{ijk}^p R_i S_j S_k, \tag{39a}$$

$$Y_p = S_p + \sum_{i=1}^{N}\sum_{j=1}^{N}\gamma_{ij}^p R_i S_j + \sum_{i=1}^{N}\sum_{j\geq i}^{N}\sum_{k\geq j}^{N}\mu_{ijk}^p S_i S_j S_k + \sum_{i=1}^{N}\sum_{j=1}^{N}\sum_{k\geq j}^{N}\nu_{ijk}^p S_i R_j R_k. \tag{39b}$$

In these equations, the new-defined variables, R_p and S_p, are respectively homogeneous to a displacement and a velocity. They are called the *normal* coordinates. The coefficients of this non-linear change of variables (a_{ij}^p, b_{ij}^p, r_{ijk}^p, u_{ijk}^p, γ_{ij}^p, μ_{ijk}^p, ν_{ijk}^p) are given in Appendix A, as well as in Touzé et al. (2004). the complete proof leading to Eqs. (39) is given in Touzé (2003).

The normal form is found by substituting for Eq. (39) in (2), and reads:

$$
\begin{aligned}
\forall\, p &= 1, \dots, N: \\
\dot{R}_p &= S_p, && \text{(40a)} \\
\dot{S}_p &= -\omega_p^2 R_p - (A^p_{ppp} + h^p_{ppp}) R_p^3 - B^p_{ppp} R_p S_p^2 \\
&\quad - R_p \left[\sum_{j>p}^{N} \left[(A^p_{jpj} + A^p_{pjj} + h^p_{pjj}) R_j^2 + B^p_{pjj} S_j^2 \right] \right. \\
&\quad \left. + \sum_{i<p} \left[(A^p_{iip} + A^p_{pii} + h^p_{iip}) R_i^2 + B^p_{pii} S_i^2 \right] \right] \\
&\quad - S_p \left[\sum_{j>p}^{N} B^p_{jpj} R_j S_j + \sum_{i<p} B^p_{iip} R_i S_i \right]. && \text{(40b)}
\end{aligned}
$$

In these last equations, new coeffients (A^p_{ijk}, B^p_{ijk}) appear: they arise from the cancellation of the quadratic terms, and write:

$$
A^p_{ijk} = \sum_{l \geq i}^{N} g^p_{il} a^l_{jk} + \sum_{l \leq i} g^p_{li} a^l_{jk}, \tag{41a}
$$

$$
B^p_{ijk} = \sum_{l \geq i}^{N} g^p_{il} b^l_{jk} + \sum_{l \leq i} g^p_{li} b^l_{jk}. \tag{41b}
$$

Equations (39)-(40) constitutes the main result for vibratory systems. The following comments are worth mentionable:

- Even though the last Eq. (40) appears longer on the page than the original one (2), the reader must be convinced that it is simpler. In particular, one can observe that a nonlinear term of the form R_i^2 or R_i^3 does not exist anymore in the equation for the pth normal coordinate R_p. These monoms are particularly important and are called *invariant-breaking* terms. Their cancellation will be related to nonlinear normal modes and invariant manifold in the next subsection.
- Thanks to the real formulation, oscillator-equations are obtained for the normal form. One can easily recover second-order oscillator equations by substituting for (40a) in (40b). This formalism allows for simple physical meaning of the different terms.
- Eq. (39) is identity-tangent: the linear results are thus recovered for small amplitudes.
- Velocity-dependent terms arise in Eqs. (40) only if quadratic nonlinearity is present in the initial problem given by (2). For example, for

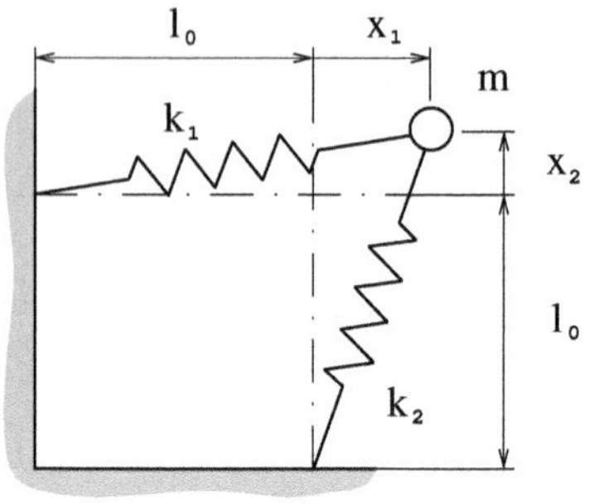

Figure 3. Schematics of the system containing quadratic and cubic non-linearities

a non-linear beam problem, for which $g^p_{ij} \equiv 0$, no velocity-dependent terms appear when considering the third-order approximation of the dynamics onto invariant manifolds.

2.5 NNMs and Normal form

Nonlinear Normal Modes (NNMs) is the core of this CISM course on Modal analysis of nonlinear systems. The main idea of defining a NNM is to try extend to notion of linear normal modes at the nonlinear stage. Linear modes offer a very interesting basis for analyzing and understanding the dynamics of linear systems. Their main property relies in the decoupling of the equations of motion, resulting in superposition theorem. This decoupling can be interpreted, on a geometrical viewpoint, in the phase space of the system, as an invariance property of the linear eigenspaces, which are two-dimensional invariant planes of the linear system.

The idea of defining a non-linear normal mode is to extend the decoupling of the linear eigenspaces exhibited at the linear stage. Letting $g^p_{ij} \equiv h^p_{ijk} \equiv 0$ in Eq. (2), and initiating a motion along the pth eigendirection results in a motion which is always contained within it. This is the *invariance* property one would be able to extend to the non-linear regime.

Let us introduce a simple two-dofs example in order to illustrate the problem. The system, composed of a mass connected to two nonlinear springs, is represented in figure 3. The dynamics of the system is described by :

$$\ddot{X}_1 + \omega_1^2 X_1 + \frac{\omega_1^2}{2}(3X_1^2 + X_2^2) + \omega_2^2 X_1 X_2 + \frac{\omega_1^2 + \omega_2^2}{2} X_1(X_1^2 + X_2^2) = 0, \tag{42a}$$

$$\ddot{X}_2 + \omega_2^2 X_2 + \frac{\omega_2^2}{2}(3X_2^2 + X_1^2) + \omega_1^2 X_1 X_2 + \frac{\omega_1^2 + \omega_2^2}{2} X_2(X_1^2 + X_2^2) = 0, \tag{42b}$$

where $X_1 = x_1/l_0$ and $X_2 = x_2/l_0$ have been nondimensionalized by the length l_0 of the springs at rest. The system is fully parameterized by the two eigenfrequencies (ω_1, ω_2). The phase space is $\mathbb{R}^4$, so that illustration of the

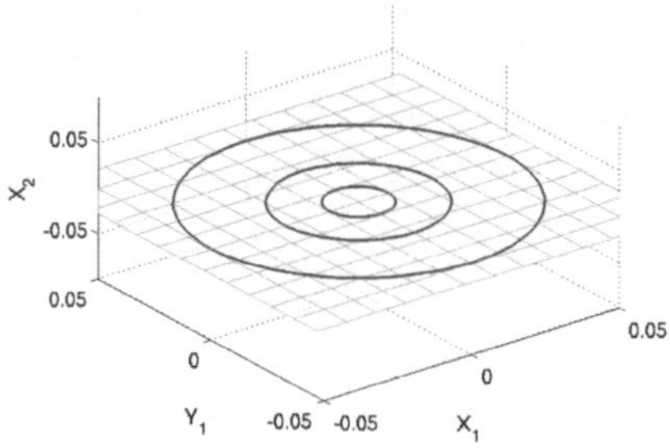

Figure 4. Trajectories (closed periodic orbits) of the *linear* system, represented in space (X_1, Y_1, X_2). Motions along the first linear mode with initial conditions $X_1(t=0) = 0.01$, 0.025 and 0.05. The linear eigenspace is represented by the horizontal plane $X_2 = 0$. The trajectories initiated along the linear mode stay within for al time t (invariance property).

complete phase space is not possible. Figure 4 shows, in a selected subspace (X_1, Y_1, X_2), with $Y_1 = \dot{X}_1$, three trajectories of the corresponding *linear* system, where quadratic and cubic terms have been cancelled. The linear modes of the system corresponds to purely vertical and purely horizontal motions of the mass. In phase space, the linear eigenspaces are the two-dimensional planes, respectively defined by $X_2 = Y_2 = 0$ for the first mode, and $X_1 = Y_1 = 0$ for the second mode. The three trajectories have been computed for three initial conditions $X_1(t=0) = 0.01$, 0.025 and 0.05 (all other coordinates equal to zero). As awaited for the linear system, the motion initiated in the first linear mode stay in this plane for all time t: this is the invariance property, that results from the decoupling of the system at the linear stage.

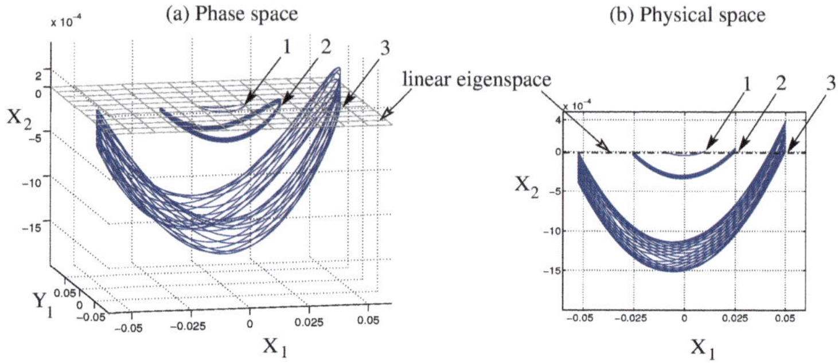

Figure 5. Trajectories of the nonlinear system, Eqs. (42). Motions initiated in the first linear mode with initial conditions $X_1(t=0) = 0.01,\ 0.025$ and 0.05 (all other coordinates to zero), illustrating the non-invariance of the linear eigenspace for the nonlinear system.

Let us consider now the complete nonlinear system. We repeat the same numerical experiment, that is to say we compute the three trajectories corresponding to an initial condition in the first linear mode with increasing amplitude $X_1(t=0) = 0.01,\ 0.025$ and 0.05. The corresponding trajectories are reported in Fig. 5. As the first one with the smallest amplitude stay in the vicinity of the linear eigenespace, it is very clear from the other two that the invariance property is not fulfilled anymore, as the trajectories are whirling around the linear eigenspace. It is worth mentionable that this behaviour only results on the presence of the invariant-breaking term X_1^2 on the second oscillator-equation (42b): as energy is fed on the first oscillator along the first linear mode, the term X_1^2 transfers a small amount of energy and excites the oscillations around the second coordinate, thus violating the invariance property.

Invariance can be recovered by selecting another set of initial conditions. Figure 6 shows the trajectories computed for (X_1, X_2)=(0.01,0) ; (0.025, 2.3*10^{-5}) and (0.05, 1.8*10^{-4}), *i.e.* the same amplitude for X_1 but with a small, selected perturbation on X_2, which allows recovering closed periodic orbits. The figure illustrates the dual definition of a NNM for conservative system, which can be viewed either as a family of periodic orbits, or as an invariant manifold, tangent at origin to the corresponding linear eigenspace.

As noted previously, the invariant-breaking terms are cancelled with the

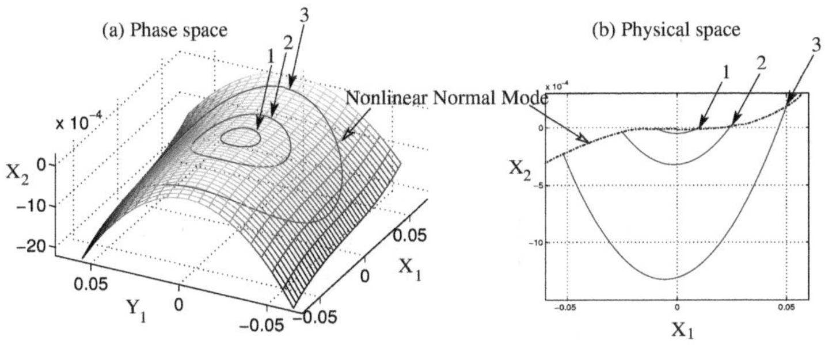

Figure 6. Trajectories (closed periodic orbits) of the nonlinear system, Eqs. (42). Motions initiated in the first nonlinear normal mode with initial conditions (X_1, X_2)=(0.01,0) ; (0.025, 2.3*10^{-5}) and (0.05, 1.8*10^{-4}). Shaded surface: invariant manifold (first NNM).

change of coordinates of the normal transformation, as long as no internal resonances are present between the eigenfrequencies. Fig. 7 presents a schematical representation of the phase space and a geometrical interpretation of the normal form transformation. As expressed with the modal coordinates, the equations of motion are written in a phase space which is spanned by the linear eigenspaces (two-dimensional planes parameterized by (X_p, Y_p)). Due to the presence of invariant-breaking terms, when the dynamics is expressed with these coordinates, linear modes are not invariant in the nonlinear regime. To recover invariance, the idea is to span the phase space with the NNMs (the invariant manifold) of the system, sketched as $\mathcal{M}_1$ and $\mathcal{M}_2$ in Fig. 7. The normal transformation is the key for expressing the new, *normal coordinates* (linked to the invariant manifolds), to the original, modal ones. The transformation is nonlinear, expressing the fact that the invariant manifolds are curved subspaces. The nonlinear transform cancels the invariant-breaking terms. The dynamics, expressed with the normal coordinates, is hence written in a curved invariant-based span of the phase space.

The invariance property is key for allowing one to operate proper truncations. Indeed, when working out with real systems emanating from the discretization of continuous structures, one is led to manipulate an a priori infinite number of modes which must be truncated for numerical treatments. The results shown here with a simple two-dofs system can easily extend to systems with numerous dofs that can be decomposed in subgroups. Due to

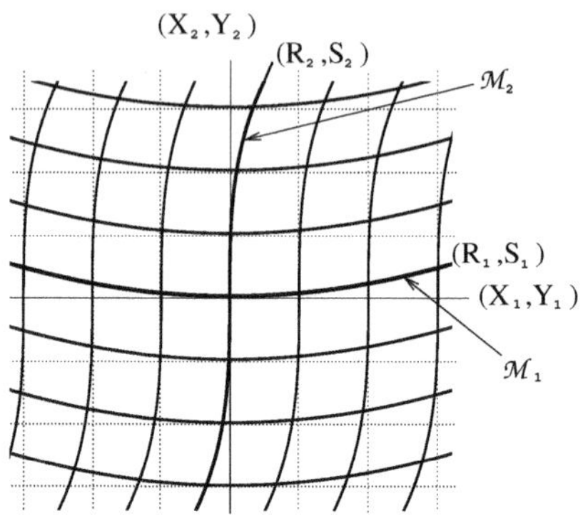

Figure 7. Schematic representation of a four-dimensional phase space for illustrating the normal transform. Two-dimensional manifolds rare represented by lines. (X_1, X_2, Y_1, Y_2): modal coordinates, cartesian grid: mesh generated by the linear eigenspace parameterization. $\mathcal{M}_1$, $\mathcal{M}_2$: invariant manifolds. (R_1, R_2, S_1, S_2): *normal* coordinates. Curved grid associated to the invariant-based span of the phase space.

the presence of invariant-breaking nonlinear terms, truncations on the dynamics expressed with the modal coordinates leads to simulate trajectories that do not exist in the complete phase space. On the other hand, once the dynamics expressed in the curved, invariant-based span, proper truncations can be realized, as the trajectories simulated with the truncated system corresponds to those of the complete system. This remark is very important and will be the key to derive accurate reduced-order models that are able *e.g.* to predict the correct type of nonlinearity for a given set of coupled oscillators.

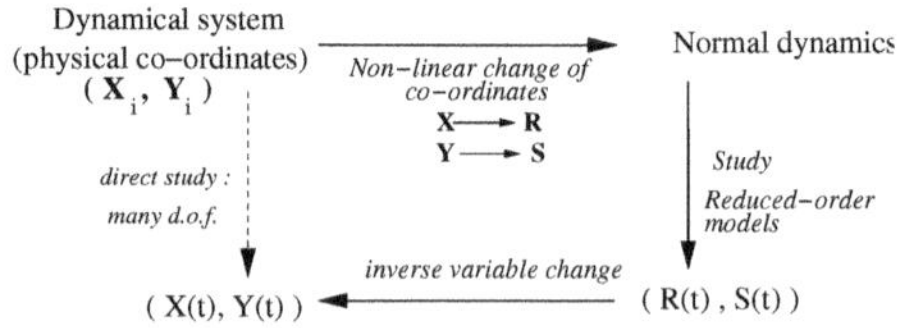

Figure 8. Illustration of the nonlinear transform for derivation of reduced-order models.

Fig. 8 illustrates the general idea for deriving accurate reduced-order models (ROMs) thanks to normal form theory. As compared to reduction methods derived by using a linear change of coordinates (*e.g.* projection onto the linear mode basis, or Proper Orthogonal Decomposition (POD) method), the idea here is to use a *nonlinear* change of coordinates. Once the system expressed in its normal form, the truncations can be realized as one is ascertained of an accurate and meaningful result thanks to the invariance property. In that respect, the method proposed here can thus be sumed up as using a more sophisiticated, nonlinear transformation, before realizing the truncations.

The next subsection considers the simplest case where the motion is reduced to a single NNM. More complicated cases involving internal resonances are handled in subsection 2.7, as well as in section 4 with applications to shells.

2.6 Single-mode motion

In order to restrict the dynamics to a single NNM, one has just to proceed as usual with the *normal* coordinates. As invariance is recovered, to study *e.g.* p^{th} NNM, all the other coordinates $k \neq p$ need to be cancelled:

$$\forall k \neq p : R_k = S_k = 0, \tag{43}$$

Susbtituting for (43) in (39) gives the geometry of the manifold in phase space (approximated at third order):

$$\forall k \neq p :$$

$$X_k = a^k_{pp} R^2_p + b^k_{pp} S^2_p + r^k_{ppp} R^3_p + u^k_{ppp} R_p S^2_p, \tag{44a}$$

$$Y_k = \gamma^k_{pp} R_p S_p + \mu^k_{ppp} S^3_p + \nu^k_{ppp} S_p R^2_p. \tag{44b}$$

This set of $2(N-1)$ equations in a phase space of dimension $2N$ defines the geometry of the invariant manifold, up to third order. Detailed comparisons with other computations led by different authors, have been realized. In particular the invariant manifold procedure proposed by Shaw and Pierre (1991) and its asymptotic development for solving out the Partial Differential Equation defining the geometry resolved in (Pesheck, 2000; Pesheck and Pierre, 1997) gives exactly the same expressions. The same comparison has also been realized with the NNM computation procedure with a multiple scale approach, as proposed by Nayfeh and Nayfeh (1994), and once again all coefficients match, evidencing the equivalence of the different methods.

The dynamics onto the pth manifold is found by substituting for (43) in

(40), and reads

$$\ddot{R}_p + \omega_p^2 R_p + (A_{ppp}^p + h_{ppp}^p) R_p^3 + B_{ppp}^p R_p \dot{R}_p^{\ 2} \ = 0 \,. \tag{45}$$

This procedure for reducing the dynamics to selected invariant subspaces, is the key for deriving out proper reduced-order models that will be able to reproduce all the qualitative and quantitative features of the complete system. This point will be key for the remainder of the presentation. Note that the normal form theory provides the most complete picture for expressing the NNMs of a system, as the nonlinear change of coordinate (39) is a complete change, from the phase space into itself. It contains the geometry of the N NNMs that are tangent at origin to the linear eigenspaces, and thus can be viewed as "extensions" of the linear modes. For recovering results obtained with other methods, *e.g.* Shaw and Pierre (1991); Nayfeh and Nayfeh (1994) from normal form theory, one has to restrict the results to particular cases. On the other hand, extending the result obtained via, *e.g.* the invariant manifold method, to more than one dof (at best N) is generally a very tedious task, see (Pesheck, 2000; Jiang, 2004).

2.7 Classification of nonlinear terms, case of internal resonance

From all the results obtained in the previous subsection, one is now able to draw out a classification of the nonlinear coupling terms appearing in the equations of motion, in order to get a clear physical understanding of their meaning on the dynamics. For that purpose, let us consider a general, two dofs system with quadratic and cubic nonlinearities:

$$\begin{aligned} &\ddot{X}_1 + \omega_1^2 X_1 + g_{11}^1 X_1^2 + g_{12}^1 X_1 X_2 + g_{22}^1 X_2^2 \\ &\qquad + h_{111}^1 X_1^3 + h_{112}^1 X_1^2 X_2 + h_{122}^1 X_1 X_2^2 + h_{222}^1 X_2^3 = 0, \end{aligned} \tag{46a}$$

$$\begin{aligned} &\ddot{X}_2 + \omega_2^2 X_2 + g_{11}^2 X_1^2 + g_{12}^2 X_1 X_2 + g_{22}^2 X_2^2 \\ &\qquad + h_{111}^2 X_1^3 + h_{112}^2 X_1^2 X_2 + h_{122}^2 X_1 X_2^2 + h_{222}^2 X_2^3 = 0, \end{aligned} \tag{46b}$$

and let us focus on the first oscillator equation. The following classification can be derived from the precedent results:

- The cubic terms X_1^3 and $X_1 X_2^2$ are *trivially resonant terms.* they can not be cancelled by the change of coordinate. Note that when no internal resonance of order two are present, *all* the quadratic terms can be cancelled by the normal transform.
- All the other terms can be cancelled by the normal transform. Among them, the terms X_2^2 and X_2^3 are *invariant-breaking terms*: they are source terms that couple equations 1 and 2, whatever a resonance relationship between the eigenfrequencies exist or not. If no internal

resonance exist, they are coined *non-resonant coupling terms*, otherwise the coupling is stronger, energy is exchanged between the modes and the terms are coined *resonant coupling terms*.

Let us consider now the case of internal resonance to explain how the precedent results must be adapted. Let us assume for illustration that a relationship $\omega_2 = 2\omega_1$ exist between the two eigenfrequencies. This is a second-order resonance relationship, which implies that quadratic terms won't be cancellable by the normal transform. To recover the resonant coupling terms that will stay in the normal form (without making the complete calculation), the following rule of thumb can be adopted. At the linear stage, the solution for X_1 and X_2 reads: $X_1 \sim \exp \pm i\omega_1 t$, $X_2 \sim \exp \pm i\omega_2 t$. The nonlinear term X_1^2 behaves as (amongst other solutions) $X_1^2 \sim \exp 2i\omega_1 t = \exp i\omega_2 t$. Hence this term can be viewed as a forcing term for the second oscillator equation, which is moreover exactly tuned at its eigenfrequency, and thus will lead to a resonance and the appearance of secular terms. This resonance is the key for explaining why a small denominator problem appear when computing the normal form, so that this term can not be removed. Following the same reasoning and considering now the monom X_1X_2, one can see that this term can produce oscillatory motions like $X_1X_2 \sim \exp i(\omega_2 - \omega_1)t = \exp i\omega_1 t$. hence this term is a forcing term acting at the resonance frequency of the first oscillator, thus it will stay in the first oscillator-equation in the normal form. As a conclusion, for that case $\omega_2 = 2\omega_1$, the normal form up to quadratic term reads:

$$\ddot{X}_1 + \omega_1^2 X_1 + g_{12}^1 X_1 X_2 = 0, \tag{47a}$$

$$\ddot{X}_2 + \omega_2^2 X_2 + g_{11}^2 X_1^2 = 0, \tag{47b}$$

At third order, the internal resonance $\omega_2 = 2\omega_1$ do not create new resonance condition, so that the cubic terms can be treated as usual. the only difficulty is to track to nonlinear coefficient of the normal form that corresponds to the monoms which have not been cancelled. These terms are easily found by looking at the formulas given in Appendix A. Then the complete normal form up to order three can easily be written.

This example shows that the treatment of internal resonance is not made too difficult with the formalism of normal form, contrary to the huge complexities involved in other methods (invariant manifold, multiple scales) to adapt their treatments to the case of internal resonance.

2.8 Damped systems

All the developments presented in the previous section have been obtained under the assumption of a conservative system. Obviously when one deals with real structures, damping is at hand and should be considered, in particular in the development of accurate reduced-order models. A simple strategy is to build the ROM without considering the damping, with the nonlinear change of coordinate presented in subsection 2.4. Then damping could be added on the normal form dynamics, *i.e.* on the reduced order model, directly. However, numerical examples have shown that this method underestimates the whole damping present in the structure (Touzé and Amabili, 2006)[3]. This underestimation is a consequence of the fact that a single NNM merges the contributions of numerous linear normal modes. When modal damping is added to each linear contribution, the decay rate of all the linear modes that are gathered to create the selected NNM are somehow added, so that the decay rate onto the manifold is not as simple as the initial decay rate postulated for the linear modal coordinates. Hence the normal form strategy must be adapted to handle the case of damped systems. The presentation now follows closely the results published in Touzé and Amabili (2006).

The starting point is now an assembly of N oscillator-equations, expressed within the modal basis, so that the linear coupling terms are diagonal. A modal damping is assumed so that the dynamics reads, $\forall p = 1...N$:

$$\ddot{X}_p+\omega_p^2 X_p+2\xi_p\omega_p\dot{X}_p+\sum_{i=1}^{N}\sum_{j\geq i}^{N} g_{ij}^p X_i X_j+\sum_{i=1}^{N}\sum_{j\geq i}^{N}\sum_{k\geq j}^{N} h_{ijk}^p X_i X_j X_k = 0. \quad (48)$$

Deriving a correct mechanical model of damping (including thermoelasticity, viscoelasticity, fluid-structure interaction...) for a large class of structure is an extremely difficult task, which also greatly depends on some specific properties of the material used. The great majority of studies on vibrations of continuous structures uses an ad-hoc viscous modal damping as the one which is here postulated. It is assumed that the modal damping introduced gives an excellent approximation of the energy losses in the considered structure, and has been finely tuned for each mode by any available method (numerical prediction or experimental fitting). Underdamped eigenmodes, corresponding to oscillatory motions, are considered, so that: $\forall p = 1...N: \ \xi_p < 1$.

Being a linear term, the modal viscous damping has an effect on the eigenvalues of the structures. For mode p, the two complex conjugated eigenvalues reads (where i is such that $i^2 = -1$):

$$\lambda_p^{\pm} \;=\; -\xi_p\omega_p \pm i\omega_p\sqrt{1-\xi_p^2} \quad (49)$$

[3]These examples will be shown in section 4.

Besides the real part of Eq. (49) which controls the decay rate of energy along the pth linear eigenspace, the imaginary part shows that the damping also have an effect on the oscillation frequency. For numerous thin metallic structures, the damping ratio ξ_p is very small for all p, so that the assumption of a lightly damped system could be considered. In that case, a first-order development of (49) shows that:

$$\lambda_p^{\pm} \approx \pm i\omega_p - \xi_p \omega_p + \mathcal{O}(\xi_p^2), \tag{50}$$

which shows that the correction on the frequency is at least a second-order effect. For computing the normal form, the general formalism can be adopted, excepting that now the eigenvalues are complex number with real and imaginary parts. At first sight, the occurrence of the real part could let us think that the trivial resonance relationships would be destroyed, as they were a simple consequence of having a purely imaginary eigenspectrum. However, the constraint that the normal form computed for the damped system is an extrapolation of the undamped system, is assumed. This implies that, by continuity, when the damping ratio ξ_p tends to zero, the new calculation must recover that obtained for the undamped system. In particular, this condition involves that small denominator of the form $1/\xi_p$ are not allowed in the calculation. Interestingly, these small denominators appears *only* for the trivially resonant terms. Hence all the calculations are led with the extra condition that when a coefficient in the normal form transform scales as $1/\xi_p$, then it must be cancelled, and the corresponding monom stay in the normal form. By doing so, only the trivially resonant terms are kept, and a continuity from undamped to damped real normal forms, is obtained.

The calculation proceeds in the same manner as in the conservative case. It is still assumed that no internal resonance are at hand[4] (the case of internal resonance being easily treated as shown in subsection 2.7). The non-linear change of co-ordinates reads:

$$\begin{aligned} X_p = R_p &+ \sum_{i=1}^{N}\sum_{j\geq i}^{N}(a_{ij}^p R_i R_j + b_{ij}^p S_i S_j) + \sum_{i=1}^{N}\sum_{j=1}^{N} c_{ij}^p R_i S_j \\ &+ \sum_{i=1}^{N}\sum_{j\geq i}^{N}\sum_{k\geq j}^{N}\left(r_{ijk}^p R_i R_j R_k + s_{ijk}^p S_i S_j S_k\right) \\ &+ \sum_{i=1}^{N}\sum_{j=1}^{N}\sum_{k\geq j}^{N}\left(t_{ijk}^p S_i R_j R_k + u_{ijk}^p R_i S_j S_k\right), \end{aligned} \tag{51a}$$

[4]Once again, for obtaining continuity with the conservative case, internal resonance relationships are defined here only for the imaginary part (frequency) of the eigenvalues.

$$Y_p = S_p + \sum_{i=1}^{N}\sum_{j\geq i}^{N}(\alpha^p_{ij}R_iR_j + \beta^p_{ij}S_iS_j) + \sum_{i=1}^{N}\sum_{j=1}^{N}\gamma^p_{ij}R_iS_j$$
$$+\sum_{i=1}^{N}\sum_{j\geq i}^{N}\sum_{k\geq j}^{N}\left(\lambda^p_{ijk}R_iR_jR_k + \mu^p_{ijk}S_iS_jS_k\right)$$
$$+\sum_{i=1}^{N}\sum_{j=1}^{N}\sum_{k\geq j}^{N}\left(\nu^p_{ijk}S_iR_jR_k + \zeta^p_{ijk}R_iS_jS_k\right) \quad (51b)$$

The normal dynamics can thus be explicitely written: $\forall p = 1...N$:

$$\dot{R}_p = S_p \quad (52a)$$
$$\dot{S}_p = -\omega_p^2 R_p - 2\xi_p\omega_p S_p - \left(h^p_{ppp} + A^p_{ppp}\right)R_p^3 - B^p_{ppp}R_pS_p^2 - C^p_{ppp}R_p^2S_p$$
$$- R_p\Bigg[\sum_{j>p}^{N}\left[(h^p_{pjj} + A^p_{pjj} + A^p_{jpj})R_j^2 + B^p_{pjj}S_j^2 + (C^p_{pjj} + C^p_{jpj})R_jS_j\right]$$
$$+\sum_{i<p}\left[(h^p_{iip} + A^p_{iip} + A^p_{pii})R_i^2 + B^p_{pii}S_i^2 + (C^p_{pii} + C^p_{ipi})R_iS_i\right]\Bigg]$$
$$- S_p\left[\sum_{j>p}^{N}(B^p_{jpj}R_jS_j + C^p_{jjp}R_j^2) + \sum_{i<p}(B^p_{iip}R_iS_i + C^p_{iip}R_i^2)\right] \quad (52b)$$

The coefficients $(A^p_{ijk},\ B^p_{ijk},\ C^p_{ijk})$ arise from the cancellation of the quadratic terms. Their expressions are:

$$A^p_{ijk} = \sum_{l\geq i}^{N} g^p_{il}a^l_{jk} + \sum_{l\leq i} g^p_{li}a^l_{jk}, \quad (53a)$$

$$B^p_{ijk} = \sum_{l\geq i}^{N} g^p_{il}b^l_{jk} + \sum_{l\leq i} g^p_{li}b^l_{jk}. \quad (53b)$$

$$C^p_{ijk} = \sum_{l\geq i}^{N} g^p_{il}c^l_{jk} + \sum_{l\leq i} g^p_{li}c^l_{jk}. \quad (53c)$$

As compared to the *conservative case*, introducing the damping in the linear operator leads to a non-linear change of co-ordinates, Eq. (39), which is now complete (in terms of the monoms (R_i, S_j)). The newly introduced coefficients: $\{c^p_{ij},\ \alpha^p_{ij},\ \beta^p_{ij},\ s^p_{ijk},\ t^p_{ijk},\ \lambda^p_{ijk},\ \zeta^p_{ijk}\}$ bring a perturbation which is at least of the order of the damping ratios $\{\xi_i\}$. More precisely, Eq. (39) may be expanded as a power series of the small perturbative terms $\{\xi_i\}$. It is then found that the coefficients that were non-zero in the conservative case (*i.e.* $\{a^p_{ij},\ b^p_{ij},\ \gamma^p_{ij},\ r^p_{ijk},\ u^p_{ijk},\ \mu^p_{ijk},\ \nu^p_{ijk}\}$) contains only even powers of the damping ratios, and the new terms, $\{c^p_{ij},\ \alpha^p_{ij},\ \beta^p_{ij},\ s^p_{ijk},\ t^p_{ijk},\ \lambda^p_{ijk},\ \zeta^p_{ijk}\}$, contains only odd powers of the damping ratios. As a consequence, the $\{A^p_{ijk},\ B^p_{ijk},\ C^p_{ijk}\}$ terms defined in Eqs. (53), can also be expanded as power series of the damping ratios. It is then found that $\{A^p_{ijk},\ B^p_{ijk}\}$ contains only even powers of the damping ratios and may be sorted according to $\mathcal{O}(\xi_i^0)$, $\mathcal{O}(\xi_i^2)$, $\mathcal{O}(\xi_i^4)$, ... So that, in the limit of a conservative systems, A^p_{ijk} and B^p_{ijk} tends to a non-zero value. On the other hand, C^p_{ijk} sorts according to odd powers terms: $\mathcal{O}(\xi_i^1)$, $\mathcal{O}(\xi_i^3)$, $\mathcal{O}(\xi_i^5)$, ... So that it is equal to zero in the conservative case. Hence this "damped" formulation allows one to see the damped normal dynamics as a perturbation of the undamped one, and could be use in great generality as the conservative results are recovered by simply cancelling all the ξ_p.

Let us comment some of the differences brought by this new formulation. To have a better picture, let us restrict ourselve to studying a single NNM, labelled p (master mode), obtained by letting $\forall k \neq p, R_k = S_k = 0$ in the previous expressions. The geometry of the pth manifold in phase space now reads:

$$\forall\, k \neq p:$$

$$X_k = a^k_{pp}R_p^2 + b^k_{pp}S_p^2 + c^k_{pp}R_pS_p + r^k_{ppp}R_p^3 + s^k_{ppp}S_p^3 + t^k_{ppp}R_p^2S_p + u^k_{ppp}R_pS_p^2, \tag{54a}$$

$$Y_k = \alpha^k_{pp}R_p^2 + \beta^k_{pp}S_p^2 + \gamma^k_{pp}R_pS_p + \lambda^k_{ppp}R_p^3 + \mu^k_{ppp}S_p^3 + \nu^k_{ppp}R_p^2S_p + \zeta^k_{ppp}R_pS_p^2. \tag{54b}$$

As compared to the conservative case, Eq. (44), new coefficients (and thus new monoms) appears. Secondly the already present coefficients, *e.g.* a^k_{pp}, b^k_{pp}, now depends on the damping values. An illustration of their dependence is shown in Fig. 9, obtained for the two-dofs system presented in subsection 2.5. The equations of motion are given by Eqs. (42), and modal damping of the form $2\xi_p\omega_p\dot{X}_p$, is added on each equation, for p=1,2. The parameters have been set to $\omega_1 = 2$, ω_2=4.5. The first NNM is selected, so that p=1 and k=2. The figure shows the variation of the quadratic coefficients of the first equation, a^2_{11}, b^2_{11} and c^2_{11}, for two different cases. In Fig. 9(a), the two damping coefficients have the same values, so that ξ_1=ξ_2=ξ, and ξ is increased from 0 to 0.4, so that a system that is more and more damped is studied. When $\xi_1=\xi_2=0$, the conservative results are recovered: c_{11}^2 is equal to zero whereas a_{11}^2, b_{11}^2 have a non-zero value. One can see that the variations may be large for some of the coefficients, so that the geometry of the manifold can be significantly changed when damping is increased. In the second case, Fig. 9(b), the damping on the first linear mode is selected at a constant, small value, $\xi_1=10^{-3}$, and ξ_2 is raised from 0 to 0.4 in order to simulate a situation where the slave mode is more and more damped, as compared to the master. Here the variations are also found to be important for large values of the damping.

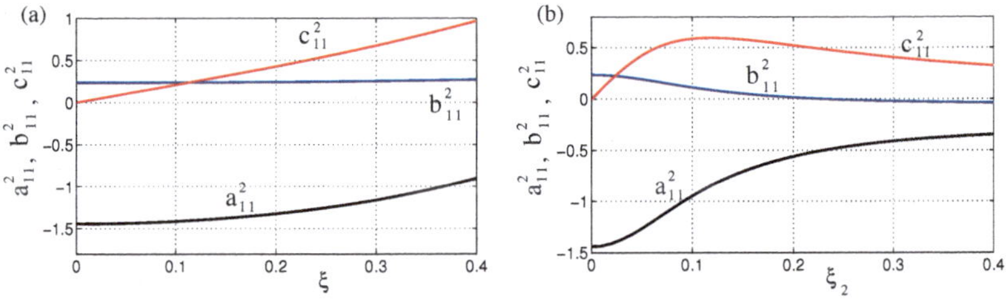

Figure 9. Variations of the coefficients a_{11}^2, b_{11}^2 and c_{11}^2, controlling the quadratic part of the geometry of the first NNM, as functions of the damping ratios, for the two-dofs system. Parameters are: $\omega_1 = 2$, $\omega_2=4.5$. (a): variations for an increasing value of the two damping ratios taken as equal: $\xi_1=\xi_2=\xi$. (b): variations for a fixed value $\xi_1=10^{-3}$ and an increasing ξ_2.

The normal dynamics for a single NNM motion when damping is included reads:

$$\ddot{R}_p+\omega_p^2 R_p+2\xi_p\omega_p\dot{R}_p+\left(h_{ppp}^p + A_{ppp}^p\right) R_p^3+B_{ppp}^p R_p\dot{R}_p^2+C_{ppp}^p R_p^2\dot{R}_p = 0. \quad (55)$$

The new coefficients appearing in this equation, as compared to Eq. (45), is C_{ppp}^p. Eq. (53c) reveals that C_{ppp}^p is constructed from all the values of the c_{pp}^k, k=1...N, that also directly influences the geometry of the manifold, Eq. (54a). The corresponding monom, $R_p^2\dot{R}_p$, can be interpreted as a nonlinear damper. Hence the whole damping for the dynamics on the pth NNM gathers a linear damping term, $2\xi_p\omega_p\dot{R}_p$, defining the energy decay rate onto the linear mode (remember that the change of coordinate is identity-tangent), and a nonlinear damping term, $C_{ppp}^p R_p^2\dot{R}_p$, allowing for a more precise definition of the global decay rate onto the manifold, taking into account all damping terms of the linear contributions that are included in the construction of the NNM. The coefficients A^p_{ppp} and B^p_{ppp}, now depends on the damping in a manner that is proportional to a^k_{pp} and b^k_{pp}, so that their variations can be inferred from Fig. 9. These coefficients are responsible for the type of nonlinearity (hardening/softening behaviour) of the NNM, so a more complete study is postponed to the next section which is entirely devoted to this problematic.

2.9 Closing remarks

The main theoretical results for deriving a normal form approach for nonlinear normal modes, has been shown in this section. The remainder of the lecture is devoted to applications of the results obtained. Some of the assumptions made in this section can be easily relaxed to extend the generality of the results. In particular, the special case of structural systems with quadratic and cubic nonlinearities, Eqs. (48), has been here assumed as it contains the great majority of applications to thin structures like beams, plates and shells, vibrating at large amplitudes. However, more general cases including for example nonlinear damping monoms of the form $\dot{X}_i X_j$ (quadratic) or $\dot{X}_i X_j X_k$, $\dot{X}_i \dot{X}_j X_k$ are of course amenable to a solution following the same steps. Another limitation could appear through the asymptotic develoments, systematically stopped at third order. Once again, this limitation can be overcome by pushing further the developments, which is not a theoretical problem, but may appear as more and more difficult, due to the complexity one has to faced, *e.g.* by pushing up to order 5. Moreover, the legitimate question is to know if the game is worth the candle. Asymptotic expansions up to order five realized for computing the NNMs with the center manifold technique by Shaw and Pierre (1993) show that the gain in accuracy brought by the fifth-order is not significative, and in some cases can be poorest than the third-order. Closer investigations on the validity limits of normal form approach by Lamarque et al. (2012), as well as thorough comparisons between asymptotics and numerical computation of NNMs by Blanc et al. (2013), appears to reach the same conclusion, that is to say that pushing further the asymptotics may lead to small improvments as compared to the computational effort needed. In the remainder of the lecture, the third-order expansion will thus always be used for applications.

3 Hardening/softening behaviour

This section is entirely devoted to the correct prediction of the type of nonlinearity (hardening/softening behaviour) for the modes of an assembly of N nonlinear oscillator equations as in Eq. (2). NNMs and normal form are used to derive a proper, easy-to-use and almost analytical method. The presentation in subsections 3.1 and 3.2 recalls the main results obtained by Touzé et al. (2004). The next subsection 3.3 with applications to shells gives the main results published by Touzé and Thomas (2006); Thomas et al. (2005), while subsection 3.4 gathers important results published by Touzé and Amabili (2006) on the influence of the damping.

3.1 Definition

In nonlinear vibrations, the oscillation frequencies of a system depend on vibration amplitude, a feature that has no counterpart in linear theory. This dependence can be of two different types. Either a *hardening* behaviour is at hand, which implies that the oscillation frequency *increases* with the amplitude, or a *softening* type nonlinearity is present, which means that the oscillation frequency *decreases* with the amplitude. This behaviour can be easily illustrated with the Duffing oscillator:

$$\ddot{X} + \omega_0^2 X + \alpha X^3 = 0. \tag{56}$$

A perturbation method (multiple scales, Poincaré-Lindstedt, Averaging, ...) can be used in order to derive the first-order relationship between the nonlinear oscillation frequency, denoted ω_{NL}, and the amplitude a of the oscillation frequency, which reads: $X(t) = a\cos(\omega_{NL} t + \phi) + \mathcal{O}(a^2)$. We obtain:

$$\omega_{NL} = \omega_0 \left(1 + \frac{3\alpha}{8\omega_0^2} a^2\right). \tag{57}$$

This equation clearly shows that a hardening behaviour is obtained when $\alpha > 0$, which also explains the origin of the vocable "hardening": when $\alpha > 0$, the nonlinear stiffness of the oscillator can be written as $\omega_0^2 X(1 + \frac{\alpha}{\omega_0^2} X^2)$, which means that the more X is large (the particle is far from the equilibrium position), the more the spring is stiff: the stiffness term has thus a "hardening behaviour". On the other hand when $\alpha < 0$, the more X is large the less the restoring force is important, and consequently the oscillation frequency decreases with vibration amplitude.

Let us consider now the case of a single nonlinear oscillator with quadratic and cubic nonlinearities. We can for example assume that the system (2) has been reduced to a single linear mode, say number p, and thus the dynamics reads:

$$\ddot{X}_p + \omega_p^2 X_p + g_{pp}^p X_p^2 + h_{ppp}^p X_p^3 = 0. \tag{58}$$

The frequency-amplitude relationship in this case writes:

$$\omega_{NL} = \omega_0 \left(1 + \tilde{\Gamma}_p a^2\right), \quad \text{with} \quad \tilde{\Gamma}_p = \frac{1}{8\omega_p^2}\left(3h_{ppp}^p - \frac{10 {g_{pp}^p}^2}{3\omega_p^2}\right). \tag{59}$$

One can see that the quadratic and cubic coefficients have opposite effects on the type of nonlinearity.

However, as it has been shown previously, linear eigenspaces are not invariant subspaces. Hence reducing a system of N linear oscillator as Eq. (2) to a single linear mode, in order to predict the type of nonlinearity of the selected mode, may lead to erroneous results. The main reason comes from the non-invariance: as the trajectories computed with Eq. (58) do not exist for the complete system, there is no reason that the type of nonlinearity match. On the other hand, if one uses the NNMs, as invariance is recovered, the type of nonlinearity can be predicted, and for the same complexity at hand. Indeed, using the formalism presented previously, reducing the dynamics to the pth NNM leads to still consider a single oscillator equation, but inside wich all the non-resonant couplings have been included. The dynamics for a single NNM is given in Eq. (45). A first-order perturbative method then leads to the following relationship:

$$\omega_{NL} = \omega_0 \left(1 + \Gamma_p a^2\right), \quad \text{with} \quad \Gamma_p = \frac{3(A^p_{ppp} + h^p_{ppp}) + \omega_p^2 B^p_{ppp}}{8\omega_p^2}. \tag{60}$$

In this equation, on can see that the influence of all the slave linear modes (used to construct the corresponding NNM) are enclosed in the A^p_{ppp} and B^p_{ppp} coefficients. Their variation is thus key to properly predict the type of nonlinearity. From Eqs. (41), and substituting the expressions of a^p_{pp} and b^p_{pp} from the nonlinear transform (see Eqs.(107) in appendix A), A^p_{ppp} and B^p_{ppp} can be explicited as:

$$A^p_{ppp} = \sum_{l=1}^{N} \frac{2\omega_p^2 - \omega_l^2}{\omega_l^2(\omega_l - 2\omega_p)(\omega_l + 2\omega_p)} (g^p_{pl} + g^p_{lp}) g^l_{pp}, \tag{61a}$$

$$B^p_{ppp} = \sum_{l=1}^{N} \frac{2}{\omega_l^2(\omega_l - 2\omega_p)(\omega_l + 2\omega_p)} (g^p_{pl} + g^p_{lp}) g^l_{pp}. \tag{61b}$$

Note that for applying these formulae, the implicit convention used throughout the notes: $g^p_{ij} = 0$ when $i > j$, must be applied.

These expressions call for the following comments:

- Considering one single mode, say p, in the original equations of motion, would lead, by substitution, to recover the type of nonlinearity given in Eq. (59). The summations, in the case of a N-dofs system, clearly shows how all the slave modes can influence the type of nonlinearity.
- The expressions in Eqs (61) shows that in case of internal resonance, a small denominator effect appears, leading to a divergence in the

expressions of a^p_{pp} and b^p_{pp}, and thus on the predicted type of non-linearity Γ_p. Even though this is in the line of all the calculations presented, a further comment is needed. Interestingly, there is only one kind of internal resonance, namely 2:1 resonance, which have an influence on the type of non-linearity. When studying the pth mode, only the lth modes, whose eigenfrequencies are such that $\omega_l = 2\omega_p$, are able to significantly change the value of T_p. Other second-order internal resonances, *e.g.* $\omega_p = 2\omega_l$, or $\omega_l + \omega_m = \omega_p$, are not able to produce a small denominator and to change the value of T_p. Finally, third-order internal resonances have no influence since only the first order correction to the backbone curve is studied (Eq. (60)).

- In case of 2:1 internal resonance, the system cannot be reduced to a single NNM. Morevover, perturbative studies shows that in the 2:1 internal resonance case, due to the presence of invariant-breaking terms, only coupled solutions exist. Thus the type of non-linearity, which is a notion associated to the backbone curve of a single oscillator, does not have anymore meaning.

The next subsection considers a two-dofs example in order to better highlight the main features of the method. Then continuous structures with an infinite number of dofs will be considered. The method of normal form for NNMs will there find a very good application, as it provides a reliable and efficient method to predict properly their type of nonlinearity.

3.2 A two dofs example

The two dofs system composed of a masse connnected to two nonlinear springs, whose equations of motions are given in Eqs. (42), is once again considered.

Considering the first linear mode leads to a dynamics governed by:

$$\ddot{X}_1 + \omega_1^2 X_1 + \frac{3\omega_1^2}{2} X_1^2 + \frac{\omega_1^2 + \omega_2^2}{2} X_1^3 = 0 \tag{62}$$

The backbone curve in this case reads:

$$\tilde{\omega}_{NL} = \omega_1 \left(1 + (-\frac{3}{4} + \frac{3\omega_2^2}{16\omega_1^2}) a^2 \right), \tag{63}$$

where $\tilde{\omega}_{NL}$ stands for the non-linear angular frequency found with a linear mode approximation, and a is the amplitude of the motion considered: $X_1 = a\cos(\tilde{\omega}_{NL} t + \beta_0) + ...$

In the parameter plane (ω_1, ω_2), regions of hardening or softening behaviour are governed by the sign of:

$$\tilde{\Gamma}_1 = -\frac{3}{4} + \frac{3\omega_2^2}{16\omega_1^2} \tag{64}$$

Considering now the first non-linear mode, which is the right approximation if one is interested in a motion non-linearly vibrating along the first physical mode, indicates that the oscillations are governed by Eq. (45), with $p = 1$. Computing the coefficients and replacing in Eq. (60) shows that the hardening or softening behaviour onto the first invariant manifold is determined by the sign of:

$$\Gamma_1 = -\frac{3}{4} + \frac{3\omega_2^2}{16\omega_1^2} + \frac{\omega_2^2(8\omega_1^2 - 3\omega_2^2)}{16\omega_1^2(\omega_2^2 - 4\omega_1^2)} \tag{65}$$

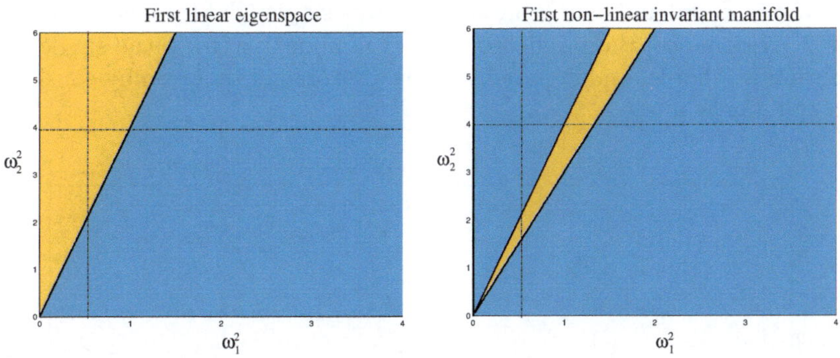

Figure 10. Regions of hardening/softening behaviour in the parameter plane (ω_1^2, ω_2^2), for the two-dofs system. Yellow: hardening behaviour, blue: softening behaviour. Left: prediction given by the sign of $\tilde{\Gamma}_1$, *i.e.* with truncation to the first linear mode. Right: prediction given by the sign of Γ_1, *i.e.* for oscillatory motions along the first NNM.

The behaviour of $\tilde{\Gamma}_1$ shows that a hardening behaviour is found when ω_2 is large as compared to ω_1, otherwise a softening behaviour is at hand. This can be easily understood by comparing the quadratic and cubic coefficients of the oscillator-equation (62). On the other hand, the prediction for hardening/softening region in parameter space with a single NNM truncation shows clearly the correction brought by considering properly the bending of the phase space caused by the presence of the second oscillator. More specifically, this effect can be drastic and can change the effective behaviour of the non-linear oscillations. Figure 10 shows the hardening and softening region in the two cases (simply given by the signs of $\tilde{\Gamma}_1$ in the linear case, and Γ_1 in the non-linear case). One can notice for example the upper-left region, which is predicted to exhibit a hardening behaviour by the linear approximation, whereas the real behaviour is soft.

In order to have a better picture of the behaviour of the type of non linearity versus the two parameters (ω_1, ω_2) of the system, Figure 11 shows two cuts in this two-dimensional parameter plane, namely for fixed ω_2=2 and ω_1 variable, then for fixed ω_1=$\sqrt{0.5}$ and ω_2 variable, *i.e.* along the two lines indicated in Fig. 10. These cuts reveals that the behaviour of $\tilde{\Gamma}_1$ is monotone, whereas Γ_1 shows a singularity, occuring for $\omega_2 = 2\omega_1$, *i.e.* at the 2:1 internal resonance between the two eigenfrequencies. As already underlined, in the vicinity of this internal resonance, the concept of the type of nonlinearity loses its meaning because the dynamics is essentially two-dimensional and cannot be reduced to a single NNM. However, far from the 2:1 resonance, the predicted type of nonlinearity from the single NNM solution is reliable, and depart from the approximation given by considering a single linear mode.

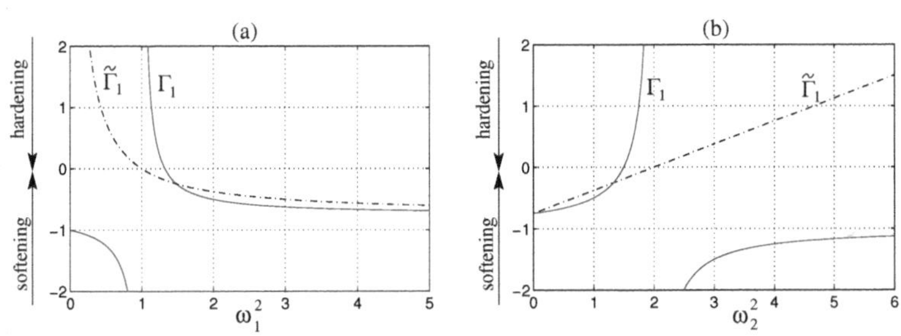

Figure 11. Type of nonlinearity for the two-dofs example, comparison between the prediction given by a single linear mode, $\tilde{\Gamma}_1$, versus the correct prediction given by a single NNM, Γ_1. (a): ω_2=2, $\omega_1^2 \in [0, 5]$, (a): ω_1=$\sqrt{0.5}$, $\omega_2^2 \in [0, 5]$

3.3 Application to shells

The method shown previously for predicting accurately the type of non-linearity, is now applied to the case of spherical-cap thin shallow shells with a varying radius of curvature R. Flat plates are known to exhibit a hardening behaviour,as it has been shown both theoretically and experimentally, see e.g. Tobias (1957); Yamaki (1961); Pandalai and Sathyamoorthy (1973); Sridhar et al. (1975); Touzé et al. (2002); Thomas et al. (2003), which means that the leading cubic coefficient h^p_{ppp} is positive. Introducing a radius of curvature R, going to infinity (perfect plate) to finite values (spherical-cap shells) introduces an asymmetry in the restoring force, due to the loss of symmetry of the neutral plane of the shell. In turn, quadratic nonlinearity appears in the equations of motions, with a magnitude proportioanl to $1/R$. The type of nonlinearity is thus awaited to vary from hardening to softening type, depending on the selected mode and the geometry. The aim of this subsection is to derive properly the prediction of this type of nonlinearity for spherical shells, with the formalism of NNMs and normal form.

The model is based on von Kármán kinematical assumptions on the strain-displacement relationship, in order to take into account moderately large vibration amplitudes. The governing equations are first recalled, then the Galerkin method is briefly reviewed in order to explain how to pass from the PDE of motion to oscillator-equations having the form of Eqs. (2), then finally the type of nonlinearity for some eigenmodes of the structure, are given. A geometrical nondimensional parameter, inversely proportional to the curvature of the shell, is introduced, in order to study how the type of nonlinearity is modified when continuously transforming a thin circular plate to a spherical-cap shell.

Von Kármán model

Von Kármán kinematical assumption relies in a clever simplification of the general 3-D strain-displacement relationship, allowing to take into account moderately large vibration amplitudes, where the coupling between in-plane and transverse motions is taken into account. The model has first been written for the static behaviour of plates (Kármán, 1910), and has then been generalized to dynamical behaviour of plates and shells (Chu and Herrmann, 1956; Efstathiades, 1971). In this subsection we follow the derivations proposed by Touzé and Thomas (2006). A spherical shell of thickness h, radius of curvature R and outer diameter $2a$, made of a homogeneous isotropic material of density ρ, Poisson ratio ν and Young's modulus E, see Fig. 12. Numerous assumptions pertaining to the derivation of von Kármán model –*e.g.* moderate rotations, in-plane inertia – are not recalled here for the sake of brevity, the interested reader is referred to Chu and Herrmann (1956); Efstathiades (1971); Touzé et al. (2002); Thomas et al. (2005); Thomas and Bilbao (2008) for a more thorough description. About the geometry of the shell, it is assumed that:

- the shell is thin: $h/a \ll 1$ and $h/R \ll 1$;
- the shell is shallow: $a/R \ll 1$;

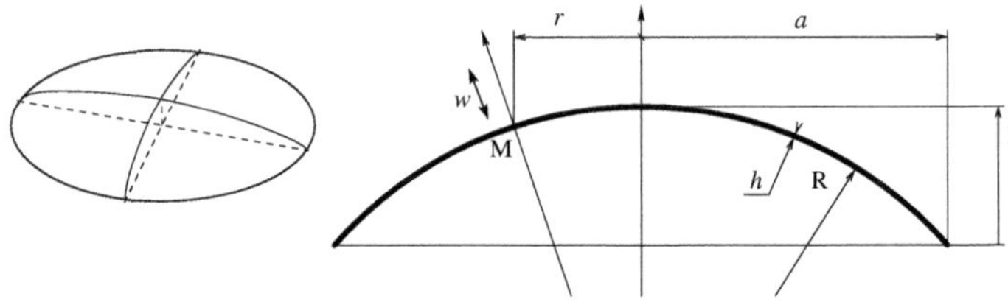

Figure 12. Geometry of the shell: three-dimensional sketch and cross section.

The equations of motion are given in terms of the transverse displacement $w(r,\theta,t)$ along the normal to the mid-surface and the Airy stress function $F(r,\theta,t)$. This is a peculiarity of the von Kármán model, which allows expression of the in-plane motions into the so-called Airy stress function F. The equations of motion reads, for all time t:

$$D\Delta\Delta w + \frac{1}{R}\Delta F + \rho h\ddot{w} = L(w,F), \tag{66a}$$

$$\Delta\Delta F - \frac{Eh}{R}\Delta w = -\frac{Eh}{2}L(w,w), \tag{66b}$$

where $D = Eh^3/12(1-\nu^2)$ is the flexural rigidity, $\ddot{w}$ is the second partial derivative of w with respect to time, Δ is the laplacian and L is a bilinear quadratic operator. With the assumption of a shallow shell fulfilled, the spatial operators are written in polar coordinates, and thus reads:

$$\Delta(\cdot) = (\cdot)_{,rr} + \frac{1}{r}(\cdot)_{,r} + \frac{1}{r^2}(\cdot)_{,\theta\theta}, \tag{67}$$

and

$$\begin{aligned} L(w,F) = w_{,rr}\left(\frac{F_{,r}}{r} + \frac{F_{,\theta\theta}}{r^2}\right) + F_{,rr}\left(\frac{w_{,r}}{r} + \frac{w_{,\theta\theta}}{r^2}\right) \\ -2\left(\frac{w_{,r\theta}}{r} - \frac{w_{,\theta}}{r^2}\right)\left(\frac{F_{,r\theta}}{r} - \frac{F_{,\theta}}{r^2}\right), \end{aligned} \tag{68}$$

Eqs (66) express the dynamics of a spherical-cap shell, without external forcing, nor damping terms. These equations are valid for plates as well, which is obtained for a radius of curvature R going to infinity. This, in turn, simplifies two terms in (66). The linear remaining terms in Eq. (66a) are the classical inertia and flexural rigidity, that one finds back in the linear Kirchhoff-Love model for vibration of plates. The linear coupling term between Eqs (66a) and (66b), proportional to $1/R$, expresses the linear coupling between transverse and in-plane motions, which exist for shells only. For plates, these terms vanish. Finally, the nonlinear coupling term reflects the nonlinear coupling between transverse and in-plane motions. One can observe that for shells, this coupling implies quadratic and cubic terms for the displacement w, whereas for plates, only cubic terms are present.

Dimensionless variables are introduced as:

$$r = a\,\bar{r}\,, \quad t = a^2\sqrt{\rho h/D}\,\bar{t}\,, \quad w = h\,\bar{w}\,, \quad F = Eh^3\,\bar{F} \tag{69}$$

Substituting the above definitions in equations of motion (66a,b) and dropping the overbars in the results, one obtains:

$$\Delta\Delta w + \varepsilon_q\,\Delta F + \ddot{w} = \varepsilon_c\,L(w,F), \tag{70a}$$

$$\Delta\Delta F - \sqrt{\kappa}\Delta w = -\frac{1}{2}L(w,w), \tag{70b}$$

where the aspect ratio κ of the shell has been introduced:

$$\kappa = \frac{a^4}{R^2\,h^2} \tag{71}$$

As it will be shown next, for a fixed value of the Poisson ratio ν, all the linear results (eigenfrequencies and mode shapes), as well as the type of non-linearity, only depend on κ, which is the only free parameter related to the geometry of the shell. The two other parameters ε_q and ε_c appearing in Eq. (70) are equal to:

$$\varepsilon_q = 12(1-\nu^2)\frac{a^2}{Rh} = 12(1-\nu^2)\sqrt{\kappa}, \quad \varepsilon_c = 12(1-\nu^2). \tag{72}$$

Their subscripts comes from the fact that they balance respectively the quadratic and the cubic terms in the non-linear ordinary differential equations governing the dynamics of the problem (see Eq. (75)).

Linear analysis

All the analysis is here performed for a free-edge boundary condition. It is derived by vanishing, at the edge $r = 1$: the membrane forces, the bending moment, the twisting moment and the transverse shear force. They are not recalled here for the sake of brevity, the interested reader can found the complete expressions in (Thomas et al., 2005; Touzé and Thomas, 2006).

The linearized equations of motion are analyzed to derive the eigenmodes and eigenfrequencies of the problem, as a function of the geometry. The eigenmodes are the solutions of:

$$\Delta\Delta\Phi + \chi\Delta\Psi - \omega^2\Phi = 0, \tag{73a}$$

$$\Delta\Delta\Psi = \Delta\Phi. \tag{73b}$$

where Φ refers to the eigenmodes of the transverse motion and Ψ to those of the membrane motion. The coefficient $\chi = 12(1-\nu^2)\kappa$ is the only parameter of the linear problem. All the study could have been realized by taking χ as the geometrical parameter, as it is sometimes done by various authors (Evensen and Ewan-Iwanowsky, 1967; Gonçalves, 1994). However, the results will be presented as functions of κ, in order to set apart the material property which appear through the Poisson ratio ν in the expression of χ. In the remainder of this study, ν is kept constant at $\nu = 0.33$.

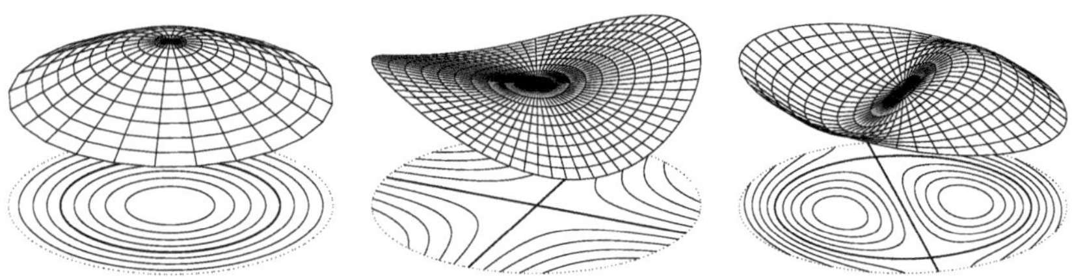

Figure 13. Three representative mode shapes of a circular spherical thin shallow shell with free edge: axisymmetric mode (0,1), purely asymmetric mode (2,0) and mixed mode (1,1).

Transverse and membrane mode shapes are numbered $\Phi_{(k,n)}$ and $\Psi_{(k,n)}$ where k is the number of nodal diameters and n the number of nodal circles. Axisymmetric modes are such that $k = 0$. For $k \geq 1$ (asymmetric modes), the associated eigenvalue has a multiplicity of two, so that for each eigenfrequency, there are two independent mode shapes, called preferential configurations or companion modes. Among these modes, *purely asymmetric modes* (such that $k \geq 2$ and $n = 0$) are distinguished from *mixed modes* (such that $k \geq 1$ and $n \geq 1$). Figure 13 shows representative pictures for each of the three distinguished family. Mode (0,1) is axisymmetric, mode (2,0) is a purely asymmetric one while (1,1) is a mixed mode.

The complete linear analysis is provided by Thomas et al. (2005). It shows that all deformed shapes, except membrane mode shapes for purely asymmetric modes, have a negligible dependence on the geometrical parameter κ. This is illustrated in Fig. 14, showing the profile for $r \in [0,\ 1]$ of two different modes: $\Phi_{(2,0)}$, the first purely asymmetric transverse mode, and $\Psi_{(2,0)}$, the first purely asymmetric in-plane mode, for large variations of the aspect ratio a/R between 0 and 0.6 (remind that the shallowness asummption implies $a/R \ll 1$). The dependence of $\Phi_{(2,0)}$ with the geometry is very slight, and this kind of behaviour is also found for all transverse modes, all in-plane modes except purely asymmetric ones.

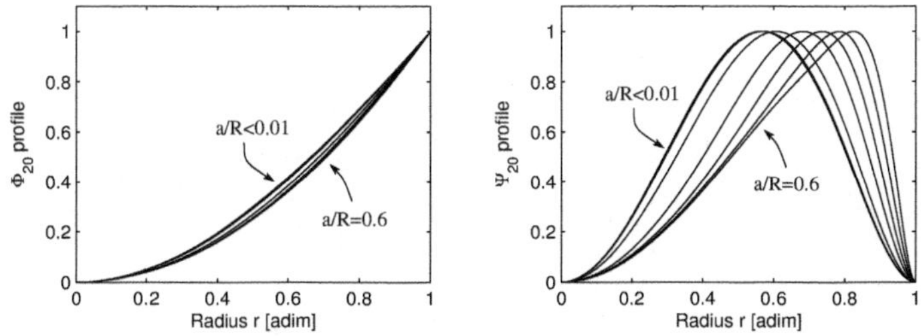

Figure 14. Profiles of theoretical asymmetric (2,0) mode shape, for several values of a/R between 0 and 0.6: (left) transverse mode and (right) membrane mode.

On the contrary, the eigenfrequencies dependence on the aspect ratio κ, represented on Fig. 15, shows a different behaviour, which leads to classify the modes into two families. The first family contains the purely asymmetric modes, since their eigenfrequencies display a slight dependence on curvature. The second family contains axisymmetric and mixed modes. They show a huge eigenfrequency dependence on curvature.

Modal expansion

The complete non-linear equations of motion (70) are projected onto the natural basis of the transverse eigenmodes. The displacement is thus written as:

$$w(r,\theta,t) = \sum_{p=1}^{+\infty} X_p(t)\,\Phi_p(r,\theta), \tag{74}$$

where the subscript p refers to a specific mode of the shell, defined by a couple $(k,\ n)$ and, if $k \neq 0$, an additional binary variable which indicates the

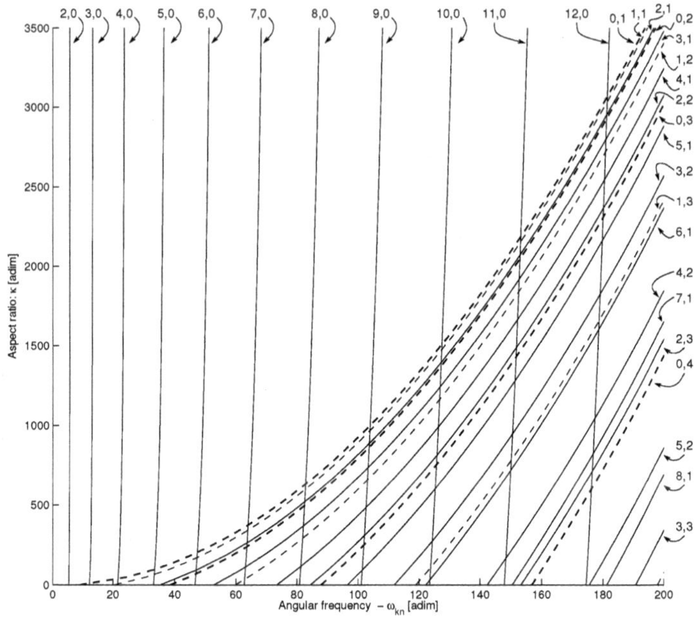

Figure 15. Dimensionless natural frequencies ω_{kn} of the shell as a function of the aspect ratio κ, for $\nu = 0.33$.

preferential configuration considered (*sine* or *cosine* companion mode). The modal displacements X_p are the unknowns, and their dynamics is governed by: $\forall p \geq 1$:

$$\ddot{X}_p + \omega_p^2 X_p + \varepsilon_q \sum_{i=1}^{+\infty} \sum_{j=1}^{+\infty} g_{ij}^p X_i X_j + \varepsilon_c \sum_{i=1}^{+\infty} \sum_{j=1}^{+\infty} \sum_{k=1}^{+\infty} h_{ijk}^p X_i X_j X_k = 0. \quad (75)$$

The expression of the non-linear coefficients are:

$$g_{ij}^p = -\iint_{\mathcal{S}_\perp} \Phi_p L(\Phi_i, \Psi_j)\, dS - \frac{1}{2} \sum_{b=1}^{+\infty} \frac{1}{\xi_b^4} \iint_{\mathcal{S}_\perp} L(\Phi_i, \Phi_j) \Upsilon_b \, dS \iint_{\mathcal{S}_\perp} \Phi_p \Delta \Upsilon_b \, dS, \quad (76)$$

$$h_{ijk}^p = \frac{1}{2} \sum_{b=1}^{+\infty} \frac{1}{\xi_b^4} \iint_{\mathcal{S}_\perp} L(\Phi_i, \Phi_j) \Upsilon_b \, dS \iint_{\mathcal{S}_\perp} \Phi_p L(\Phi_k, \Upsilon_b)\, dS. \quad (77)$$

The Υ_n, as well as its associated zero ξ_n, are defined in (Thomas et al., 2005). $\mathcal{S}_\perp$ is the domain defined by $(r, \theta) \in [0\ 1] \times [0\ 2\pi[$.

Now that the PDE has been reduced to nonlinear oscillator equations, the formalism of NNMs and normal form can be applied to derive the type of nonlinearity for each mode of the shell. The type of nonlinearity is dictated by the sign of:

$$T_p = \frac{1}{8\omega_p^2}\left[3(A^p_{ppp} + \varepsilon_c h^p_{ppp}) + \omega_p^2 B^p_{ppp}\right], \tag{78}$$

where the expressions for A^p_{ppp} and B^p_{ppp} are given as in Eq. (61):

$$A^p_{ppp} = \varepsilon_q^2 \sum_{l=1}^{+\infty} \frac{2\omega_p^2 - \omega_l^2}{\omega_l^2(\omega_l^2 - 4\omega_p^2)}(g^p_{pl} + g^p_{lp})g^l_{pp}, \tag{79}$$

$$B^p_{ppp} = \varepsilon_q^2 \sum_{l=1}^{+\infty} \frac{2}{\omega_l^2(\omega_l^2 - 4\omega_p^2)}(g^p_{pl} + g^p_{lp})g^l_{pp}. \tag{80}$$

Before showing the results, two important comments are worth mentionable:

- The nonlinear coefficients g^p_{ij} and h^p_{ijk} shows a very slight dependence on the curvature of the shell. This is the consequence of the slight dependence of the mode shapes with the aspect ratio, as the nonlinear coefficients are computed from integrals involving the mode shape functions (Eqs (76-77)). Hence the main effect of the shell's geometry on the trend of non-linearity is described by the relative variations of the eigenfrequencies, shown on Fig. 15.
- As it appears in Eqs (79-80), when studying the trend of non-linearity of the pth mode, one has to keep all the l modes such that $g^l_{pp} \neq 0$, and $g^p_{pl} \neq 0$ or $g^p_{lp} \neq 0$. As shown by Thomas et al. (2005), a number of coefficients $\{g^p_{ij}\}_{p,i,j\geq 1}$ are equal to zero due to the rotational symmetry of the structure. The conditions for these quadratic coefficients to be non-zero are expressed in terms of the number of nodal diameters k_l and k_p of the l and p modes. They read:
  (i) $g^l_{pp} \neq 0$ if $k_l \in \{2k_p, 0\}$.
  (ii) $g^p_{pl} \neq 0$ or $g^p_{lp} \neq 0$ if $k_p \in \{k_l + k_p, |k_l - k_p|\}$.
  These rules show that two classes of modes have to be retained when studying the type of non-linearity of the pth mode: axisymmetric ($k_l = 0$) as well as asymmetric modes having twice the number of nodal diameters ($k_l = 2k_p$). No other mode has an influence on the type of non-linearity.

In the remainder of the study, N will refer to the number of modes retained in this specific subset composed of the pertinent ones with respect to the type of non-linearity.

From circular plates to spherical-cap shells

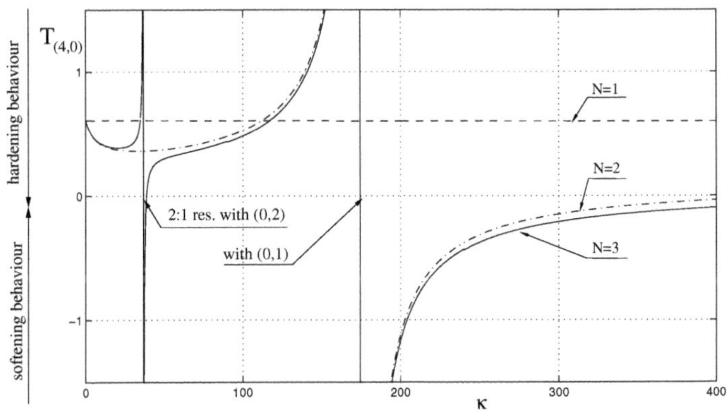

Figure 16. Type of nonlinearity for mode (4,0) as function of the geometrical parameter κ. N=1: system truncate to linear mode (4,0). N=2: truncation including (4,0) and (0,1). N=3: (4,0), (0,1) and (0,2).

The type of nonlinearity is now computed for three different modes of the shell, representing each of the three families. A purely asymmetric mode, (4,0), is first selected. The variation of $T_{(4,0)}$ as function of κ is shown in Fig. 16. According to the rules underlined for truncation, only axisymmetric modes and asymmetric modes with eight nodal diameters, must be taken into account. For κ=0, the shell is a perfect circular plate. In this case the leading cubic coefficient is positive, so that a hardening behaviour is at hand. The dashed line ($N = 1$) shows the prediction given by considering only the linear mode for computing the type of nonlinearity. The correct prediction is severely affected by the presence of 2:1 internal resonances, creating discontinuities. For mode (4,0), only two 2:1 internal resonances are possible with the modes that could interact to influence the type of nonlinearity: with mode (0,2) at $\kappa = 36.91$, and with mode (0,1) at $\kappa = 174.1$. In conclusion for this purely asymmetric mode, one can underline the fundamental importance of axisymmetric modes for an accurate prediction of the type of nonlinearity. Secondly, hardening behaviour is observed until the 2:1 resonance with mode (0,1), where softening behaviour settles down. Finally, The type of non-linearity tends to zero as κ tends to infinity.

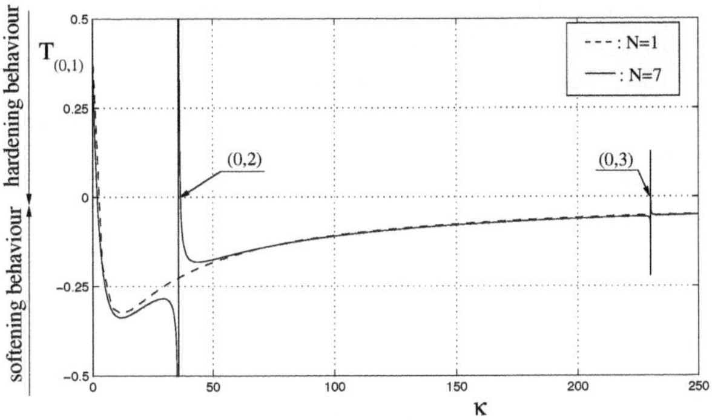

Figure 17. Type of nonlinearity for mode (0,1) as function of the geometrical parameter κ. N=1: truncation containing only mode (0,1). N=7: converged result with inclusion of the seventh first axisymmetric modes, from (0,1) to (0,7).

The case of an axisymmetric mode is now considered with mode (0,1). The coupling rules **(i)** and **(ii)**, indicate that only axisymmetric modes have to be kept. The main difference with the previous case is the behaviour of the eigenfrequencies with respect to κ. As it can be seen on Fig. 15, axisymmetric eigenfrequencies increase with curvature. Hence, an infinity of 2:1 internal resonances are now possible, with all the other axisymmetric modes.

The result of computation is shown on Fig. 17. It can be seen that the effect of the geometry –the increase of κ– is much more pronounced than for the asymmetric mode: the inital hardening behaviour ($\kappa = 0$) becomes softening at $\kappa = 1.93$, and not because of a 2:1 internal resonance. Two resonances, leading to a change in behaviour, are then shown: at $\kappa = 35.97$, where the following relationship is fulfilled: $2\omega_{(0,1)} = \omega_{(0,2)} = 43.21$. Then at $\kappa = 230.1$, where 2:1 resonance occurs with mode (0,3). These 2:1 resonances lead to a return to hardening behaviour. However, it occurs on a very little interval, which is already negligible for the resonance with (0,2), and completely negligible for (0,3). The next 2:1 resonances (with (0,4) at $\kappa = 756.9$, with (0,5) at $\kappa = 1871.5$ ...) occur on intervals which are always smaller and thus are not shown.

Single-mode prediction is also shown on Fig. 17. Although the 2:1

resonances are missed, the general behaviour is correctly predicted: change from hardening to softening due to curvature is found at $\kappa = 1.95$ instead of $\kappa = 1.93$, and the asymptotic behaviour, which becomes neutral when κ tends to infinity, is recovered. These results show that for the specific case of the fundamental axisymmetric mode, the single-mode approximation, used in the precedent studies (Grossman et al., 1969; Yasuda and Kushida, 1984; Sathyamoorthy, 1994; Varadan and Pandalai, 1978), predicts the essential features, in spite of a too severe truncation.

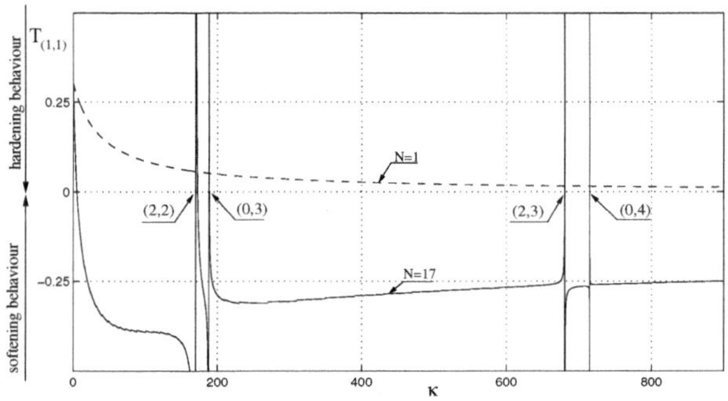

Figure 18. Type of nonlinearity for mode (1,1) as function of the geometrical parameter κ. N=1: truncation including only linear mode (1,1). N=17: converged result including modes (1,1), (2,0) to (2,4) (with both configurations), (0,1) to (0,5).

The case of a mixed mode, namely (1,1) is presented on Fig. 18. As for the axisymmetric modes, the effect of geometry is important and leads to a change of behaviour for a very small value of the aspect ratio: $\kappa = 5.3$. Then 2:1 internal resonances occurs, with modes (2,2), (0,3), (2,3), (0,4), ... Their number is unlimited, as for the axisymmetric case. The change of behaviour occurs on very small intervals. The first one, due to 2:1 resonance with mode (2,2), is hardly negligible, and the others have no chance to be experimentally measurable. It can be thus conclude that except on a very small interval ($\kappa \in [0\,,\ 5.3]$), mode (1,1) behaves in a softening way.

The single-mode approximation is also shown on Fig. 18. It predicts a hardening behaviour which becomes neutral when κ tends to infinity. The converged result is obtained for $N = 17$ modes, namely : (1,1); (2,0) to (2,4); (0,1) to (0,5), and shows that, contrary to the precedent cases, coefficient $T_{(1,1)}$ tends to a finite value when κ tends to infinity. Hence the behaviour remains softening and does not becomes neutral for large values of the aspect ratio.

3.4 Influence of the damping

This subsection is devoted to studying the influence of the damping on the type of nonlinearity. This problematic could appear unusual or ill-posed, because the type of nonlinearity is generally defined for conservative systems through the backbone curve. However, a more general view is needed to tackle the cases of forced responses and/or dissipative free oscillations, where the oscillation frequency also depend on vibration amplitude.

As it has been shown in subsection 2.8, the normal form theory with inclusion of damping allows extending the results continuously from conservative to dissipative systems. Moreover, a clear dependence of the coefficients on the damping ratios, has been underlined. The aim of this subsection is thus to show how this dependence can influence the type of nonlinearity.

For that purpose, let us consider the two-dofs system, Eqs. (42), with inclusion of damping terms. The motions on the first NNM are governed by the dynamical reduced equation:

$$\ddot{R}_1+\omega_1^2 R_1+2\xi_1\omega_1\dot{R}_1+\left(h_{111}^1+A_{111}^1\right)R_1^3+B_{111}^1 R_1\dot{R}_1^2+C_{111}^1 R_1^2\dot{R}_1=0. \quad (81)$$

and the type of nonlinearity is given by the sign of Γ_1:

$$\Gamma_1=\frac{3(A_{111}^1+h_{111}^1)+\omega_1^2 B_{111}^1}{8\omega_1^2} \quad (82)$$

As a consequence of the particular behaviour of the A_{111}^1 and B_{111}^1 with increasing values of ξ_2, the type of non-linearity may change with increasing damping. Figure 19 shows, for the two-dofs example with parameter values $\omega_1=3$, $\omega_2=5.4$, and $\xi_1=0.001$, that when ξ_2 increases (simulating the presence of a slave mode which is more and more damped) the type of nonlinearity of the first mode may be affected and change from hardening to softening behaviour. In this case, it happens for $\xi_2=0.081$, so that the ratio of the two modal damping is equal to: $\xi_2/\xi_1=81$.

Another case is studied in Fig. 20, where now the two linear modal damping coefficients ξ_1 and ξ_2 vary of the same quantity, so that the ratio ξ_2/ξ_1 is kept constant. This more realistic case could for example simulate a structure whose global damping is raised. Here again, it is also observed that a global increase of the amount of damping have a significant effect on the type of non-linearity. From these two examples, it is concluded that the damping tends to enhance and favours the softening behaviour.

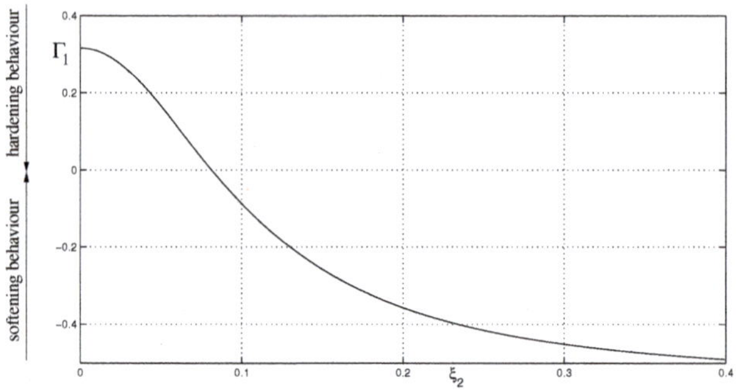

Figure 19. Type of non-linearity Γ_1, defined by Eq. (82), for increasing values of ξ_2. The behaviour turns from hardening to softening type for $\xi_2 = 0.081$. Other selected values are: $\omega_1 = 3$, $\omega_2 = 5.4$, and $\xi_1 = 0.001$.

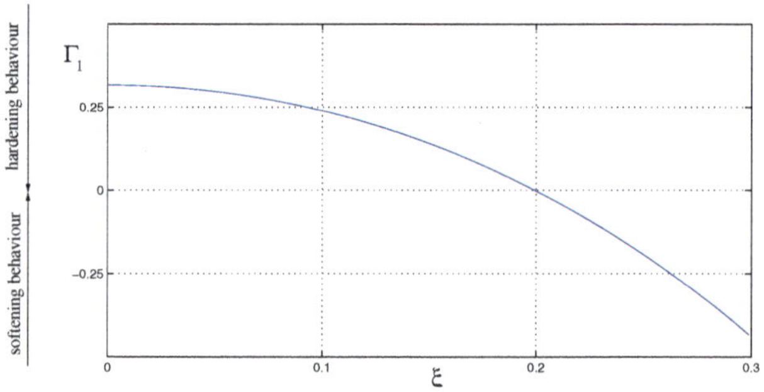

Figure 20. Type of non-linearity Γ_1 for increasing values of the global damping in the system. The two modal damping values here are equal : $\xi_1 = \xi_2 = \xi$. Other selected values are: $\omega_1 = 3$, $\omega_2 = 5.4$.

This particular effect of the damping on the type of non-linearity can thus significantly change predictions based on the undamped system. Let us now observe how the type of nonlinearity is modified on a section of the map of nonlinearity as function of the two parameters (ω_1, ω_2), shown in Fig. 10. More precisely, the line ω_2=2 and varying ω_1, already shown in Fig. 11(a), is now considered, for a fixed value of ξ_1=0.001, and increasing values of ξ_2. The results are given in Fig. 21.

When damping is not considered, the type of non-linearity Γ_1 displays a discontinuity at the internal resonance value where $\omega_1 = 1$, ie. where 2:1 resonance occurs : $\omega_2 = 2\omega_1$. At this discontinuity point, the behaviour changes abruptly from softening to hardening type. The discontinuity is due to the presence of internal resonance which leads to small denominators in the solution. As already argued, in a small intervall near the 2:1 internal resonance point, single-mode solutions do not exist anymore, and the concept of the type of non-linearity loses its meaning.

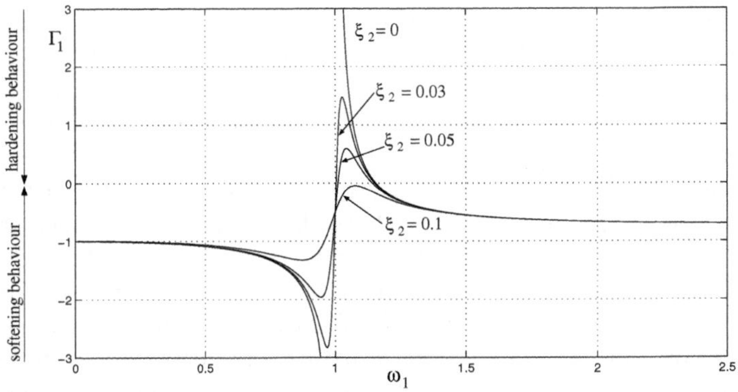

Figure 21. Type of non-linearity for different values of ξ_2, illustrating the fact that the discontinuity at the 2:1 internal resonance is smoothened by increasing the damping of the slave oscillator. $\omega_2 = 2$ and $\xi_1 = 0.001$.

Here, it is shown that taking into account the whole damping of the structure smoothens the discontinuity. For increasing values of ξ_2, Figure 21 shows that the region of hardening behaviour after the 2:1 internal resonance decreases, and can even disappear, which happens here for $\xi_2 = 0.1$. This is a reflection of the fact that increasing ξ_2 changes the second eigenvalue $\lambda_2^{\pm} = -\xi_2\omega_2 \pm i\omega_2\sqrt{1-\xi_2^2}$, so that the eigenvalues of the damped system stem apart from the resonance condition so that the discontinuity due to the small denomitar smoothens.

From this study, it can be concluded that a careful prediction of the type of non-linearity must include the damping in the analysis. Examples on continuous structures (imperfect plates dans spherical shells), are shown by Touzé et al. (2008b), underlining that the results obtained here with the two-dofs system generalize.

3.5 Closing remarks

A general strategy has been proposed for deriving the type of nonlinearity of an assembly of nonlinear oscillator equations, based on normal form theory and reduction to a single NNM. With this method, the same complexity is at hand as compared to predictions given by using a single linear mode, as a single oscillator equation is used for the prediction, so that analytical results are derived without resorting to time-consuming numerical simulations including the N oscillator equations. As shown by several authors, see *e.g.* Nayfeh et al. (1992); Nayfeh (1995); Nayfeh and Lacarbonara (1997); Touzé et al. (2004), using a single LNM can lead to erroneous results in the prediction. The reason is the non invariance of the linear eigenspace, as shown in the previous section. The effect of the other modes on the type of nonlinearity, has been underlined, in particular the presence of 2:1 internal resonance.

This section has revealed that using NNMs for reduced-order modeling appears as an appealing method. Indeed, one is at least ascertained to predict the correct type of nonlinearity with the reduced model. For the last section of this lecture, the focus is on the derivation of ROMs for structural systems in forced vibration. The case of harmonic forcing, in the vicinity of one eigenfrequency, is studied. This will allows to demonstrate the ability of ROMs based on NNMs to properly recover a complete bifurcation diagram.

4 Reduced-order models for resonantly forced response

The goal of this section is to use the reduced-order modelling strategy based on NNMs and normal form theory, in order to compute the harmonically forced response of thin structures, vibrating at large amplitudes and excited in the vicinity of one of its eigenfrequency. Applications to shells will be specifically shown, and a comparison with the Proper Orthogonal Decomposition (POD) method will highlight the benefit of using NNMs in this case. In the first subsection, the derivation of the ROM is briefly reviewed and the advantage of using the normal form method including the damping is shown on a two-dofs example. The presentation in subsection 4.1 recalls some of the results published by Touzé and Amabili (2006). Subsection 4.2 selects one of the examples shown in (Touzé et al., 2008a) for illustration. Subsection 4.3 with application to a closed circular cylindrical shell is taken from Touzé and Amabili (2006), while the comparison with the POD method in the last subsection is published in (Amabili and Touzé, 2007).

4.1 Derivation of the reduced-order model

The previous developments have clearly highlighted the general method for deriving the ROM, shown schematically in Fig. 8. From a Partial Differential Equation, the first step is a projection onto the basis of the linear eigenmodes. The second step is to use the normal transform in order to express the dynamical equations in an invariant-based span of the phase space. The last step consists in truncating the system by keeping the resonnant NNMs. All the examples used consider the case of a harmonic external forcing, the frequency of which is in the vicinity of an eigenfrequency of the selected structure. In the simplest case, when no internal resonance is present among the eigenfrequencies of the shell, only the NNM whose eigenfrequency is near the excitation, has to be kept in the truncation, so that a very simple ROM, consisting of a single oscillator-equation, is derived. This case will be illustrated in subsection 4.2 with a doubly-curved panel. If internal resonance exists, then all the NNMs, the frequencies of which are contained in the internal resonance, must be kept in the truncation. This case will be illustrated in subsection 4.3, where the ROM of an asymmetric mode of a closed circular cylindrical shell, is considered. Due to the rotational symmetry of the circular shell, asymmetric modes appears by pairs of companion modes, in the same manner as what has been observed in circular plates and shells in subsection 3.3. Hence a 1:1 internal resonance is present and two NNMs must be retained in the truncation.

Two main assumptions are thus retained in the derivation of the ROM:

- A third-order asymptotic expansion is used for the normal transform. Hence all the results are a priori accurate up to order three. In particular, the proposed method shows the best results for systems containing quadratic and cubic nonlinearities, as the correction brought by the quadratic term on the cubic nonlinearities is well taken into account. For systems containing only cubic nonlinearity, the improvement as compared to linear mode truncation, is important for the transformation only, but not for the dynamical solutions. When considering large amplitude vibrations, this third-order limitation can become problematic in terms of accuracy of solutions.
- Application of the proposed ROMs to real situations leads to consider external forces applied to the structure. On the mathematical viewpoint, external forces must be taken into account in the normal form computation, as proposed for example by Elphick et al. (1987). However, it overshoots the mark of the present study, since the formulation must turn to definitions of *time-dependent* invariant manifolds. In the mechanical context, time-dependent manifolds have been computed *e.g.* by Jiang et al. (2005), requiring a huge computational effort,

since the numerical procedure must be repeated for each forcing frequency. Moreover, a consequence of the numerical procedure is that the results are no more expressed under a differential formulation, which renders parametric studies numerically expensive.

In this study, the ROM will be obtained by adding the external force directly to the normal form. The main advantage is that the calculation derived in section 2 is intrinsical to the structure, whereas rigorous computations including the external force must be done for each type of forcing studied. Secondly, the perturbation brought by the external force onto the normal form is at least a second-order effect (Iooss and Adelmeyer, 1998). Hence, this first approximation will be used to derive simple ROMs, and the results presented in the next sections shows that qualitative and quantitative results are generally obtained.

Before applicating the method to thin shells, a justification for using the normal form with damping is in order. Indeed, the same reasoning made for the external forcing could also have justified a more simple treatment of the damping terms. The ROMS could have been built for the system without damping, using the formulas exposed in subsection 2.4, and then damping and external forces could have been added directly to the normal form. This method will be coined the "conservative NNM" method in the following, and will be compared to the second strategy, termed "damped NNM" formalism, where the damping is included in the normal transform, as shown in subsection 2.8. For that purpose, the two-dofs example consisting of the mass connected to the two nonlinear springs, is used for illustration. The equations of motion, given in Eqs. (42), are completed by adding damping and forcing terms, and reads:

$$\ddot{X}_1 + \omega_1^2 X_1 + 2\xi_1\omega_1\dot{X}_1 + \frac{\omega_1^2}{2}(3X_1^2 + X_2^2) + \omega_2^2 X_1 X_2 + \frac{\omega_1^2 + \omega_2^2}{2} X_1(X_1^2 + X_2^2) = F_1\cos(\Omega t), \tag{83a}$$

$$\ddot{X}_2 + \omega_2^2 X_2 + 2\xi_2\omega_2\dot{X}_2 + \frac{\omega_2^2}{2}(3X_2^2 + X_1^2) + \omega_1^2 X_1 X_2 + \frac{\omega_1^2 + \omega_2^2}{2} X_2(X_1^2 + X_2^2) = 0. \tag{83b}$$

The forcing is considered on the first mode only, and the excitation frequency is such that $\Omega \approx \omega_1$.

Three different truncations, having the same complexity (a single nonlinear oscillator equation), are used as reduced-order models. The first one is the most simple, and consists in keeping only the linear mode by imposing

$X_2 = 0$:

$$\ddot{X}_1 + \omega_1^2 X_1 + 2\xi_1\omega_1\dot{X}_1 + \frac{3\omega_1^2}{2}X_1^2 + \frac{\omega_1^2+\omega_2^2}{2}X_1^3 = F_1\cos(\Omega t) \tag{84}$$

The second ROM, following the "conservative NNM" method, is found by subsituting in Eqs. (45) and (61) the coefficients by their expressions. It reads:

$$\ddot{R}_1 + \omega_1^2 R_1 + 2\xi_1\omega_1\dot{R}_1 + \left[\frac{\omega_2^2 - 2\omega_1^2}{2} + \frac{(2\omega_1^2-\omega_2^2)}{2(\omega_2^2-4\omega_1^2)}\omega_2^2\right]R_1^3 + \left[-3 + \frac{\omega_2^2}{\omega_2^2-4\omega_1^2}\right]\dot{R}_1^2 R_1 = F_1\cos(\Omega t). \tag{85}$$

Finally, the third ROM is found by applying the "damped NNM" strategy. It is given by Eq. (81), which is here recalled for p=1:

$$\ddot{R}_1 + \omega_p^2 R_1 + 2\xi_1\omega_1\dot{R}_1 + \left(h_{111}^1 + A_{111}^1\right)R_1^3 + B_{111}^1 R_1\dot{R}_1^2 + C_{111}^1 R_1^2\dot{R}_1 = F_1\cos(\Omega t), \tag{86}$$

where A_{111}^1, B_{111}^1 and C_{111}^1 have not been replaced by their complete expressions for that particular problem as their expressions are now too lengthy. Note however that, as compared to (85), these coefficients now depends explicitely on the damping. In Eq. (85) the only term producing an energy loss is $2\xi_1\omega_1\dot{R}_1$, whereas in (86), the additional term $C_{111}^1 R_1^2\dot{R}_1$ is also dissipative, and depends on ξ_1 and ξ_2, so that a better approximation of the whole damping in the system is awaited for the second ROM based on "damped NNMs".

The performance of the three ROMs are compared on Fig. 22, where the parameter values of the system have been set to ω_1=2, ω_2=4.5, ξ_1=0.001, ξ_2=0.01, so as to simulate the presence of a damped mode and its influence on the frequency-response curve of the first mode. The forcing is such that $\Omega \sim \omega_1$, and the amplitude is set as F_1=5.10^{-4} for that first simulation. The solutions of each model is obtained by continuation of periodic orbits, and the software AUTO is used for that purpose. The reference solution is of course obtained with the complete system, Eqs. (83). One can see that the linear mode truncation predicts an incorrect hardening behaviour, in the line of the results already obtained in the precedent section, and is thus not reliable. On the other hand, the two ROMS based on NMMs predict the correct type of nonlinearity. The ROM constructed with the "conservative NNM" method underestimates the damping in the system, and gives a maximum amplitude which is slightly larger that the reference solution. This means that taking into account only $2\xi_1\omega_1\dot{R}_1$ as dissipative

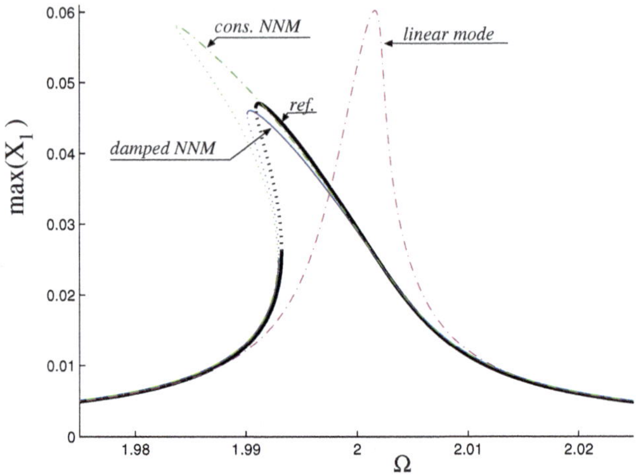

Figure 22. Frequency-response curves (maximum of the first coordinate X_1 versus excitation frequency Ω). Thick solid line : reference solution. Thin blue solid line :"damped NNM", Eq. (86). Dash-dotted green line : "conservative NNM", Eq. (85). Magenta dash-dotted thin line: linear mode truncation, Eq. (84). Selected values: ω_1=2, ω_2=4.5, ξ_1=0.001, ξ_2=0.01, F_1=5.10^{-4}.

term in the ROM is an incorrect estimation of the whole damping present in the system. The ROM constructed with the "damped NNM" method gives a very satisfactory result. The slight differences observed with the reference solution can be attributed to the two main assumptions made in the construction of the ROM, *i.e.* the third-order asymptotic development and the time-independent approximation of the manifold.

These trends are enhanced when the amplitude of the forcing is increased to F_1=1.10^{-3}, as shown in Fig. 23. In this case, the two ROMs: linear and "conservative NNM" give unaceptable result. The ROM constructed with the "damped NNM" method shows a correct behaviour, even though the slight discrepancies already observed for F_1=5.10^{-4} are now clearly observable.

In the remainder of this section, the reduced-order modeling strategy is applied to thin shells including a priori an infinite number of dofs. Truncations to one or two NNMs are realized. As numerous modes, with increasing values of their modal damping factor, will be gathered in a single NNM, the "damped NNM" strategy is systematically used. The examples

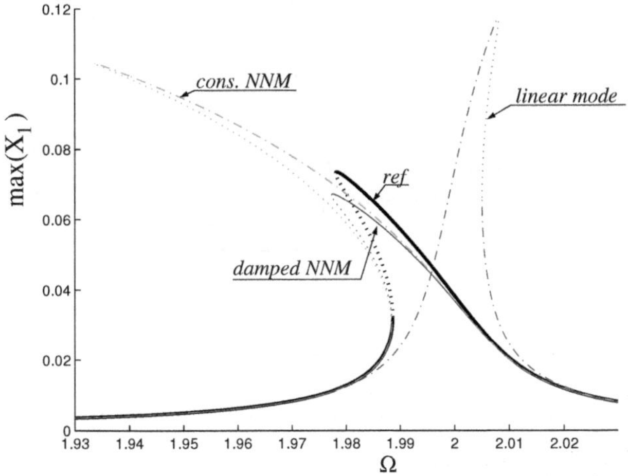

Figure 23. Frequency-response curves (maximum of the first coordinate X_1 versus excitation frequency Ω). Same as Fig. 22, except the amplitude of the forcing which have been increased: F_1=1.10^{-3}.

shown herein underline the importance of having a correct estimation of the damping onto the manifold.

4.2 Application : the case of a doubly-curved panel with in-plane inertia

A hyperbolic paraboloid panel with rectangular base, referred to as the HP panel in the following, is first selected. Frequency response curves, in the vicinity of its fundamental mode, will be computed for different ROMs. A sketch of the shell is shown in Fig. 24. The curvilinear coordinate system is denoted by (O, x_1, x_2, z), with the origin O at one edge of the panel. R_1 and R_2 (assumed to be independent of x_1 and x_2) are the principal radii of curvature, a and b are curvilinear length, and h is the thickness. The radii of curvature are such that $R_1 = -R_2$. For the numerical results, dimensions of the panel are selected as: $a = b = 0.1$ m, $R_x = -R_y = 1$ m, and thickness $h = 1$ mm. The material is linear elastic with Young's modulus $E = 206.10^9$ Pa, density $\rho = 7800$ kg.m^{-3} and Poisson's ratio $\nu = 0.3$.

Donnell's non-linear shell theory is used for expressing the kinematics of the shell. This modelisation is of larger extent than the von Kármán model presented in subsection 3.3. Indeed, Donnell non-linear shell theory do not

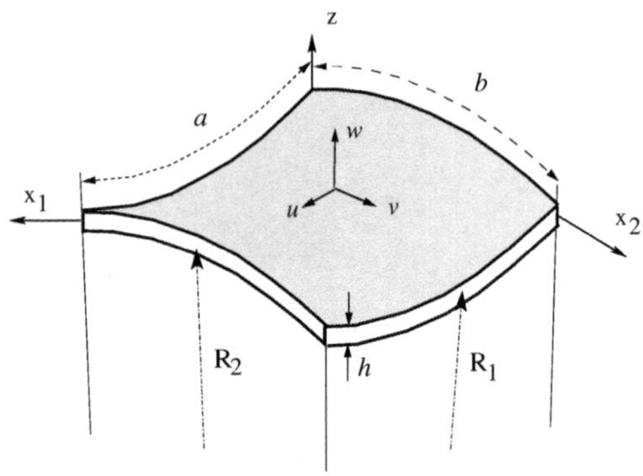

Figure 24. Geometry and coordinate systems for the selected shells.

neglect the in-plane inertia, and can thus be used with confidence for non-shallow shell and on a wider range of displacement amplitudes (and/or a larger range of midfrequencies). The membrane displacements are denoted by u and v, and the normal displacement is w. Boundary conditions are simply supported.

The drawback of retaining in-plane inertia is that additional degrees of freedom must be taken into the expansion as a consequence that the simplified Donnell's shallow-shell formulation cannot be used. Secondly, the computation of the eigenmodes can become more difficult. For these reasons, ad-hoc expansion functions are here used for discretizing the problem. The basis functions used here are :

$$\phi^{(u)}_{m,n}(x_1,x_2) = \cos(m\pi x_1/a)\sin(n\pi x_2/b), \tag{87a}$$
$$\phi^{(v)}_{m,n}(x_1,x_2) = \sin(m\pi x_1/a)\cos(n\pi x_2/b), \tag{87b}$$
$$\phi^{(w)}_{m,n}(x_1,x_2) = \sin(m\pi x_1/a)\sin(n\pi x_2/b). \tag{87c}$$

The three displacements are denoted as: w for transverse motions, u and v

for in-plane motions. They are expanded as:

$$u(x_1, x_2, t) = \sum_{m,n=1}^{M_u, N_u} u_{mn}(t)\phi_{mn}^{(u)}(x_1, x_2), \tag{88a}$$

$$v(x_1, x_2, t) = \sum_{m,n=1}^{M_v, N_v} v_{mn}(t)\phi_{mn}^{(v)}(x_1, x_2), \tag{88b}$$

$$w(x_1, x_2, t) = \sum_{m,n=1}^{M_w, N_w} w_{mn}(t)\phi_{mn}^{(w)}(x_1, x_2). \tag{88c}$$

The number of basis functions in each direction is free and governed by the integers M_u, N_u, M_v, N_v, M_w, and N_w.

Let $\mathbf{q}$ be the vector of generalized coordinates, gathering together all the unknown functions of time introduced by the expansions given in Eqs. (88):

$$\mathbf{q} = [u_{m,n},\ v_{m,n},\ w_{m,n}]^T, \quad m = 1, ...M_u,\ M_v,\ M_w,\ n = 1, ...N_u,\ N_v,\ N_w. \tag{89}$$

In the remainder, P refers to the dimension of $\mathbf{q}$, *i.e.* the number of generalized coordinates used for discretizing the shell. The generic element of $\mathbf{q}$ is denoted by q_p. Finally, in the three considered cases, the result of the discretization gives a set of coupled non-linear oscillator equations to solve. They writes:

$$\ddot{q}_p + 2\zeta_p\omega_p\dot{q}_p + \sum_{i=1}^{P} z_i^p q_i + \sum_{i,j=1}^{P} z_{i,j}^p q_i q_j + \sum_{i,j,k=1}^{P} z_{i,j,k}^p q_i q_j q_k = f_p \cos(\omega t). \tag{90}$$

Modal damping in Eq. (90) is considered in the classical form $2\zeta_p\omega_p\dot{q}_p$, and $\mathbf{f} = [f_1\ ...\ f_P]^T$ is the vector of the projected external forcing considered.

The reference solution is obtained by simulating Eqs. (90). For deriving the bifurcation diagram for the panel (or the frequency response curve in forced vibration), a numerical solution is obtained by using a pseudo-arclength continuation method implemented in the software AUTO (Doedel et al., 2002). The convergence of the solutions with respect to the number P of generalized coordinates retained has already been done in previous studies. It has been shown that, for the HP panel, $P = 22$ basis functions were needed for obtaining convergence (Amabili, 2005). As a consequence of this large value, computation time associated with the numerical simulations with AUTO for obtaining a single frequency-response curve are large.

The first idea for reducing the size of the system is to use the linear normal modes (LNMs). Let $\mathbf{L} = [z_i^p]_{p,i}$ be the linear part of Eq. (90), and $\mathbf{P}$ the matrix of eigenvectors (numerically computed) of $\mathbf{L}$ such that: $\mathbf{P}^{-1}\mathbf{L}\mathbf{P} = \mathbf{\Lambda}$, with $\mathbf{\Lambda} = \text{diag}\left[\omega_p^2\right]$, and ω_p the eigenfrequencies of the structure. A linear change of coordinates is computed, $\mathbf{q} = \mathbf{P}\mathbf{X}$, where $\mathbf{X} = [X_1 \ldots X_P]^T$ is, by definition, the vector of modal coordinates. Application of $\mathbf{P}$ makes the linear part diagonal, so that the dynamics can now be expressed in the eigenmodes basis, and reads, $\forall p = 1, \ldots, P$:

$$\ddot{X}_p + 2\zeta_p\omega_p\dot{X}_p + \omega_p^2 X_p + \sum_{i,j=1}^{P} g_{ij}^p X_i X_j + \sum_{i,j,k=1}^{P} h_{ijk}^p X_i X_j X_k = F_p \cos(\omega t). \tag{91}$$

The application of $\mathbf{P}$ let the viscous damping unchanged, and $\mathbf{F} = \mathbf{P}^{-1}\mathbf{f} = [F_1 \ldots F_P]^T$ is the new vector of modal forces. The quadratic and cubic non-linear coupling coefficients $\{g_{ij}^p\}$ and $\{h_{ijk}^p\}$ are computed from the $\{z_{i,j}^p\}$ and $\{z_{i,j,k}^p\}$ appearing in Eq (90) with matrix operations involving $\mathbf{P}$. The dimension of $\mathbf{X}$ is P, but truncation can now be realized by keeping any number of LNMs. Let P_{LNM} be the dimension of the truncation operated in $\mathbf{X}$. Convergence studies will be realized by increasing P_{LNM} from 1 to P. Since the LNMs possesses some interesting properties (in particular orthogonality), it is awaited to obtain convergence for $P_{LNM} \leq P$.

Finally, from Eqs. (91), the NNM method with normal form is applied to obtain a more severely reduced order model. For selecting the number of NNMs for building the ROMs, one needs to know the eigenfrequencies in order to test the presence of internal resonance before proceeding. With dimensions and material properties as chosen, the fundamental frequency is 488.1 Hz. Nondimensionalizing the frequencies by the first one, it results that the list of frequencies are, for the six first: 1, 5.18, 5.18, 9.01, 46.22, 78.05. The fundamental mode (simply supported boundary condition) appears as a "breathing mode" without nodal lines. Then a degenerate eigenvalue with multiplicity two is observed, corresponding to two modes with a nodal line either on the x_1 or x_2 direction. In particular no simple internal resonance exists between the first mode and the next. Hence the simplest model, including a single NNM, is selected and should be able to catch the main features of the dynamics of the resonant response.

The response of the HP panel to harmonic excitation in the vicinity of the first eigenfrequency ω_1 is numerically computed. The convergence of the solution has been carefully studied by Amabili (2005) for an excitation amplitude $\tilde{f}$ of 4.37 N applied at the center of the panel. It has been shown that 22 basis functions were necessary to obtain convergence. More precisely, the generalized coordinates retained for this reference solution are: $w_{1,1}$, $w_{1,3}$, $w_{3,1}$, $w_{3,3}$, $u_{1,1}$, $u_{3,1}$, $u_{1,3}$, $u_{3,3}$, $u_{1,5}$, $u_{5,1}$, $u_{3,5}$, $u_{5,3}$, $u_{5,5}$, $v_{1,1}$, $v_{3,1}$, $v_{1,3}$, $v_{3,3}$, $v_{1,5}$, $v_{5,1}$, $v_{3,5}$, $v_{5,3}$, $v_{5,5}$. The damping parameter ζ_p has been set to 0.004 for each mode: $\forall p = 1...22, \zeta_p = 0.004$.

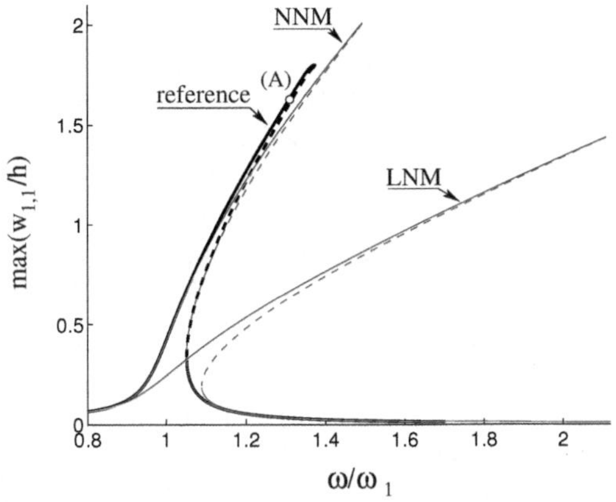

Figure 25. Frequency-response curve for the HP panel, harmonically excited in the vicinity of the first eigenfrequency ω_1. The reference solution is compared to the solution given by keeping a single linear mode (LNM) or a single NNM. The excitation amplitude is $\tilde{f} = 4.37$ N. Point (A), with $\omega = 1.3\omega_1$, is used for time integration, see Fig. 27.

Figure 25 shows the frequency-response curve for the reference solution, with 22 basis functions. As a single NNM ROM is awaited to give good results, it is decided to compare two ROMs having the same complexity (a single nonlinear oscillator equation). The first one is obtained by keeping in the truncation only the first LNM ($P_{LNM} = 1$). Eqs (91) are restricted to the first one :

$$\ddot{X}_1 + 2\zeta_1\omega_1\dot{X}_1 + \omega_1^2 X_1 + g_{11}^1 X_1^2 + h_{111}^1 X_1^3 = F_1 \cos(\omega t). \tag{92}$$

Branches of solution are numerically obtained by continuation with AUTO, then the original coordinates are recovered via: $\mathbf{q} = \mathbf{PX}$, where, in $\mathbf{X}$, only the first coordinate X_1 is different from zero.

The second reduced-order model is obtained by keeping the first NNM: Eqs. (52) are truncated by letting $R_p = 0$, $\forall p = 2...22$. The dynamics onto the invariant manifold is then governed by :

$$\ddot{R}_1 + 2\zeta_1\omega_1\dot{R}_1 + \omega_1^2 R_1 + \left(h_{111}^1 + A_{111}^1\right) R_1^3 + B_{111}^1 R_1 \dot{R}_1^2 + C_{111}^1 R_1^2 \dot{R}_1 = F_1 \cos(\omega t). \tag{93}$$

Eq. (93) is solved numerically with AUTO, then one uses Eqs. (39) to come back to the modal coordinates, and finally the matrix of eigenvectors **P** allows reconstitution of the amplitudes in the basis of selected projection functions. Thanks to the non-linear nature of the change of variable (39), all the modal amplitudes are non-zero.

Figure 25 shows the main coordinate $w_{1,1}$, having the most significant response. One can observe that the nonlinearity is of the hardening type, and that the amplitude of the response, of the order of two times the thickness, is large. For the ROMs, it is observed that whereas reduction to a single linear mode gives poor result, reduction to a single NNM give a satisfactory result, with a slight overestimation of the maximum vibration amplitude.

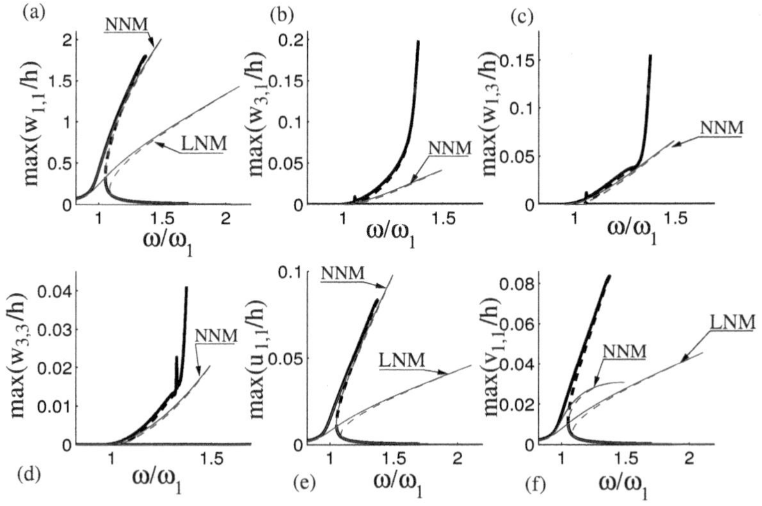

Figure 26. Maximum amplitude of the response of 6 generalized coordinates versus excitation frequency, for an excitation amplitude of $\tilde{f} = 4.37$ N. Reference solution (thick line) is compared to the reduction to a single linear mode (LNM) and a single non-linear mode (NNM). (a): maximum of $w_{1,1}$. (b): maximum of $w_{3,1}$. (c): maximum of $w_{1,3}$. (d): maximum of $w_{3,3}$. (e): maximum of $u_{1,1}$. (f): maximum of $v_{1,1}$.

Moreover, as shown in Figure 26, the reduced model composed of a single NNM, thanks to the non-linear change of coordinate, allows recovering all the other coordinates that are not directly excited. Fig. 26 shows the six main coordinates, *i.e.* the first four coordinates in transverse direction,

$w_{1,1}$, $w_{3,1}$, $w_{1,3}$ and $w_{3,3}$, as well as the first two longitudinal coordinates $u_{1,1}$ and $v_{1,1}$. It is observed that with the NNM ROM, energy is recovered in all the coordinates, with a good approximation of the original amplitudes. On the other hand, for the model composed of a single linear mode, non-zero amplitudes are recovered only on $w_{1,1}$, $u_{1,1}$ and $v_{1,1}$, as these three coordinates are linearly coupled to create the first eigenmode described by X_1 which is simulated. But a vanishing response is found with this LNM ROM for $w_{3,1}$, $w_{1,3}$ and $w_{3,3}$.

This first result emphasizes the main characteristic of the NNM ROM : the geometrical complexity due to the curvature of the invariant manifold, is first computed in the non-linear change of coordinates. Once the dynamics reduced to the manifold, a single oscillator equation is sufficient to recover the dynamics. Then, coming back to the original coordinates allows recovering energy onto the slave modes thanks to the non-linear projection.

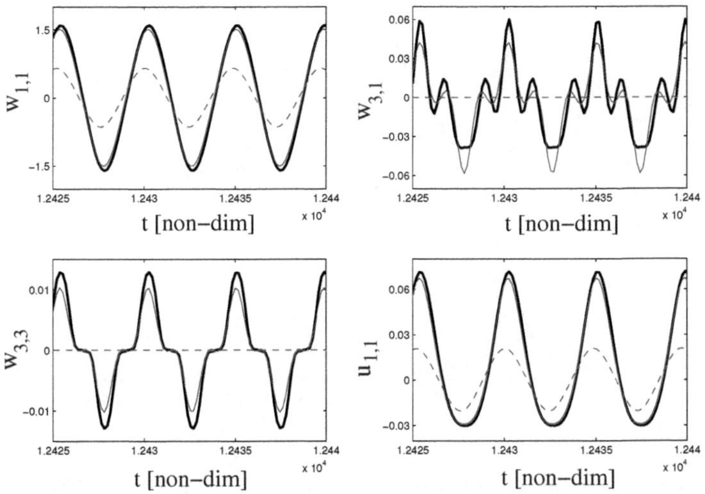

Figure 27. Time-domain response of 4 generalized coordinates of the HP panel, excitation frequency $\omega = 1.3\omega_1$, amplitude $\tilde{f} = 4.37$ N. Reference solution (thick line) is compared to the NNM solution (thin line), and the LNM solution (dashed line).

The time solutions for the four most significant coordinates is shown in Fig. 27. Once again, the reference solution is compared to the two reduced models composed of a single linear and non-linear mode. Time integrations have been performed for $\tilde{f} = 4.37$ N and $\omega = 1.3\omega_1$ (Point (A) on Fig. 25). Whereas the reduction to a single linear mode is not acceptable, the solutions provided by a single NNM are very good. Despite the fact that only one oscillator-equation is simulated, a variety of complex signals are recovered thanks to the non-linear change of coordinates.

The convergence of the solution with an increasing number of LNMs is shown in Fig. 28 for the excitation amplitude of 4.37 N. It is found that the convergence is very slow : 15 LNMs are necessary to obtain an acceptable solution. The solution with 11 LNMs is qualitatively different from the converged solution with a strange loop appearing in the frequency response, and is thus not acceptable. Hence a very slow convergence with respect to increasing P_{LNM} is found, and using the linear normal modes is not very favourable as compared to the projection functions used. On the

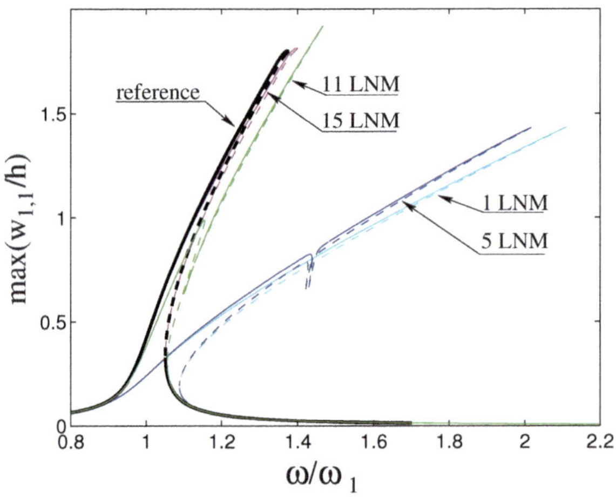

Figure 28. Maximum amplitude response of $w_{1,1}$ versus excitation frequency for $\tilde{f} = 4.37$ N, showing the convergence of the solution when increasing P_{LNM}. Reference solution (22 basis functions) is compared to truncations with 1 linear mode, 5 LNM, 11 LNM and 15 LNM.

other hand, it has been found that increasing the number of NNMs kept in the truncation in Eqs (52) do not change anything in the solution : the added NNMs have been found to stay with constant neglectable amplitude, and the same solution is found as the one obtained with a single NNM. This is a logical consequence of the invariance property of the NNMs. Hence the solution with a single NNM seems to be the best ROM possible. The only way to improve the results found here is not in increasing the number of NNMs, but in overshooting the two limitations of the present approximation used for generating the NNMs.

Finally, the robustness of the ROMs with respect to increasing the amplitude of the forcing, is studied. Fig. 29 shows the results obtained for a lower excitation amplitude: $\tilde{f} = 2.84$ N, and for a larger one: $\tilde{f} = 6.62$ N. For $\tilde{f} = 2.84$ N, the result given by the NNM ROM is almost perfectly coincident with the reference solution obtained with 22 basis function, whereas the model with a single linear mode give unacceptable results. For the larger amplitude, $\tilde{f} = 6.62$ N, the result deteriorates for the NNM-reduced model, which is not able to catch the saturation loop found by the reference solution at the top of the frequency-response curve. The observation of the other coordinates (not shown for the sake of brevity) shows that this loop reflects the fact that most of the energy is, at this point, absorbed by the higher modes, the amplitude of which significantly and abruptly increase. More precisely, it appears that an internal resonance appears between nonlinear frequencies of the system. Indeed, as the frequencies vary with amplitude, they can fulfill a resonance relationship for a given vibration amplitude even though the linear frequencies are not commensurate. The phenomena encountered here for this level of vibration appears to be of this type with an important increase of energy exchanged from the fundamentally excited modes to higher modes. Hence the reduced model should be changed to catch this new phenomenon, but this appears to be over the scope of the present study.

As a conclusion on the HP panel, the dynamics has been reduced from 22 dofs to a single NNM. Results shows that the reduction, computed with an asymptotic expansion to approach the invariant manifold, gives very good results for vibration amplitudes up to 1.5 times the panel thickness h. Beyond this value, the two approximations used for generating the ROM do not hold anymore. On the other hand, using truncations with LNMs did not allow substantial improvement as compared to the selected basis functions used for discretizing the problem.

4.3 Application : the case of a closed circular cylindrical shell

A water-filled perfect circular cylindrical shell, simply supported, and harmonically excited in the neighbourhood of the fundamental frequency, is selected in order to derive a NNM-based ROM for a continuous structure. A detailed discussion on the model can be found in (Amabili, 2003; Amabili

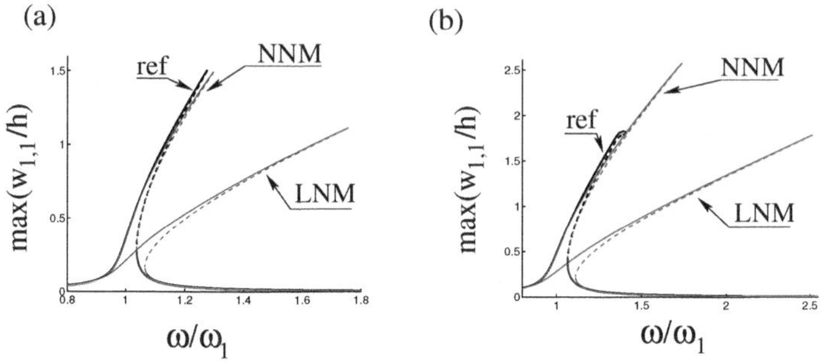

Figure 29. Frequency-response curve for (a): $\tilde{f} = 2.84$ N, and (b): $\tilde{f} = 6.62$ N. Reference solution (ref) is compared to truncations with a single linear mode (LNM) and a single non-linear mode (NNM).

et al., 2003). Donnell's non-linear shallow-shell theory is used to take into account large amplitude motions, so that in-plane inertia, transverse shear deformation and rotary inertia are neglected. Note that Donnell's shallow-shell theory is equivalent to von Kármán model already used in subsection 3.3, so that the equations of motion have the same form. For the transverse deflection $w(x, \theta, t)$, it reads:

$$D\nabla^4 w + ch\dot{w} + \rho h\ddot{w} = f - p + \frac{1}{R}\frac{\partial^2 F}{\partial x^2} + \frac{1}{R^2}\left[\frac{\partial^2 F}{\partial \theta^2}\frac{\partial^2 w}{\partial x^2} - 2\frac{\partial^2 F}{\partial x \partial \theta}\frac{\partial^2 w}{\partial x \partial \theta} + \frac{\partial^2 F}{\partial x^2}\frac{\partial^2 w}{\partial \theta^2}\right], \tag{94}$$

where D is the flexural rigidity, E Young's modulus, ν Poisson's ratio, h the shell thickness, R the mean shell radius, ρ the mass density, c the coefficient of viscous damping, p the radial pressure applied to the surface of the shell by the contained fluid, and f is a point excitation, located at $(\tilde{\theta}, \tilde{x})$:

$$f = \tilde{f}\delta(R\theta - R\tilde{\theta})\delta(x - \tilde{x})\cos(\omega t). \tag{95}$$

F is the usual Airy stress function, which satisfies the following compatibility equation:

$$\frac{1}{Eh}\nabla^4 F = -\frac{1}{R}\frac{\partial^2 w}{\partial x^2} + \left[\left(\frac{\partial^2 w}{R\partial x \partial \theta}\right)^2 - \frac{\partial^2 w}{\partial x^2}\frac{\partial^2 w}{R^2 \partial \theta^2}\right]. \tag{96}$$

A circumferentially closed circular cylindrical shell of length L is considered. Mathematical expressions of boundary conditions are given in (Amabili et al., 2003, 2006). The contained fluid is assumed to be incompressible, inviscid and irrotational. The expression of p, which describes the fluid-structure interaction, is given by Amabili (2003).

The PDE of motion is discretized by projection onto the natural modes basis. The reference solution, whose convergence has been carefully verified in (Amabili, 2003; Pellicano et al., 2002), is computed by keeping 16 natural modes. The transverse deflection is thus expanded via:

$$w(x,\theta,t) = \sum_{\substack{m=1\\k=1}}^{3} \left[A_{m,kn}(t)\cos(kn\theta) + B_{m,kn}(t)\sin(kn\theta)\right]\sin(\lambda_m x) + \sum_{m=1}^{4} A_{(2m-1),0}(t)\sin(\lambda_{(2m-1)}x), \tag{97}$$

where n is the number of circumferential waves, m the number of longitudinal half-waves (for symmetry reasons, only odd values are retained), $\lambda_m = m\pi/L$; $A_{m,n}(t)$ and $B_{m,n}(t)$ are the generalized coordinates. By use of the Galerkin method, 16 second-order differential equations are obtained. They are in the form of the general equations used as the starting point of this study, Eq. (2). The following correspondence between modal coordinates is used: $A_{1,n}$ and $B_{1,n}$ are X_1 and X_2, $A_{3,n}$ and $B_{3,n}$ are X_3 and X_4, $A_{1,2n}$ and $B_{1,2n}$ are X_5 and X_6, $A_{3,2n}$ and $B_{3,2n}$ are X_7 and X_8, $A_{1,3n}$ and $B_{1,3n}$ are X_9 and X_{10}, $A_{3,3n}$ and $B_{3,3n}$ are X_{11} and X_{12}, asymmetric modes $A_{1,0}$, $A_{3,0}$, $A_{5,0}$ and $A_{7,0}$ are X_{13} to X_{16}. Finally, modal damping is postulated.

The reference solution is obtained for the following shell: L = 520 mm, R = 149.4 mm, h = 0.519 mm, $E = 2.06.10^{11}$ Pa, ρ = 7800 kg.m^{-3}, ρ_F = 1000 kg.m^{-3} (water-filled shell), and ν =0.3. The excitation frequency ω is set in the vicinity of the fundamental mode (n = 5, m = 1), whose eigenfrequency is 79.21 Hz. Modal damping $\xi_{1,n} = 0.0017$ is assumed. The harmonic point excitation has a magnitude of 3 N and is located at $\tilde{x} = L/2$ and $\tilde{\theta} = 0$. Finally, the displacements are normalized with respect to the thickness h, and the time with respect to the period of the first eigenfrequency $\omega_{1,n}$. The frequency-response curves are numerically obtained with the software AUTO.

The response of the shell in the vicinity of an asymmetric mode is investigated. As a consequence of the rotational symmetry displayed by the shell, asymmetric modes appears by pairs, and 1:1 internal resonance exists between each pair of companion modes. Hence, the minimal model which could capture accurately the dynamics is composed of two NNMs. The ROM is build by applying the non-linear change of co-ordinates, Eq. (39), to the dynamical systems, so that after this operation, the dynamics is written in terms of the new coordinates (R_p, S_p) that are the continuation of the linear ones. Thanks to the invariance property, the truncation can now be done without losing important informations. Two couples of master coordinates, (R_1, S_1) and (R_2, S_2) are selected, which are related to $(A_{1,n}, \dot{A}_{1,n})$ and $(B_{1,n}, \dot{B}_{1,n})$. All other normal coordinates (R_p, S_p), for $p \geq 3$, are set to zero. The normal dynamics with these two master coordinates derives from Eq. (52): one has just to write this system for $p = 1, 2$. The dynamics onto this four-dimensional invariant manifold is thus governed by:

$$\begin{aligned}\ddot{R}_1 + \omega_1^2 R_1 + 2\xi_1\omega_1\dot{R}_1 &+ (A_{111}^1 + h_{111}^1)R_1^3 + B_{111}^1 R_1 \dot{R}_1^2 \\ &+ (A_{212}^1 + A_{122}^1 + h_{122}^1)R_1 R_2^2 + B_{122}^1 R_1 \dot{R}_2^2 + B_{212}^1 R_2 \dot{R}_1 \dot{R}_2 \\ &+ C_{111}^1 R_1^2 \dot{R}_1 + (C_{122}^1 + C_{212}^1) R_1 R_2 \dot{R}_2 + C_{221}^1 R_2^2 \dot{R}_1 = f\cos(\omega t)\end{aligned} \tag{98a}$$

$$\begin{aligned}\ddot{R}_2 + \omega_2^2 R_2 + 2\xi_2\omega_2\dot{R}_2 &+ (A_{222}^2 + h_{222}^2)R_2^3 + B_{222}^2 R_2 \dot{R}_2^2 \\ &+ (A_{112}^2 + A_{211}^2 + h_{112}^2)R_2 R_1^2 + B_{211}^2 R_2 \dot{R}_1^2 + B_{112}^2 R_1 \dot{R}_1 \dot{R}_2 \\ &+ C_{222}^2 R_2^2 \dot{R}_2 + (C_{121}^2 + C_{211}^2) R_1 R_2 \dot{R}_1 + C_{112}^2 R_1^2 \dot{R}_2 = 0\end{aligned} \tag{98b}$$

where the coefficients A_{ijk}^p, B_{ijk}^p and C_{ijk}^p are given by Eq. (53). In case of low-order internal resonance, the dynamical monoms corresponding to resonant terms should normally be added into the normal form, Eq. (52), which were derived for the case of no internal resonance. However, the case considered here (a perfect shell) does not produce new terms because of the perfect symmetry of the initial problem. For example, one could have find a monom like $R_1^2 R_2$ in the first equation as it is a resonant term. However, this dynamical term is not present in the original equation because $h_{112}^1 = 0$, so it is not present in the normal form.

For comparison, the ROM obtained with the *"conservative NNM"* formulation is also computed. It can be obtained from Eq. (98) by setting $C_{ijk}^p = 0$, and A_{ijk}^p, B_{ijk}^p to their values obtained for $\xi_i = 0$, $\forall i$. Frequency-response curves are numerically obtained with AUTO for the three following models: reference solution with 16 degrees-of-freedom, and the two ROMs corresponding to "damped" and "conservative" NNMs. For these simulation, the original modal coordinates are simply recovered by using Eq. (39).

Figure 30 shows the frequency-response curves for the driven mode $A_{1,n}$ and its companion mode $B_{1,n}$. The full system simulation with 16 dofs is presented with a thick line. The dynamical response is composed of two main branches of solution. The first branch correspond to single-mode motion where only the driven mode, $A_{1,n}$, is present in the response ($B_{1,n}$=0).

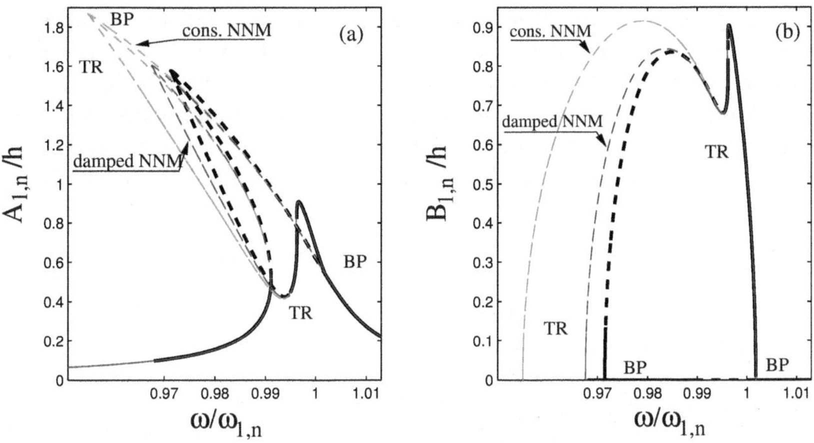

Figure 30. Maximum amplitude of transverse shell vibration versus excitation frequency. (a): driven mode $A_{1,n}$. (b): companion mode $B_{1,n}$. Thick line: reference solution (unstable states (dashed thick line) and bifurcation points are indicated: BP : pitchfork, TR: Neimarck-Sacker (torus) bifuration). Conservative and damped NNM ROMs are represented with thin line.

This branch present a softening-type nonlinearity. Due to the presence of the companion mode in 1:1 internal resonance with the driven mode, the main branch loses stability at the Bifurcation point (BP), where a secondary branch of coupled solutions emerge. Along this secondary branch, the coupling between the two resonant modes is strong and energy is exchanged inbetwwen them, resulting in important values for $B_{1,n}$. Moreover, this secondary branch encounters an instability with the presence of two Neimarck-Sacker bifurcation points, denoted as TR (like "torus" bifurcation) on the figure. Inbetween these two Neimarck-Sacker bifurcations, the harmonic solutions are not stable anymore, and a quasiperiodic response is observed.

Let us now observe how the ROMs build with NNMs are able to recover all the fine details of this bifurcation diagram. The *"conservative NNM"* case is plotted with a dash-dotted line. One can see that all the dynamical features of the original system are recovered: the two branches are found as well as the nature of the bifurcations and the stability. This result was awaited since it is a fundamental property of the normal form to recover the essential dynamical properties, thus the qualitative behaviour (number and nature of bifurcations) will always be predicted by the ROM. As already mentioned in the 2-dof example, a higher value of the maximum amplitude is found, showing that the damping has been underestimated. This is corrected with the *"damped NNM"* ROM, which gives a very good result, although one may argue that the softening effect is a little bit overestimated.

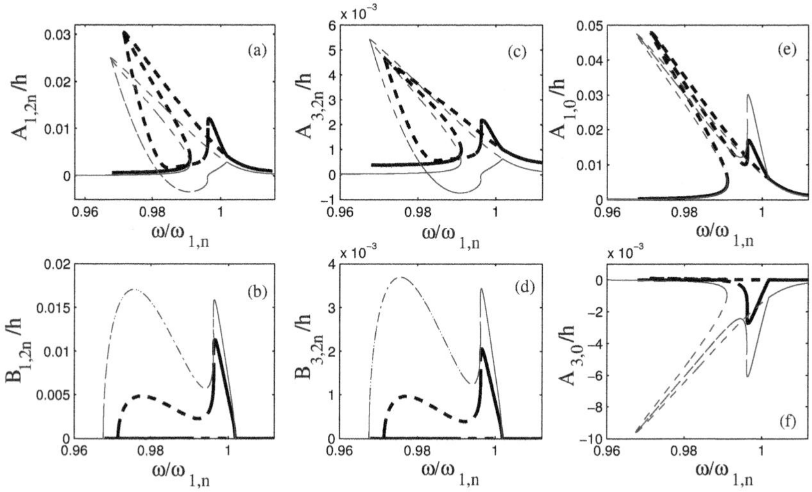

Figure 31. Maximum amplitude of transverse shell vibration versus excitation frequency for six slave modal coordinates: (a) and (b): driven and companion of asymmetric (1,2n) mode; (c) and (d): driven and companion of asymmetric (3,2n) mode; (e) and (f): axisymmetric (1,0) and (3,0) modes. Thick line: reference solution. Thin line : ROM computed from the two-dof "damped NNM" master coordinates.

Recovering the original modal coordinates with Eq. (39) shows that, thanks to the curvature of the invariant manifold in phase space, slight contributions are present onto all others linear modal coordinates. For comparison, the sixth most important modal amplitudes are represented in Fig. 31, for the full-order model, and the *"damped NNM"* ROM. This figure shows that the reduced model with two equations allows to recover all the modal amplitudes with good accuracy.

This example shows that the method can be easily used for reducing the non-linear dynamics of geometrically non-linear structures. The main advantage relies in the quickness of the method: computing all the coefficients appearing in Eqs. (39, 53, 52) is immediate on a standard PC for this 16-dof full-order system. As two approximations are used to produce the ROM (namely a time-invariant manifold is used, and it is computed by a third-order asymptotic development), it is awaited that the results can deteriorate for very large amplitude motions. However, this example shows that up to two times the thickness of the shell, the ROM is robust.

4.4 Comparison with the Proper Orthogonal Decomposition method

This last section is devoted to a comparison between reduced-order modeling as proposed with NNM via normal form theory, and the more popular Proper Orthogonal Decomposition (POD) method. For that purpose, the case of the circular cylindrical shell, filled with water and harmonically excited in the vicinity of one of its fundamental frequency (as in the previous subsection), is selected. A brief presentation of the POD method is first carried out, then some results are presented in order to illustrate the advantages and drawbacks of each method.

The POD method optimally extracts the spatial information necessary to characterize the spatio-temporal complexity and inherent dimension of a system, from a set of temporal snapshots of the response, gathered from either numerical simulations or experimental data. This point is important in the comparison since one needs to have at hand a set of solutions of the system, in order to build the POD ROM from that set. This can be seen as an advantage in an experimental context, where the POD method is routinely used to extract *e.g.* coherent structures from measurements. However in a theoretical context, this appears as a drawback, since one is obliged to run some simulation to have a set of results in order to build the POD ROM. Moreover, the question of the choice of the dynamical solutions to compute, from which the ROM is built, is very important, and a great care must be taken in this choice.

In the present context, the temporal responses are obtained via direct simulation with the conventional Galerkin solution. The question of the selected dynamical solutions for these direct solutions is postponed to the results to better explain the differences one can obtain in choosing *e.g.* either a regular or a chaotic solution. The Proper Orthogonal Modes (POMs) obtained by the POD method are denoted $\Psi_i(r,\theta)$. Once they are obtained, a Galerkin approach will be used in order to derive oscillator equations for the ROM, via:

$$w(r,\theta,t) = \sum_{i=1}^{P_{POD}} a_i(t)\Psi_i(r,\theta). \tag{99}$$

In this expression, $a_i(t)$ are the proper orthogonal coordinates and P_{POD} is the number of POMs (dof) used to build the POD model. It is awaited to obtain P_{POD} significantly smaller than P_{LNM}, the number of linear modes necessary for the conventional Galerkin method to converge. The question is here to compare P_{POD} with P_{NNM}.

Let us now explain briefly the method used to find out the POD modes $\Psi_i(r,\theta)$. The interested reader is referred to general introduction on the POD method to have a complete description, see *e.g.* Berkooz et al. (1993); Feeny and Kappagantu (1997); Kerschen and Golinval (2002); Kerschen et al. (2005), here only the main steps are recalled. The displacement field $w(r,\theta,t)$ is first divided into its time-mean value $\bar{w}(r,\theta)$ and the zero-mean response $\tilde{w}(r,\theta,t) = w(r,\theta,t) - \bar{w}$. In the POD method, the POMs are obtained by minimizing the objective function defined by:

$$\tilde{\lambda} =< (\Psi_i(r,\theta) - \tilde{w}(r,\theta,t))^2 >, \quad \forall (r,\theta) \in \Omega \tag{100}$$

where Ω denotes the space domain, $<>$ the time-averaging operation and $\Psi_i(r,\theta)$ the generic POD mode. For minimizing the objective function, a collection of dynamical states are needed. They are collected from a direct simulation. For example, having realized a temporal solution of the complete problem for a given external force, resulting in a given dynamical behaviour, one is led with a discrete collection of events $\tilde{w}(r,\theta,t_n)$, for n=0 to N_t, the number of time steps of the dynamical simulation. Let us denote by $\tilde{w}_n$ the so-called temporal snapshots of the computed solution $\tilde{w}(r,\theta,t_n)$. Then, the time-averaging operation of a series of N_t snapshots writes:

$$< \tilde{w}(r,\theta,t) >= \frac{1}{N_t}\sum_{n=1}^{N_t} \tilde{w}_n \tag{101}$$

Minimizing of the objective function (100) is obtained, after some mathematics, by solving the following eigenvalue problem:

$$\int_\Omega < \tilde{w}(r,\theta,t)\tilde{w}(r',\theta',t) > \Psi(r',\theta')\, r'\mathrm{d}r'\mathrm{d}\theta' = \lambda\Psi(r,\theta), \tag{102}$$

where $< \tilde{w}(r,\theta,t)\tilde{w}(r',\theta',t) >$ is the time-averaged spatial autocorrelation function.

A Galerkin projection scheme for determining POMs semi-analytically, and in parallel to approximate the solution of the PDE, is now presented.

This technique has been developped initially by Sarkar and Païdoussis (2004) for beams conveying fluid and extended to circular cylindrical shells by Amabili et al. (2003, 2006); Amabili and Touzé (2007). The main idea is to express the POD modes as linear combination of the eigenmodes and the structures, hence allowing for:

- A physical interpretation of the POMs
- An analytical expression of the dynamical system governing the dynamics of the problem in the POD basis.

The generic POM is projected on the eigenmodes $\Phi_i(r,\theta)$ of the shell as:

$$\Psi(r,\theta) = \sum_{i=1}^{P_{LNM}} \alpha_i \Phi_i(r,\theta), \tag{103}$$

where α_i are unknown coefficients. Then, the following eigenvalue problem is finally obtained:

$$\underline{\underline{A}}\,\underline{\alpha} = \lambda \underline{\underline{B}}\,\underline{\alpha}, \tag{104}$$

where

$$A_{ij} = \tau_i \tau_j < \tilde{q}_i(t)\tilde{q}_j(t) >, \quad B_{ij} = \tau_i \delta_{ij}, \quad \tau_i = \int_\Omega \Phi_i^2(r,\theta) r \mathrm{d}r \mathrm{d}\theta, \tag{105}$$

δ_{ij} is the Kronecker symbol, $\tilde{q}_i(t) = q_i(t) - \bar{q}_i$ is the zero-mean response of the ith generalized coordinate, with $\bar{q}_i$ being its mean. The norm of the basis functions τ_i in the present case is $\pi RL/2$ for asymmetric modes and πRL for axisymmetric modes. In Eq. (104), $\underline{\underline{A}}$ and $\underline{\underline{B}}$ are symmetric and positive definite matrices of dimension $P_{LNM} \times P_{LNM}$, and $\underline{\alpha}$ is a vector containing the P_{LNM} unknown coefficients of the POMs. The eigenvectors $\underline{\alpha}$ corresponding to the largest eigenvalues (known as dominant POMs) in Eq. (104) can now be inserted in Eq. (103) that gives a basis for the approximate solution of the PDE using the Galerkin approach; this will be referred to as the POD-Galerkin scheme hereafter.

Recombination of modal expressions given in Eqs (103), (99) and (97) allows deriving the accurate expansion of the POD mode onto the basis of the linear eigenmodes of the shell as:

$$\begin{aligned} w(r,\theta,t) &= \sum_{i=1}^{P_{POD}} a_i(t) \sum_{j=1}^{P_{LNM}} \alpha_{ji} \Phi_j(r,\theta) \\ &= \sum_{i=1}^{P_{POD}} \sum_{m=1}^{M} \sum_{n=0}^{N} \left[\alpha_{m,n,i}\cos(n\theta) + \beta_{m,n,i}\sin(n\theta)\right] sin(\lambda_m r), \end{aligned} \tag{106}$$

where on the right-hand side two different symbols, $\alpha_{m,n,i}$ and $\beta_{m,n,i}$, have been introduced to differentiate the coefficients of the POMs for cosine and sine terms in θ and are given by the corresponding α_{ji}. Eq. (106) is used to solve the equations of motion of the shell, given in the beginning of the previous subsection, Eqs. (94) and (96), with the Galerkin method to find the equations of motion of the ROM. Moreover, Eq. (106) has still the same shape over the shell surface as Eq. (97); therefore, the fluid-structure interaction can be treated with the same approach used for the Galerkin method. This is not surprising, because the POD modes have been projected on the eigenmodes.

The coefficients $\alpha_{m,n,i}$ and $\beta_{m,n,i}$ are also meaningful in order to get a physical interpretation of the POMs in phase space. Indeed, the POD method can be seen as a geometrical method that span the phase space with orthogonal modes, as enforced by the linear eigenvalue problem that defines the POMs, Eq. (104). These orthogonal modes are defined by the direction of the phase space where most of the information is present, which also explains why the choice of the data used for computing the POMs is crucial. In the POD-Galerkin scheme developed here, the POMs are expressed in the basis of the linear eigenmodes of the selected structure, so that the coefficients $\alpha_{m,n,i}$ and $\beta_{m,n,i}$ can be directly read as the deviation between the POMs and the linear eigenspaces.

The comparison between the ROMs either based on NNMs, or on POD, is now addressed. For that purpose, the case of the circular cylindrical shell filled with water, excited harmonically in the vicinity of its fundamental mode, is considered. The performances of the ROMs are assessed by comparing their ability to recover the full bifurcation diagram found in the previous subsection, Fig. 30. Fig. 32 shows the comparison between the reference model with 16 linear modes, as already shown, together with the ROM composed of two NNMs (also shown in the previous section), and finally the result given by considering three POD modes.

The selection of the temporal response used for building the POD-based ROM is fully documented in (Amabili et al., 2003), illustrating the difficulties encountered for choosing a proper case allowing to recover the response shown in Fig. 32. In particular, Fig. 32 shows two points, b and c, that have been used in the process. For point b, the frequency is $\omega = 0.995\omega_1$, the response is on the second branch, which means that the coupling between the two configurations is activated. However, the point b is selected just before the Neimarck-Sacker bifurcation, so that the response is harmonic. On the other hand, point c, for $\omega = 0.991\omega_1$, is also on the coupled solutions branch, but in the quasiperiodic regime. If one uses point b for building the

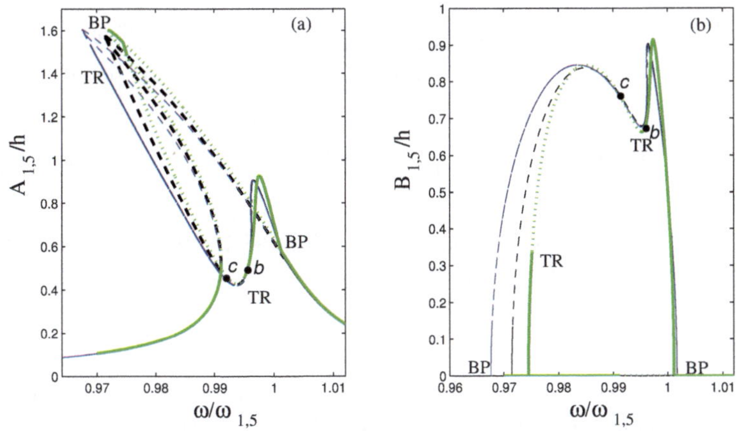

Figure 32. Maximum amplitude of transverse shell vibration versus excitation frequency. (a): driven mode $A_{1,n}$. (b): companion mode $B_{1,n}$. Thick line: reference solution (unstable states (dashed thick line) and bifurcation points are indicated: BP : pitchfork, TR: Neimarck-Sacker (torus) bifuration). Blue thin line: ROM obtained by considering two NNMs. Green thick line: ROM obtained by considering three POMs. Points b and c denotes the dynamical responses used for building the POD ROM.

POD model, results are not satisfactory at all: the solution branches moves off the reference solution (Amabili et al., 2003). Secondly, it has then been found impossible to recover the Neimarck-Sacker bifurcation point in that case, so that the quasiperiodic regime is not predicted by the POD based ROM.

For the POD ROM, the best result has been obtained by using point c, in the quasiperiodic regime. The reason is that the quasiperiodic orbits shows larger variations in phase space. Hence the available information for building the ROM is more important. For reproducing the complete bifurcation diagram, the best solution has been obtained by retaining three POMs, it is shown as a green line in Fig. 32. One can observe that the complete bifurcation diagram has been fairly recovered, with all bifurcation points and special regimes found.

Comparing now the NNM and the POD-based ROMS, one can observe that both are able to recover the bifurcation diagram. However, three POMs are necessary, versus only two NNMs. Retaining only the first two POMs

i^{th} POM	$\alpha_{1,5,i}$	$\beta_{1,5,i}$	$\alpha_{1,10,i}$	$\beta_{1,10,i}$	$\alpha_{1,0,i}$	$\alpha_{3,0,i}$
1	1	0	0.000213	0	0.000043	8.8e-6
2	0	1	0	-0.00029	0	0
3	0.000123	0	-0.1847	0	-0.9641	0.1855

Table 1. Coefficients $\alpha_{m,n,i}$ and $\beta_{m,n,i}$ defined in Eq. (106), for the first three POD modes obtained with point c.

gives incorrect results: in that case, the first bifurcation point allowing to go from the uncoupled, first branch of solution; to the coupled, secondary branch, is not detected, so that 1:1 resonant motions are not predicted ! On the other hand, the NNM-based ROM gives very satisfactorily results with only two modes, and with direct application from the PDE of motions, without having to work out with a difficult selection of a correct database to construct the ROM.

A geometrical interpretation, in the phase space, allows to understand why three POMs are necessary to reconstruct the correct solution, whereas only two NNMs are enough. The coefficients $\alpha_{m,n,i}$ and $\beta_{m,n,i}$ for the first three POMs, obtained by using point c (quasiperiodic state) for creating the snapshots database, are given in Table 1. One can see that $\alpha_{1,5,1}$=1, which means that the first POD mode is very near the first linear eigenspace, with a very slight deviation given by the small, but non-zero, values of $\alpha_{1,10,1}$ and $\alpha_{1,0,1}$. The second line shows that the second POD mode is very near the second linear eigenspace, with a very slight correction along $B_{1,10}$. Finally the third line shows that the third POM is essentially colinear to the first axisymmetric mode (1,0), with some slight corrections.

Figure 33 shows a cut-off (Poincaré section) in the phase space of the complete problem, which is of dimension 33 (16 linear modes with displacement and velocity, plus the forcing term). The plane selected for the section is $A_{(1,5)}$, $A_{(1,0)}$: driven mode and first axisymmetric mode. The clouds of points are obtained from the direct simulation obtained with the full model, at points b (harmonic coupled solution) and c (coupled quasiperiodic motion). As awaited, the quasiperiodic motion occupies a larger volume in phase space. The most important appears in the non-negligible motions along the axisymmetric mode. The fundamental importance of retaining axisymmetric modes for analyzing nonlinear vibrations of asymmetric modes of shells has already been underlined in many studies (Amabili et al., 1998, 1999; Amabili and Païdoussis, 2003). It is the result of a non-resonant,

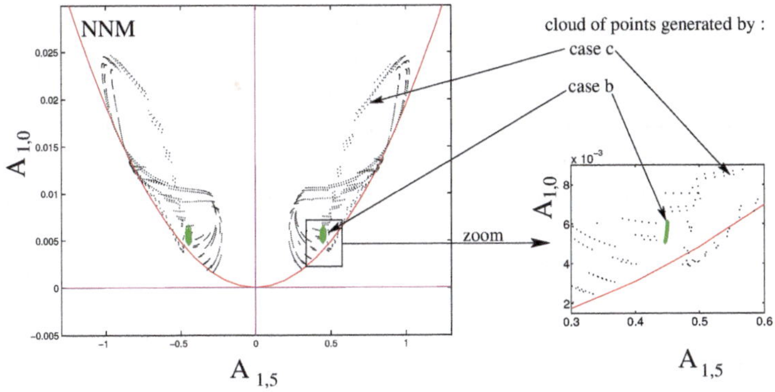

Figure 33. Poincaré section in $(A_{(1,5)},\ A_{(1,0)})$. Cloud of points: section of the orbits with the Poincaré plane, cases b and c. POD differs very little from the original axis, whereas the cut of the 4-dimensional invariant manifold (curved hyperbolic line) goes right in-between the points (NNMs).

invariant-breaking term between these two modes. Consequently, the response of $A_{(1,0)}$, though non-resonant, is not negligible, as ascertained by Figs. 31(e) and 33. The generating axis of the POD and NNM-based ROMS are also shown in Fig. 33. As already underlined, the POMs differ very slightly from the linear modes. On the other hand, the NNM method allows, thanks to normal form theory, to include the non-resonant coupling terms *into* the change of coordinates. The cut of the NNM is shown in Fig. 33, one can see that it takes naturally into account this non-resonant coupling term into the geometry of the manifold, so that the subspaces are very near the points of the reference simulation. This explains why the POD method need to take into account the third POM: otherwise the axisymmetric coupling would have been missed, hence resulting in an incorrect ROM. On the other hand, The NNM-based ROM is accurate with two NNMs only, as the bending of the phase space is taken into account in the nonlinear change of coordinates.

As a conclusion on this section, the examples with shell models have shown the ability of the NNM-based ROMs to produce accurate models for predicting the resonant forced response of thin structures with moderately large vibration amplitudes. The last subsection where comparison with the POD method has been drawn out, underlined two important features of the

NNM method:

- the method is directly applicable on the resulting dynamical system, as a supplementary (nonlinear) change of coordinates. There is no need to compute, as in the POD method, a reference solution, in order to be able to build the ROM. This is a great advantage, more especially as the choice of correct solutions is a tricky problem that need experience and cannot be realized blindly, see *e.g.* Amabili et al. (2003).
- thanks to the invariant (curved) based span of the phase space, the results obtained are of better agreement, with less NNMs than POMs. This is the advantage of using a representation basis that have a physical meaning, and is able to properly catch the non-resonant couplings, that are important for the geometry, but not for the dynamics. This underlines again a key idea claiming for using NNMS for reduced-order modeling. There is an "apparent" complexity of the dynamics which is only due to the non-resonant coupling terms, which creates the curvature of the invariant manifolds. This complexity is said to be "apparent" since it can be cancelled through a nonlinear change of coordinates. Once this complexity embedded in the geometry, a better reduction is obtained, and the complexity due to the resonant terms –that are really key for the dynamical properties of the solutions– can be more easily studied.

These conclusions need however to be moderated by the two following points:

- As proposed, the NNM-based ROM relies on two important assumptions: a third-order asymptotic method is used, and a time-invariant manifold is adopted. These assumptions have the great advantage to produce simple ROMS that are derived quickly without complicated numerical computations. However, increasing the amplitudes of vibration, the NNM-based ROM can lose its accuracy due to these two assumptions.
- The NNM method relies on a local theory, whereas the POD method is global (in the sense of the phase space). Hence the POD method can be more robust to large variations of the parameters. This is for example illustrated by Amabili and Touzé (2007).

Bibliography

M. Amabili. Theory and experiments for large-amplitude vibrations of empty and fluid-filled circular cylindrical shell with imperfections. *Journal of Sound and Vibration*, 262(4):921–975, 2003.

M. Amabili. Non-linear vibrations of doubly-curved shallow shells. *International Journal of Non-linear Mechanics*, 40(5):683–710, 2005.

M. Amabili and M. P. Païdoussis. Review of studies on geometrically nonlinear vibrations and dynamics of circular cylindrical shells and panels, with and without fluid-structure interaction. *ASME Applied Mechanical Review*, 56(4):349–381, 2003.

M. Amabili and C. Touzé. Reduced-order models for non-linear vibrations of fluid-filled circular cylindrical shells: comparison of pod and asymptotic non-linear normal modes methods. *Journal of Fluids and Structures*, 23 (6):885–903, 2007.

M. Amabili, F. Pellicano, and M. P. Païdoussis. Non-linear vibrations of simply supported, circular cylindrical shells, coupled to quiescent fluid. *Journal of Fluids and Structures*, 12:883–918, 1998.

M. Amabili, F. Pellicano, and M. P. Païdoussis. Non-linear dynamics and stability of circular cylindrical shells containing flowing fluid, part II: large-amplitude vibrations without flow. *Journal of Sound and Vibration*, 228(5):1103–1124, 1999.

M. Amabili, A. Sarkar, and M. P. Païdoussis. Reduced-order models for nonlinear vibrations of cylindrical shells via the proper orthogonal decomposition method. *Journal of Fluids and Structures*, 18(2):227–250, 2003.

M. Amabili, A. Sarkar, and M. P. Païdoussis. Chaotic vibrations of circular cylindrical shells: Galerkin versus reduced-order models via the proper orthogonal decomposition method. *Journal of Sound and Vibration*, 290 (3-5):736–762, 2006.

G. Berkooz, P. Holmes, and J.L. Lumley. The proper orthogonal decomposition in the analysis of turbulent flows. *Annual review of Fluid Mechanics*, 25:539–575, 1993.

F. Blanc, C. Touzé, J.-F. Mercier, and A.-S. Bonnet Ben-Dhia. On the numerical computation of nonlinear normal modes for reduced-order modelling of conservative vibratory systems. *Mechanical Systems and Signal Processing*, 36(2):520–539, 2013.

H-N Chu and G. Herrmann. Influence of large amplitudes on free flexural vibrations of rectangular elastic plates. *Journal of Applied Mechanics*, 23:532–540, 1956.

E.J. Doedel, R. Paffenroth, A.R. Champneys, T.F. Fairgrieve, Y.A. Kuznetsov, B.E. Oldeman, B. Sandstede, and X. Wang. Auto 2000: Continuation and bifurcation software for ordinary differential equations. Technical report, Concordia University, 2002.

H. Dulac. Solutions d'un système d'équations différentielles dans le voisinage de valeurs singulières. *Bulletin de la Société Mathématique de France*, 40:324–383, 1912.

G. J. Efstathiades. A new approach to the large-deflection vibrations of imperfect circular disks using Galerkin's procedure. *Journal of Sound and Vibration*, 16(2):231–253, 1971.

C. Elphick, G. Iooss, and E. Tirapegui. Normal form reduction for time-periodically driven differential equations. *Physics Letters A*, 120(9):459–463, 1987.

H.A. Evensen and R.M. Ewan-Iwanowsky. Dynamic response and stability of shallow spherical shells subject to time-dependent loading. *AIAA Journal*, 5(5):969–976, 1967.

B.F. Feeny and R. Kappagantu. On the physical interpretation of proper orthogonal modes in vibrations. *Journal of Sound and Vibration*, 211 (4):607–616, 1997.

P. B. Gonçalves. Axisymmetric vibrations of imperfect shallow spherical caps under pressure loading. *Journal of Sound and Vibration*, 174(2): 249–260, 1994.

P. L. Grossman, B. Koplik, and Y-Y. Yu. Nonlinear vibrations of shallow spherical shells. *ASME Journal of Applied Mechanics*, 39E:451–458, 1969.

J. Guckenheimer and P. Holmes. *Nonlinear oscillations, dynamical systems and bifurcations of vector fields.* Springer-Verlag, New-York, 1983.

G. Iooss and M. Adelmeyer. *Topics in bifurcation theory.* World scientific, New-York, 1998. second edition.

L. Jézéquel and C. H. Lamarque. Analysis of non-linear dynamical systems by the normal form theory. *Journal of Sound and Vibration*, 149(3): 429–459, 1991.

D. Jiang. *Nonlinear modal analysis based on invariant manifolds. Application to rotating blade systems.* PhD thesis, University of Michigan, 2004.

D. Jiang, C. Pierre, and S. Shaw. Nonlinear normal modes for vibratory systems under harmonic excitation. *Journal of Sound and Vibration*, 288 (4-5):791–812, 2005.

Th.von Kármán. Festigkeitsprobleme im maschinenbau. *Encyklopdie der Mathematischen Wissenschaften*, 4(4):311–385, 1910.

G. Kerschen and J.C. Golinval. Physical interpretation of the proper orthogonal modes using the singular value decomposition. *Journal of Sound and Vibration*, 249(5):849–865, 2002.

G. Kerschen, J.C. Golinval, A.F. Vakakis, and L.A. Bergman. The method of proper orthogonal decomposition for dynamical characterization and order reduction of mechanical systems: an overview. *Nonlinear Dynamics*, 41:147–169, 2005.

C.-H. Lamarque, C. Touzé, and O. Thomas. An upper bound for validity limits of asymptotic analytical approaches based on normal form theory. *Nonlinear Dynamics*, 70(3):1931–1949, 2012.

P. Manneville. *Instabilités, chaos et turbulence.* Editions de l'Ecole Polytechnique, Palaiseau, 2004.

P. Manneville. *Dissipative structures and weak turbulence.* Academic Press, 1990.

A. H. Nayfeh. On direct methods for constructing nonlinear normal modes of continuous systems. *Journal of Vibration and Control*, 1:389–430, 1995.

A. H. Nayfeh and W. Lacarbonara. On the discretization of distributed-parameter systems with quadratic and cubic non-linearities. *Nonlinear Dynamics*, 13:203–220, 1997.

A. H. Nayfeh and S. A. Nayfeh. On nonlinear normal modes of continuous systems. *Trans. ASME/Journal of Vibration and Acoustics*, 116:129–136, 1994.

A. H. Nayfeh, J. F. Nayfeh, and D. T. Mook. On methods for continuous systems with quadratic and cubic nonlinearities. *Nonlinear Dynamics*, 3:145–162, 1992.

K. A. V. Pandalai and M. Sathyamoorthy. On the modal equations of large amplitude flexural vibration of beams, plates, rings and shells. *International Journal of Non-linear Mechanics*, 8(3):213–218, 1973.

F. Pellicano, M. Amabili, and M. P. Païdoussis. Effect of the geometry on the non-linear vibration of circular cylindrical shells. *International Journal of Non-linear Mechanics*, 37:1181–1198, 2002.

E. Pesheck. *Reduced-order modeling of nonlinear structural systems using nonlinear normal modes and invariant manifolds.* PhD thesis, University of Michigan, 2000.

E. Pesheck and C. Pierre. A global methodology for the modal reduction of large non-linear systems containing quadratic and cubic non-linearities. In *Proc. of ASME DETC'97*, 1997.

H. Poincaré. *Les méthodes nouvelles de la mécanique céleste.* Gauthiers-Villars, Paris, 1892.

A. Sarkar and M. P. Païdoussis. A cantilever conveying fluid: coherent modes versus beam modes. *International Journal of Non-Linear Mechanics*, 39:467481, 2004.

M. Sathyamoorthy. Vibrations of moderatly thick shallow spherical shells at large amplitude. *Journal of Sound and Vibration*, 172(1):63–70, 1994.

S. W. Shaw and C. Pierre. Non-linear normal modes and invariant manifolds. *Journal of Sound and Vibration*, 150(1):170–173, 1991.

S. W. Shaw and C. Pierre. Normal modes for non-linear vibratory systems. *Journal of Sound and Vibration*, 164(1):85–124, 1993.

S. Sridhar, D. T. Mook, and A. H. Nayfeh. Non-linear resonances in the forced responses of plates, part I: symmetric responses of circular plates. *Journal of Sound and Vibration*, 41(3):359–373, 1975.

O. Thomas and S. Bilbao. Geometrically nonlinear flexural vibrations of plates: In-plane boundary conditions and some symmetry properties. *Journal of Sound and Vibration*, 315(3):569–590, 2008.

O. Thomas, C. Touzé, and A. Chaigne. Asymmetric non-linear forced vibrations of free-edge circular plates, part II: experiments. *Journal of Sound and Vibration*, 265(5):1075–1101, 2003.

O. Thomas, C. Touzé, and A. Chaigne. Non-linear vibrations of free-edge thin spherical shells: modal interaction rules and 1:1:2 internal resonance. *International Journal of Solids and Structures*, 42(11-12):3339–3373, 2005.

S. A. Tobias. Free undamped non-linear vibrations of imperfect circular disks. *Proc. of the Instn. of Mech. Eng.*, 171:691–700, 1957.

C. Touzé. A normal form approach for non-linear normal modes. Technical report, Publications du LMA, numéro 156, (ISSN: 1159-0947, ISBN: 2-909669-20-3), 2003.

C. Touzé and M. Amabili. Non-linear normal modes for damped geometrically non-linear systems: application to reduced-order modeling of harmonically forced structures. *Journal of Sound and Vibration*, 298(4-5): 958–981, 2006.

C. Touzé and O. Thomas. Non-linear behaviour of free-edge shallow spherical shells: effect of the geometry. *International Journal of Non-linear Mechanics*, 41(5):678–692, 2006.

C. Touzé, O. Thomas, and A. Chaigne. Asymmetric non-linear forced vibrations of free-edge circular plates, part I: theory. *Journal of Sound and Vibration*, 258(4):649–676, 2002.

C. Touzé, O. Thomas, and A. Chaigne. Hardening/softening behaviour in non-linear oscillations of structural systems using non-linear normal modes. *Journal of Sound and Vibration*, 273(1-2):77–101, 2004.

C. Touzé, M. Amabili, and O. Thomas. Reduced-order models for large-amplitude vibrations of shells including in-plane inertia. *Computer Methods in Applied Mechanics and Engineering*, 197(21-24):2030–2045, 2008a. URL `http://dx.doi.org/10.1016/j.cma.2008.01.002`.

C. Touzé, C. Camier, G. Favraud, and O. Thomas. Effect of imperfections and damping on the type of non-linearity of circular plates and shallow spherical shells. *Mathematical Problems in Engineering*, 2008:Article ID 678307, 2008b. doi:10.1155/2008/678307.

T. K. Varadan and K. A. V. Pandalai. Non-linear flexural oscillations of orthotropic shallow spherical shells. *Computers and Structures*, 9:417–425, 1978.

S. Wiggins. *Introduction to applied nonlinear dynamical systems and chaos.* Springer-Verlag, New-York, 2003. Second edition.

N. Yamaki. Influence of large amplitudes on flexural vibrations of elastic plates. *ZAMM*, 41(12):501–510, 1961.

K. Yasuda and G. Kushida. Nonlinear forced oscillations of a shallow spherical shell. *Bull. JSME*, 27(232):2233–2240, 1984.

A Expressions of the coefficients for the normal transform in the conservative case

Expressions of the coefficients (a^p_{ij}, b^p_{ij}, r^p_{ijk}, u^p_{ijk}, γ^p_{ij}, μ^p_{ijk}, ν^p_{ijk}) used in Eqs. (39), which allows one to go from the modal variables (associated with the orthogonal linear grid) to the normal coordinates (associated to the invariant manifold and the curved grid) are given here.

A.1 Quadratic coefficients

- The following expressions are obtained, $\forall p = 1, ..., N$:

$$\forall\, i = 1, ..., N\,, \forall\, j \geq i, ..., N:$$

$$a^p_{ij} = \frac{\omega_i^2 + \omega_j^2 - \omega_p^2}{D_{ijp}} g^p_{ij} \tag{107a}$$

$$b^p_{ij} = \frac{2}{D_{ijp}} g^p_{ij} \tag{107b}$$

$$c^p_{ij} = 0 \tag{107c}$$

$$\alpha^p_{ij} = 0 \tag{107d}$$

$$\beta^p_{ij} = 0 \tag{107e}$$

$$\gamma^p_{ii} = \frac{2}{4\omega_i^2 - \omega_p^2} g^p_{ii} \tag{107f}$$

$$\forall\, i = 1, ..., N\,, \forall\, j > i, ..., N:$$

$$\gamma^p_{ij} = \frac{\omega_j^2 - \omega_i^2 - \omega_p^2}{D_{ijp}} g^p_{ij} \tag{107g}$$

$$\gamma^p_{ji} = \frac{\omega_i^2 - \omega_j^2 - \omega_p^2}{D_{ijp}} g^p_{ij} \tag{107h}$$

where $D_{ijp} = (\omega_i + \omega_j - \omega_p)(\omega_i + \omega_j + \omega_p)(-\omega_j + \omega_i + \omega_p)(\omega_i - \omega_j - \omega_p)$.

A.2 Cubic coefficients

- The following coefficients, which correspond to the trivially resonant terms, are equal to zero:

$$\begin{aligned}
\forall p = 1, ..., N: \quad & u^p_{ppp} = r^p_{ppp} = \mu^p_{ppp} = \nu^p_{ppp} = 0 \\
\forall j > p...N: \quad & r^p_{pjj} = u^p_{pjj} = u^p_{jpj} = 0 \\
& \mu^p_{pjj} = \nu^p_{pjj} = \nu^p_{jpj} = 0 \\
\forall i < p: \quad & r^p_{iip} = u^p_{iip} = u^p_{pii} = 0 \\
& \mu^p_{iip} = \nu^p_{iip} = \nu^p_{pii} = 0
\end{aligned} \tag{108a}$$

• The non-zero coefficients are now given, $\forall p = 1, ..., N$:
$\forall\, i = 1, ..., N,\ i \neq p:$

$$\begin{aligned}
r^p_{iii} &= \frac{1}{D^{(1)}_{ip}} \left[(7\omega_i^2 - \omega_p^2)(h^p_{iii} + A^p_{iii}) + 2\omega_i^4 B^p_{iii} \right] \\
u^p_{iii} &= \frac{1}{D^{(1)}_{ip}} \left[6h^p_{iii} + 6A^p_{iii} + (3\omega_i^2 - \omega_p^2) B^p_{iii} \right] \\
\mu^p_{iii} &= u^p_{iii} \\
\nu^p_{iii} &= \frac{1}{D^{(1)}_{ip}} \left[(9\omega_i^2 - 3\omega_p^2)(h^p_{iii} + A^p_{iii}) + 2\omega_p^2 \omega_i^2 B^p_{iii} \right]
\end{aligned}$$

where $D^{(1)}_{ip} = (\omega_p^2 - \omega_i^2)(\omega_p^2 - 9\omega_i^2)$.

$\forall\, i = 1, ..., N-1,\ i \neq p,\ \forall j > i, ..., N:$

$$
\begin{aligned}
r^p_{ijj} =& \frac{\omega_i^2(\omega_i^2 - 2\omega_j^2 - 2\omega_p^2) + (\omega_p^2 - 4\omega_j^2)(\omega_p^2 - 2\omega_j^2)}{(\omega_i^2 - \omega_p^2) D^{(1)}_{ijp}} \left[A^p_{ijj} + A^p_{jij} + h^p_{ijj}\right] \\
&- \frac{2\omega_j^4(\omega_p^2 - 4\omega_j^2 + 3\omega_i^2)}{(\omega_i^2 - \omega_p^2) D^{(1)}_{ijp}} B^p_{ijj} + \frac{2\omega_i^2\omega_j^2}{D^{(1)}_{ijp}} B^p_{jij} \\
u^p_{ijj} =& \frac{\omega_i^2(\omega_i^2 - 2\omega_j^2 - 2\omega_p^2) + (\omega_p^2 - 4\omega_j^2)(\omega_p^2 - 2\omega_j^2)}{(\omega_i^2 - \omega_p^2) D^{(1)}_{ijp}} B^p_{ijj} \\
&+ \frac{8\omega_j^2 - 6\omega_i^2 - 2\omega_p^2}{(\omega_i^2 - \omega_p^2) D^{(1)}_{ijp}} [A^p_{ijj} + A^p_{jij} + h_{ijj}] - \frac{2\omega_i^2}{D^{(1)}_{ijp}} B^p_{jij} \\
u^p_{jij} =& \frac{1}{D^{(1)}_{ijp}} \left[4\,(A^p_{jij} + A^p_{ijj} + h^p_{ijj}) - 4\,{w_j}^2 B^p_{ijj} + (4\,{w_j}^2 - {w_p}^2 + {w_i}^2) B^p_{jij}\right] \\
\nu^p_{ijj} =& \frac{6\omega_i^2\omega_j^2 + 2\omega_j^2\omega_p^2 + 2\omega_p^2\omega_i^2 - 8\omega_j^4 - \omega_p^4 - \omega_i^4}{(\omega_p^2 - \omega_i^2) D^{(1)}_{ijp}} \left[A^p_{ijj} + A^p_{jij} + h^p_{ijj}\right] \\
&+ \frac{2\omega_j^4(3\omega_p^2 + \omega_i^2 - 4\omega_j^2)}{(\omega_p^2 - \omega_i^2) D^{(1)}_{ijp}} B^p_{ijj} - \frac{\omega_j^2(-\omega_p^2 + 4\omega_j^2 - \omega_i^2)}{D^{(1)}_{ijp}} B^p_{jij} \\
\nu^p_{jij} =& \frac{8\omega_j^2 - 2\omega_i^2 - 2\omega_p^2}{D^{(1)}_{ijp}} \left[A^p_{ijj} + A^p_{jij} + h^p_{ijj}\right] \\
&+ \frac{2\omega_j^2\omega_p^2 - 8\omega_j^4 + 2\omega_j^2\omega_i^2}{D^{(1)}_{ijp}} B^p_{ijj} + \frac{\omega_p^2\omega_i^2 + 4\omega_j^2\omega_i^2 - \omega_i^4}{D^{(1)}_{ijp}} B^p_{jij} \\
\mu^p_{ijj} =& \frac{6\omega_i^2\omega_j^2 + 2\omega_j^2\omega_p^2 + 2\omega_p^2\omega_i^2 - 8\omega_j^4 - \omega_p^4 - \omega_i^4}{(\omega_p^2 - \omega_i^2) D^{(1)}_{ijp}} B^p_{ijj} \\
&+ \frac{6\omega_p^2 + 2\omega_i^2 - 8\omega_j^2}{(\omega_p^2 - \omega_i^2) D^{(1)}_{ijp}} \left[A^p_{ijj} + A^p_{jij} + h^p_{ijj}\right] + \frac{4\omega_j^2 - \omega_i^2 - \omega_p^2}{D^{(1)}_{ijp}} B^p_{jij}
\end{aligned}
$$

where $D^{(1)}_{ijp} = (\omega_p + \omega_i - 2\omega_j)(\omega_p + \omega_i + 2\omega_j)(-\omega_p + \omega_i + 2\omega_j)(-\omega_p + \omega_i - 2\omega_j)$.

$\forall\ i = 1, ..., N-1,\ \forall j > i, ..., N,\ j \neq p$:

$$
\begin{aligned}
r^p_{iij} =& \frac{2\omega_i^2(4\omega_i^2 - 3\omega_p^2 - \omega_j^2) + (\omega_p - \omega_j)^2(\omega_p + \omega_j)^2}{(\omega_p^2 - \omega_j^2)D^{(2)}_{ijp}} \left[-A^p_{iij} - A^p_{jii} - h^p_{iij}\right] \\
& - \frac{2\omega_i^4(4\omega_i^2 - \omega_p^2 - 3\omega_j^2)}{(\omega_p^2 - \omega_j^2)D^{(2)}_{ijp}} B^p_{jii} - \frac{2\omega_i^2\omega_j^2}{D^{(2)}_{ijp}} B^p_{iij} \\
u^p_{jii} =& - \frac{2\omega_i^2(4\omega_i^2 - 3\omega_p^2 - \omega_j^2) + (\omega_p - \omega_j)^2(\omega_p + \omega_j)^2}{(\omega_p^2 - \omega_j^2)D^{(2)}_{ijp}} B^p_{jii} \\
& + \frac{8\omega_i^2 - 2\omega_p^2 - 6\omega_j^2}{(\omega_p^2 - \omega_j^2)D^{(2)}_{ijp}} \left[-A^p_{jii} - A^p_{iij} - h^p_{iij}\right] - \frac{2\omega_j^2}{D^{(2)}_{ijp}} B^p_{iij} \\
u^p_{iij} =& \frac{1}{D^{(2)}_{ijp}} \left[4(A^p_{jii} + A^p_{iij} + h^p_{iij}) - 4\omega_i^2 B^p_{jii} - (\omega_p^2 - \omega_j^2 - 4\omega_i^2) B^p_{iij}\right] \\
\nu^p_{jii} =& \frac{8\omega_i^4 + \omega_j^4 + \omega_p^4 - 2\omega_p^2\omega_i^2 - 6\omega_i^2\omega_j^2 - 2\omega_j^2\omega_p^2}{(\omega_j^2 - \omega_p^2)D^{(2)}_{ijp}} \left[A^p_{jii} + A^p_{iij} + h^p_{iij}\right] \\
& + \frac{8\omega_i^6 - 2\omega_j^2\omega_i^4 - 6\omega_p^2\omega_i^4}{(\omega_j^2 - \omega_p^2)D^{(2)}_{ijp}} B^p_{jii} + \frac{\omega_i^2(\omega_j^2 - 4\omega_i^2 + \omega_p^2)}{D^{(2)}_{ijp}} B^p_{iij} \\
\nu^p_{iij} =& \frac{8\omega_i^2 - 2\omega_p^2 - 2\omega_j^2}{D^{(2)}_{ijp}} \left[A^p_{jii} + A^p_{iij} + h^p_{iij}\right] \\
& + \frac{2\omega_i^2\omega_j^2 - 8\omega_i^4 + 2\omega_i^2\omega_p^2}{D^{(2)}_{ijp}} B^p_{jii} + \frac{-\omega_j^4 + 4\omega_i^2\omega_j^2 + \omega_j^2\omega_p^2}{D^{(2)}_{ijp}} B^p_{iij} \\
\mu^p_{iij} =& \frac{8\omega_i^2 - 2\omega_j^2 - 6\omega_p^2}{(\omega_j^2 - \omega_p^2)D^{(2)}_{ijp}} \left[A^p_{iij} + A^p_{jii} + h^p_{iij}\right] + \frac{4\omega_i^2 - \omega_j^2 - \omega_p^2}{D^{(2)}_{ijp}} B^p_{iij} \\
& - \frac{6\omega_i^2\omega_j^2 + 2\omega_j^2\omega_p^2 + 2\omega_i^2\omega_p^2 - 8\omega_i^4 - \omega_p^4 - \omega_j^4}{(\omega_j^2 - \omega_p^2)D^{(2)}_{ijp}} B^p_{jii}
\end{aligned}
$$

where $D^{(2)}_{ijp} = (\omega_p + 2\omega_i - \omega_j)(\omega_p + 2\omega_i + \omega_j)(-\omega_p + 2\omega_i + \omega_j)(-\omega_p + 2\omega_i - \omega_j)$.

$\forall\ i = 1, ..., N-2,\ \forall j > i, ..., N-1,\ \forall k > j, ..., N$:

$$
\begin{array}{rcll}
u^{p}_{ijk} & = & \frac{1}{D^{(3)}_{ijkp}}[& (\quad 2\omega_{j}^{4}-6\omega_{i}^{4}-4\omega_{k}^{2}\omega_{p}^{2}+4\omega_{i}^{2}\omega_{p}^{2}+4\omega_{i}^{2}\omega_{k}^{2}+2\omega_{k}^{4} \\
 & & & \quad +4\omega_{i}^{2}\omega_{j}^{2}+2\omega_{p}^{4}-4\omega_{k}^{2}\omega_{j}^{2}-4\omega_{j}^{2}\omega_{p}^{2})(h^{p}_{ijk}+A^{p}_{ijk}+A^{p}_{kij}+A^{p}_{jik}) \\
 & & & +(\quad -\omega_{k}^{2}\omega_{j}^{4}-\omega_{i}^{2}\omega_{k}^{4}-2\omega_{j}^{2}\omega_{i}^{2}\omega_{p}^{2}+3\omega_{j}^{2}\omega_{p}^{4}-\omega_{k}^{4}\omega_{j}^{2}-2\omega_{i}^{2}\omega_{k}^{2}\omega_{p}^{2} \\
 & & & \quad +3\omega_{i}^{2}\omega_{p}^{4}-\omega_{j}^{4}\omega_{i}^{2}-\omega_{i}^{4}\omega_{k}^{2}+3\omega_{k}^{2}\omega_{p}^{4}+\omega_{i}^{6}-\omega_{p}^{6}-3\omega_{k}^{4}\omega_{p}^{2}+\omega_{k}^{6}-3\omega_{j}^{4}\omega_{p}^{2} \\
 & & & \quad +10\omega_{i}^{2}\omega_{k}^{2}\omega_{j}^{2}+\omega_{j}^{6}-2\omega_{k}^{2}\omega_{j}^{2}\omega_{p}^{2}-3\omega_{i}^{4}\omega_{p}^{2}-\omega_{j}^{2}\omega_{i}^{4})B^{p}_{ijk} \\
 & & & +(\quad -2\omega_{j}^{4}\omega_{i}^{2}+4\omega_{i}^{4}\omega_{p}^{2}+4\omega_{j}^{2}\omega_{i}^{4}+6\omega_{i}^{2}\omega_{k}^{4}-4\omega_{i}^{4}\omega_{k}^{2} \\
 & & & \quad +4\omega_{j}^{2}\omega_{i}^{2}\omega_{p}^{2}-2\omega_{i}^{2}\omega_{p}^{4}-2\omega_{i}^{6}-4\omega_{i}^{2}\omega_{k}^{2}\omega_{p}^{2}-4\omega_{i}^{2}\omega_{k}^{2}\omega_{j}^{2})B^{p}_{jik} \\
 & & & +(\quad -2\omega_{i}^{2}\omega_{p}^{4}+6\omega_{j}^{4}\omega_{i}^{2}+4\omega_{i}^{2}\omega_{k}^{2}\omega_{p}^{2}-4\omega_{i}^{2}\omega_{k}^{2}\omega_{j}^{2}-4\omega_{j}^{2}\omega_{i}^{2}\omega_{p}^{2} \\
 & & & \quad -4\omega_{j}^{2}\omega_{i}^{4}-2\omega_{i}^{2}\omega_{k}^{4}+4\omega_{i}^{4}\omega_{p}^{2}-2\omega_{i}^{6}+4\omega_{i}^{4}\omega_{k}^{2})B^{p}_{kij}] \\
u^{p}_{jik} & = & \frac{1}{D^{(3)}_{ijkp}}[& (\quad -2\omega_{p}^{4}+6\omega_{j}^{4}-2\omega_{i}^{4}-4\omega_{k}^{2}\omega_{j}^{2}+4\omega_{k}^{2}\omega_{p}^{2} \\
 & & & \quad -2\omega_{k}^{4}+4\omega_{i}^{2}\omega_{k}^{2}-4\omega_{i}^{2}\omega_{j}^{2}+4\omega_{i}^{2}\omega_{p}^{2}-4\omega_{j}^{2}\omega_{p}^{2})(h^{p}_{ijk}+A^{p}_{ijk}+A^{p}_{kij}+A^{p}_{jik}) \\
 & & & +(\quad 2\omega_{j}^{6}+4\omega_{k}^{2}\omega_{j}^{2}\omega_{p}^{2}+4\omega_{i}^{2}\omega_{k}^{2}\omega_{j}^{2}+2\omega_{j}^{2}\omega_{i}^{4}+4\omega_{k}^{2}\omega_{j}^{4} \\
 & & & \quad -4\omega_{j}^{2}\omega_{i}^{2}\omega_{p}^{2}+2\omega_{j}^{2}\omega_{p}^{4}-4\omega_{j}^{4}\omega_{i}^{2}-4\omega_{j}^{4}\omega_{p}^{2}-6\omega_{k}^{4}\omega_{j}^{2})B^{p}_{ijk} \\
 & & & +(\quad 3\omega_{j}^{4}\omega_{p}^{2}+\omega_{k}^{2}\omega_{j}^{4}-3\omega_{k}^{2}\omega_{p}^{4}+\omega_{i}^{2}\omega_{k}^{4}+\omega_{k}^{4}\omega_{j}^{2}+3\omega_{i}^{4}\omega_{p}^{2} \\
 & & & \quad +2\omega_{i}^{2}\omega_{k}^{2}\omega_{p}^{2}+\omega_{j}^{4}\omega_{i}^{2}+\omega_{i}^{4}\omega_{k}^{2}-\omega_{i}^{6}+3\omega_{k}^{4}\omega_{p}^{2}-\omega_{j}^{6}-\omega_{k}^{6}+2\omega_{j}^{2}\omega_{i}^{2}\omega_{p}^{2}+\omega_{p}^{6} \\
 & & & \quad +2\omega_{k}^{2}\omega_{j}^{2}\omega_{p}^{2}-3\omega_{j}^{2}\omega_{p}^{4}-10\omega_{i}^{2}\omega_{k}^{2}\omega_{j}^{2}-3\omega_{i}^{2}\omega_{p}^{4}+\omega_{j}^{2}\omega_{i}^{4})B^{p}_{jik} \\
 & & & +(\quad 4\omega_{i}^{2}\omega_{k}^{2}\omega_{j}^{2}+4\omega_{j}^{4}\omega_{i}^{2}+2\omega_{j}^{2}\omega_{p}^{4}-6\omega_{j}^{2}\omega_{i}^{4}+2\omega_{k}^{4}\omega_{j}^{2}-4\omega_{j}^{4}\omega_{p}^{2} \\
 & & & \quad +2\omega_{j}^{6}-4\omega_{k}^{2}\omega_{j}^{4}-4\omega_{k}^{2}\omega_{j}^{2}\omega_{p}^{2}+4\omega_{j}^{2}\omega_{i}^{2}\omega_{p}^{2})B^{p}_{kij}] \\
u^{p}_{kij} & = & \frac{1}{D^{(3)}_{ijkp}}[& (\quad 2\omega_{p}^{4}+2\omega_{j}^{4}+4\omega_{i}^{2}\omega_{k}^{2}-6\omega_{k}^{4}+4\omega_{k}^{2}\omega_{p}^{2}-4\omega_{i}^{2}\omega_{p}^{2} \\
 & & & \quad +4\omega_{k}^{2}\omega_{j}^{2}-4\omega_{i}^{2}\omega_{j}^{2}-4\omega_{j}^{2}\omega_{p}^{2}+2\omega_{i}^{4})(h^{p}_{ijk}+A^{p}_{ijk}+A^{p}_{kij}+A^{p}_{jik}) \\
 & & & +(\quad -4\omega_{i}^{2}\omega_{k}^{2}\omega_{j}^{2}-2\omega_{k}^{6}-2\omega_{i}^{4}\omega_{k}^{2}-4\omega_{k}^{2}\omega_{j}^{2}\omega_{p}^{2}+4\omega_{i}^{2}\omega_{k}^{2}\omega_{p}^{2} \\
 & & & \quad +6\omega_{k}^{2}\omega_{j}^{4}-2\omega_{k}^{2}\omega_{p}^{4}+4\omega_{i}^{2}\omega_{k}^{4}-4\omega_{k}^{4}\omega_{j}^{2}+4\omega_{k}^{4}\omega_{p}^{2})B^{p}_{ijk} \\
 & & & +(\quad -2\omega_{k}^{6}-4\omega_{i}^{2}\omega_{k}^{4}+4\omega_{k}^{2}\omega_{j}^{2}\omega_{p}^{2}-4\omega_{i}^{2}\omega_{k}^{2}\omega_{p}^{2}-2\omega_{k}^{2}\omega_{p}^{4}+6\omega_{i}^{4}\omega_{k}^{2} \\
 & & & \quad -4\omega_{i}^{2}\omega_{k}^{2}\omega_{j}^{2}+4\omega_{k}^{4}\omega_{j}^{2}-2\omega_{k}^{2}\omega_{j}^{4}+4\omega_{k}^{4}\omega_{p}^{2})B^{p}_{jik} \\
 & & & +(\quad -\omega_{k}^{2}\omega_{j}^{4}-\omega_{i}^{2}\omega_{k}^{4}-2\omega_{j}^{2}\omega_{i}^{2}\omega_{p}^{2}+3\omega_{j}^{2}\omega_{p}^{4}-\omega_{k}^{4}\omega_{j}^{2}-2\omega_{i}^{2}\omega_{k}^{2}\omega_{p}^{2} \\
 & & & \quad +3\omega_{i}^{2}\omega_{p}^{4}-\omega_{j}^{4}\omega_{i}^{2}-\omega_{i}^{4}\omega_{k}^{2}+3\omega_{k}^{2}\omega_{p}^{4}+\omega_{i}^{6}-\omega_{p}^{6}-3\omega_{k}^{4}\omega_{p}^{2}+\omega_{k}^{6} \\
 & & & \quad -3\omega_{j}^{4}\omega_{p}^{2}+10\omega_{i}^{2}\omega_{k}^{2}\omega_{j}^{2}+\omega_{j}^{6}-2\omega_{k}^{2}\omega_{j}^{2}\omega_{p}^{2}-3\omega_{i}^{4}\omega_{p}^{2}-\omega_{j}^{2}\omega_{i}^{4})B^{p}_{kij}] \\
r^{p}_{ijk} & = & \frac{1}{D^{(3)}_{ijkp}}[& (\quad -\omega_{k}^{2}\omega_{j}^{4}-\omega_{i}^{2}\omega_{k}^{4}-2\omega_{j}^{2}\omega_{i}^{2}\omega_{p}^{2}+3\omega_{j}^{2}\omega_{p}^{4}-\omega_{k}^{4}\omega_{j}^{2} \\
 & & & \quad -2\omega_{i}^{2}\omega_{k}^{2}\omega_{p}^{2}+3\omega_{i}^{2}\omega_{p}^{4}-\omega_{j}^{4}\omega_{i}^{2}-\omega_{i}^{4}\omega_{k}^{2}+3\omega_{k}^{2}\omega_{p}^{4}+\omega_{i}^{6}-\omega_{p}^{6}-3\omega_{k}^{4}\omega_{p}^{2}+\omega_{k}^{6} \\
 & & & \quad -3\omega_{j}^{4}\omega_{p}^{2}+10\omega_{i}^{2}\omega_{k}^{2}\omega_{j}^{2}+\omega_{j}^{6}-2\omega_{k}^{2}\omega_{j}^{2}\omega_{p}^{2}-3\omega_{i}^{4}\omega_{p}^{2}-\omega_{j}^{2}\omega_{i}^{4})(h^{p}_{ijk}+A^{p}_{ijk}+A^{p}_{kij}+ \\
 & & & +(\quad 4\omega_{i}^{2}\omega_{k}^{2}\omega_{j}^{4}-4\omega_{k}^{2}\omega_{j}^{4}\omega_{p}^{2}+4\omega_{i}^{2}\omega_{k}^{4}\omega_{j}^{2}-4\omega_{j}^{2}\omega_{k}^{4}\omega_{p}^{2}-6\omega_{i}^{4}\omega_{k}^{2}\omega_{j}^{2}+2\omega_{k}^{2}\omega_{j}^{2}\omega_{p}^{4} \\
 & & & \quad +4\omega_{i}^{2}\omega_{k}^{2}\omega_{j}^{2}\omega_{p}^{2}-4\omega_{j}^{4}\omega_{k}^{4}+2\omega_{k}^{2}\omega_{j}^{6}+2\omega_{j}^{2}\omega_{k}^{6})B^{p}_{ijk} \\
 & & & +(\quad 4\omega_{i}^{2}\omega_{k}^{4}\omega_{j}^{2}+4\omega_{i}^{2}\omega_{k}^{2}\omega_{j}^{2}\omega_{p}^{2}+2\omega_{i}^{2}\omega_{k}^{2}\omega_{p}^{4}+2\omega_{i}^{2}\omega_{k}^{6}+2\omega_{i}^{6}\omega_{k}^{2}-4\omega_{i}^{4}\omega_{k}^{4} \\
 & & & \quad -6\omega_{i}^{2}\omega_{k}^{2}\omega_{j}^{4}-4\omega_{i}^{2}\omega_{k}^{4}\omega_{p}^{2}-4\omega_{i}^{4}\omega_{k}^{2}\omega_{p}^{2}+4\omega_{i}^{4}\omega_{k}^{2}\omega_{j}^{2})B^{p}_{jik} \\
 & & & +(\quad -4\omega_{i}^{4}\omega_{j}^{4}+2\omega_{i}^{2}\omega_{j}^{2}\omega_{p}^{4}-6\omega_{i}^{2}\omega_{k}^{4}\omega_{j}^{2}+2\omega_{i}^{2}\omega_{j}^{6}+2\omega_{i}^{6}\omega_{j}^{2}-4\omega_{i}^{2}\omega_{j}^{4}\omega_{p}^{2} \\
 & & & \quad +4\omega_{i}^{2}\omega_{k}^{2}\omega_{j}^{2}\omega_{p}^{2}+4\omega_{i}^{2}\omega_{k}^{2}\omega_{j}^{4}-4\omega_{i}^{4}\omega_{j}^{2}\omega_{p}^{2}+4\omega_{i}^{4}\omega_{k}^{2}\omega_{j}^{2})B^{p}_{kij}] \\
\mu^{p}_{ijk} & = & \frac{1}{D^{(3)}_{ijkp}}[& (\quad -4\omega_{i}^{2}\omega_{j}^{2}+4\omega_{i}^{2}\omega_{p}^{2}+12\omega_{i}^{2}\omega_{k}^{2}-4\omega_{k}^{2}\omega_{j}^{2}+4\omega_{k}^{2}\omega_{p}^{2}+2\omega_{p}^{4} \\
 & & & \quad +10\omega_{j}^{4}-6\omega_{i}^{4}-6\omega_{k}^{4}-12\omega_{j}^{2}\omega_{p}^{2})(h^{p}_{ijk}+A^{p}_{ijk}+A^{p}_{kij}+A^{p}_{jik}) \\
 & & & +(\quad 10\omega_{i}^{2}\omega_{k}^{2}\omega_{j}^{2}-2\omega_{k}^{2}\omega_{j}^{2}\omega_{p}^{2}+2\omega_{i}^{2}\omega_{k}^{2}\omega_{p}^{2}-6\omega_{j}^{2}\omega_{i}^{2}\omega_{p}^{2}+9\omega_{k}^{2}\omega_{j}^{4} \\
 & & & \quad +3\omega_{j}^{6}-11\omega_{k}^{4}\omega_{j}^{2}+\omega_{k}^{2}\omega_{p}^{4}+3\omega_{i}^{2}\omega_{k}^{4}+\omega_{k}^{4}\omega_{p}^{2}-\omega_{k}^{6}-3\omega_{i}^{4}\omega_{p}^{2} \\
 & & & \quad +3\omega_{i}^{2}\omega_{p}^{4}+\omega_{i}^{6}84-\omega_{p}^{6}+5\omega_{j}^{2}\omega_{p}^{4}-7\omega_{j}^{4}\omega_{p}^{2}-5\omega_{j}^{4}\omega_{i}^{2}+\omega_{j}^{2}\omega_{i}^{4}-3\omega_{i}^{4}\omega_{k}^{2})B^{p}_{ijk} \\
 & & & +(\quad -18\omega_{i}^{2}\omega_{k}^{2}\omega_{j}^{2}-3\omega_{j}^{2}\omega_{p}^{4}+5\omega_{k}^{4}\omega_{j}^{2}+5\omega_{j}^{2}\omega_{i}^{4}-\omega_{k}^{2}\omega_{j}^{4}-\omega_{j}^{4}\omega_{i}^{2}+7\omega_{i}^{4}\omega_{p}^{2} \\
 & & & \quad +3\omega_{i}^{2}\omega_{k}^{4}+\omega_{p}^{6}-\omega_{j}^{6}-3\omega_{k}^{6}-3\omega_{i}^{6}+3\omega_{j}^{4}\omega_{p}^{2}+7\omega_{k}^{4}\omega_{p}^{2}+3\omega_{i}^{4}\omega_{k}^{2}+6\omega_{k}^{2}\omega_{j}^{2}\omega_{p}^{2} \\
 & & & \quad +6\omega_{j}^{2}\omega_{i}^{2}\omega_{p}^{2}-5\omega_{k}^{2}\omega_{p}^{4}-6\omega_{i}^{2}\omega_{k}^{2}\omega_{p}^{2}-5\omega_{i}^{2}\omega_{p}^{4})B^{p}_{jik} \\
 & & & +(\quad \omega_{k}^{4}\omega_{j}^{2}+\omega_{k}^{6}-2\omega_{j}^{2}\omega_{i}^{2}\omega_{p}^{2}+10\omega_{i}^{2}\omega_{k}^{2}\omega_{j}^{2}+3\omega_{k}^{2}\omega_{p}^{4}-6\omega_{k}^{2}\omega_{j}^{2}\omega_{p}^{2}-3\omega_{i}^{2}\omega_{k}^{4} \\
 & & & \quad +2\omega_{i}^{2}\omega_{k}^{2}\omega_{p}^{2}+5\omega_{j}^{2}\omega_{p}^{4}+3\omega_{i}^{4}\omega_{k}^{2}-11\omega_{j}^{2}\omega_{i}^{4}-3\omega_{k}^{4}\omega_{p}^{2}+\omega_{i}^{4}\omega_{p}^{2}+3\omega_{j}^{6} \\
 & & & \quad +\omega_{i}^{2}\omega_{p}^{4}-\omega_{i}^{6}-\omega_{p}^{6}+9\omega_{j}^{4}\omega_{i}^{2}-7\omega_{j}^{4}\omega_{p}^{2}-5\omega_{k}^{2}\omega_{j}^{4})B^{p}_{kij}]
\end{array}
$$

$$
\begin{aligned}
\nu^p_{kij} = \frac{1}{D^{(3)}_{ijkp}}\Big[\; (\; & -2\omega_j^2\omega_i^2\omega_p^2 + \omega_j^4\omega_p^2 - \omega_j^6 + 3\omega_k^2\omega_j^4 + \omega_k^6 - 7\omega_i^4\omega_p^2 - 11\omega_j^4\omega_i^2 + 2\omega_k^2\omega_j^2\omega_p^2 \\
& +9\omega_j^2\omega_i^4 + 3\omega_i^6 + 5\omega_i^2\omega_p^4 + \omega_j^2\omega_p^4 - 5\omega_i^4\omega_k^2 + 10\omega_i^2\omega_k^2\omega_j^2 - 6\omega_i^2\omega_k^2\omega_p^2 + 3\omega_k^2\omega_p^4 \\
& -3\omega_k^4\omega_p^2 - \omega_p^6 + \omega_i^2\omega_k^4 - 3\omega_k^4\omega_j^2)(h^p_{ijk} + A^p_{ijk} + A^p_{kij} + A^p_{jik}) \\
+(\; & 3\omega_i^4\omega_p^4 + \omega_i^8 - 6\omega_i^4\omega_k^2\omega_p^2 - 5\omega_i^4\omega_k^4 + 6\omega_i^2\omega_k^2\omega_j^2\omega_p^2 + 5\omega_i^2\omega_j^2\omega_p^4 - 7\omega_i^2\omega_j^4\omega_p^2 \\
& +18\omega_i^4\omega_k^2\omega_j^2 + 3\omega_i^2\omega_k^6 - 3\omega_i^2\omega_k^2\omega_j^4 - 7\omega_i^2\omega_k^4\omega_p^2 - 6\omega_i^4\omega_j^2\omega_p^2 + \omega_i^6\omega_k^2 - 3\omega_i^2\omega_k^4\omega_j^2 \\
& +5\omega_i^2\omega_k^2\omega_p^4 + \omega_i^6\omega_j^2 - 5\omega_i^4\omega_j^4 - 3\omega_i^6\omega_p^2 - \omega_i^2\omega_p^6 + 3\omega_i^2\omega_j^6)B^p_{jik} \\
+(\; & 3\omega_k^2\omega_j^6 + 6\omega_i^2\omega_j^4\omega_p^2 + 5\omega_i^4\omega_j^4 + 7\omega_i^4\omega_j^2\omega_p^2 - 9\omega_i^4\omega_k^2\omega_j^2 - \omega_i^2\omega_j^6 + \omega_p^6\omega_j^2 - \omega_j^2\omega_k^4\omega_p^2 \\
& -3\omega_p^4\omega_j^4 + 3\omega_j^6\omega_p^2 + 11\omega_i^2\omega_k^4\omega_j^2 - 3\omega_i^6\omega_j^2 - 5\omega_i^2\omega_j^2\omega_p^4 - 2\omega_k^2\omega_j^4\omega_p^2 - \omega_k^2\omega_j^2\omega_p^4 \\
& +2\omega_i^2\omega_k^2\omega_j^2\omega_p^2 + \omega_j^2\omega_k^6 - 3\omega_j^4\omega_k^4 - 10\omega_i^2\omega_k^2\omega_j^4 - \omega_j^8)B^p_{ijk} \\
+(\; & -4\omega_i^4\omega_k^2\omega_j^2 - 6\omega_i^2\omega_k^4\omega_j^2 - 4\omega_i^4\omega_j^4 - 12\omega_i^4\omega_j^2\omega_p^2 + 2\omega_i^2\omega_j^2\omega_p^4 + 4\omega_i^2\omega_j^4\omega_p^2 \\
& +12\omega_i^2\omega_k^2\omega_j^4 + 10\omega_i^6\omega_j^2 - 6\omega_i^2\omega_j^6 + 4\omega_i^2\omega_k^2\omega_j^2\omega_p^2)B^p_{kij}\Big]
\end{aligned}
$$

$$
\begin{aligned}
\nu^p_{jik} = \frac{1}{D^{(3)}_{ijkp}}\Big[\; (\; & \omega_i^2\omega_k^4 + \omega_i^4\omega_p^2 + 3\omega_j^2\omega_i^4 + \omega_k^4\omega_p^2 - \omega_p^6 - 3\omega_j^4\omega_p^2 + 3\omega_j^2\omega_p^4 + 2\omega_j^2\omega_i^2\omega_p^2 + \omega_j^6 \\
& +3\omega_k^4\omega_j^2 + \omega_i^2\omega_p^4 + 2\omega_k^2\omega_j^2\omega_p^2 - \omega_k^6 + 2\omega_i^2\omega_k^2\omega_j^2 - 10\omega_i^2\omega_k^2\omega_p^2 - 3\omega_j^4\omega_i^2 \\
& +\omega_i^4\omega_k^2 - 3\omega_k^2\omega_j^4 - \omega_i^6 + \omega_k^2\omega_p^4)(h^p_{ijk} + A^p_{ijk} + A^p_{kij} + A^p_{jik}) \\
+(\; & -3\omega_i^4\omega_k^4 - \omega_i^2\omega_k^2\omega_p^4 - 2\omega_i^2\omega_k^4\omega_j^2 - 2\omega_i^2\omega_k^4\omega_p^2 - \omega_i^4\omega_k^2\omega_p^2 + \omega_i^6\omega_k^2 \\
& +\omega_k^2\omega_j^6 - 2\omega_j^2\omega_k^4\omega_p^2 - \omega_i^4\omega_k^2\omega_j^2 + 3\omega_i^2\omega_k^6 - \omega_k^2\omega_j^2\omega_p^4 - 3\omega_p^4\omega_k^4 + 3\omega_k^6\omega_p^2 - \omega_k^2\omega_j^4\omega_p^2 \\
& -\omega_k^8 + 10\omega_i^2\omega_k^2\omega_j^2\omega_p^2 + 3\omega_j^2\omega_k^6 - 3\omega_j^4\omega_k^4 - \omega_i^2\omega_k^2\omega_j^4 + \omega_k^2\omega_p^6)B^p_{ijk} \\
+(\; & -4\omega_i^2\omega_k^2\omega_j^2\omega_p^2 + 4\omega_i^2\omega_k^4\omega_j^2 + 4\omega_i^4\omega_k^2\omega_j^2 - 2\omega_i^2\omega_k^6 + 6\omega_i^2\omega_k^2\omega_p^4 - 4\omega_i^4\omega_k^2\omega_p^2 \\
& -2\omega_i^6\omega_k^2 + 4\omega_i^4\omega_k^4 - 2\omega_i^2\omega_k^2\omega_j^4 - 4\omega_i^2\omega_k^4\omega_p^2)B^p_{jik} \\
+(\; & -2\omega_i^4\omega_j^2\omega_p^2 + 3\omega_i^6\omega_k^2 - 3\omega_i^4\omega_k^4 + 3\omega_i^6\omega_p^2 - 3\omega_i^4\omega_p^4 - \omega_i^2\omega_j^2\omega_p^4 - 2\omega_i^4\omega_k^2\omega_p^2 \\
& -\omega_i^8 - \omega_i^2\omega_k^4\omega_p^2 + \omega_i^2\omega_j^6 - 2\omega_i^4\omega_k^2\omega_j^2 - \omega_i^2\omega_k^2\omega_p^4 + 3\omega_i^6\omega_j^2 + 10\omega_i^2\omega_k^2\omega_j^2\omega_p^2 \\
& +\omega_i^2\omega_k^6 + \omega_i^2\omega_p^6 - \omega_i^2\omega_k^2\omega_j^4 - 3\omega_i^4\omega_j^4 - \omega_i^2\omega_j^4\omega_p^2 - \omega_i^2\omega_k^4\omega_j^2)B^p_{kij}\Big]
\end{aligned}
$$

$$
\begin{aligned}
\nu^p_{ijk} = \frac{1}{D^{(3)}_{ijkp}}\Big[\; (\; & -11\omega_k^2\omega_j^4 - 3\omega_i^4\omega_p^2 + 9\omega_k^4\omega_j^2 - \omega_p^6 + \omega_j^4\omega_p^2 - 7\omega_k^4\omega_p^2 - \omega_j^6 + \omega_i^4\omega_k^2 \\
& +2\omega_j^2\omega_i^2\omega_p^2 - 3\omega_j^2\omega_i^4 + 3\omega_k^6 - 6\omega_i^2\omega_k^2\omega_p^2 + \omega_j^2\omega_p^4 + 3\omega_i^2\omega_p^4 + 5\omega_k^2\omega_p^4 - 5\omega_i^2\omega_k^4 \\
& +\omega_i^6 + 3\omega_j^4\omega_i^2 - 2\omega_k^2\omega_j^2\omega_p^2 + 10\omega_i^2\omega_k^2\omega_j^2)(h^p_{ijk} + A^p_{ijk} + A^p_{kij} + A^p_{jik}) \\
+(\; & 12\omega_i^2\omega_k^2\omega_j^4 - 6\omega_k^2\omega_j^6 - 4\omega_i^2\omega_k^4\omega_j^2 + 4\omega_i^2\omega_k^2\omega_j^2\omega_p^2 + 4\omega_k^2\omega_j^4\omega_p^2 \\
& +2\omega_k^2\omega_j^2\omega_p^4 - 4\omega_j^4\omega_k^4 - 12\omega_j^2\omega_k^4\omega_p^2 - 6\omega_i^4\omega_k^2\omega_j^2 + 10\omega_j^2\omega_k^6)B^p_{ijk} \\
+(\; & 3\omega_k^2\omega_j^6 + 6\omega_i^2\omega_k^2\omega_j^2\omega_p^2 + \omega_j^2\omega_k^6 - 3\omega_i^2\omega_k^2\omega_j^4 - \omega_k^2\omega_p^6 - 3\omega_i^4\omega_k^2\omega_j^2 \\
& +18\omega_i^2\omega_k^4\omega_j^2 - 6\omega_i^2\omega_k^4\omega_p^2 - 5\omega_j^4\omega_k^4 + 3\omega_i^6\omega_k^2 + \omega_i^2\omega_k^6 + 5\omega_i^2\omega_k^2\omega_p^4 - 6\omega_j^2\omega_k^4\omega_p^2 \\
& +\omega_k^8 + 5\omega_k^2\omega_j^2\omega_p^4 - 3\omega_k^6\omega_p^2 + 3\omega_p^4\omega_k^4 - 7\omega_i^4\omega_k^2\omega_p^2 - 5\omega_i^4\omega_k^4 - 7\omega_k^2\omega_j^4\omega_p^2)B^p_{jik} \\
+(\; & -\omega_i^4\omega_j^2\omega_p^2 - \omega_j^8 - 5\omega_k^2\omega_j^2\omega_p^4 + 7\omega_j^2\omega_k^4\omega_p^2 + 5\omega_j^4\omega_k^4 + 3\omega_j^6\omega_p^2 - \omega_k^2\omega_j^6 \\
& -3\omega_p^4\omega_j^4 + \omega_i^6\omega_j^2 + 2\omega_i^2\omega_k^2\omega_j^2\omega_p^2 + 11\omega_i^4\omega_k^2\omega_j^2 + \omega_p^6\omega_j^2 - 10\omega_i^2\omega_k^2\omega_j^4 - 3\omega_j^2\omega_k^6 \\
& -3\omega_i^4\omega_j^4 - 9\omega_i^2\omega_k^4\omega_j^2 + 3\omega_i^2\omega_j^6 + 6\omega_k^2\omega_j^4\omega_p^2 - \omega_i^2\omega_j^2\omega_p^4 - 2\omega_i^2\omega_j^4\omega_p^2)B^p_{kij}\Big]
\end{aligned}
$$

where $D^{(3)}_{ijkp} = (\omega_k + \omega_i - \omega_p - \omega_j)(\omega_k + \omega_i - \omega_p + \omega_j)(-\omega_k + \omega_i + \omega_p + \omega_j)(-\omega_k + \omega_i + \omega_p - \omega_j)(\omega_k + \omega_i + \omega_p - \omega_j)(\omega_k + \omega_i + \omega_p + \omega_j)(-\omega_k + \omega_i - \omega_p + \omega_j)(-\omega_k + \omega_i - \omega_p - \omega_j)$.

Nonlinear normal modes in damped - forced systems

O.V.Gendelman

Faculty of Mechanical Engineering, Technion - Israel Institute of Technology, Haifa, Israel

Abstract In the previous chapters of this volume, the nonlinear normal modes (NNMs) were defined and explored in conservative systems. In current chapter, the notion is extended for dynamical systems with viscous damping, as well as for the systems with external forcing. We start with a discussion about normal modes in linear damped systems. Then, we discuss an effect of the damping on the NNMs in nonlinear models. Finally, it is demonstrated that the problem of targeted energy transfer (TET) may be efficiently treated with the help of the NNM concept with account of possible time scale separation – both in autonomous and in forced systems.

1 Introduction

This volume is devoted to the nonlinear modal analysis. In a conservative case, despite lack of the superposition, one can conveniently define the nonlinear normal modes (NNMs) and efficiently use them for certain problems of the nonlinear analysis. Different approaches and notions were developed; some of these ideas and related results are reviewed in this volume in Chapters 1 and 2. Needless to say, the NNMs are only particular solutions and cannot completely represent the dynamics even in simple low-dimensional benchmark models. Still, some important dynamical responses may be revealed this way. Besides, one should mention that quite often the NNMs are *the only* analytic or semi-analytic solutions available for even moderately nonlinear system – all the rest of dynamics may be obtained only by numeric brutal force.

However, if the system has damping (even linear damping, to say nothing of the nonlinear one), then the things become much more complicated. One can perceive the NNMs as analytic extensions of the linear normal modes – of course, if such extension is possible in specific system. In linear damped system, however, the traditional normal modes generically cease to exist, as it is demonstrated below in Section 1. So, if one sees the NNMs as

G. Kerschen (Ed.), *Modal Analysis of Nonlinear Mechanical Systems*, CISM International Centre for Mechanical Sciences DOI 10.1007/ 978-3-7091-1791-0_4

analytic continuation of something in the linear case, the damping poses a problem - there is nothing to continue analytically! In the same time, as we mention below in Section 1, there is a way to construct certain form of the modal description for the linear damped system. These modes are complex and form 2N-dimensional modal space for the system with N degrees of freedom. It might be that these "modes" can serve as a basis for the theory of analytic continuation in the damped case. Thus, one might be able to define the damped NNMs rigorously. To the best of the author's knowledge, such theory does not exist today.

There exists another major complication related to the definition of the NNMs in the damped case. The NNMs are known to undergo numerous and sometimes complicated bifurcations [(1)]. In Section 2 it is demonstrated that under the damped dynamics the passage through the bifurcation generically destroys the NNM - even as it is stable in the conservative case. So, common definitions of the NNM stability [(1), (2)] are hardly applicable for the damped case; alternative approaches do not seem to exist - again, to the best of the author's knowledge.

Given all this information, very reasonable question arises – why one should bother himself with the problem of the NNMs in damped systems? The answer is presented in Sections 3 and 4. It terns out that in certain dynamical systems with damping the "damped" NNM can be defined *asymptotically* as an attractor of dynamic flow at certain time scale. This definition can be put to good use, for instance, for the problem of targeted energy transfer (TET)[(3), (4)]. Section 3 is devoted to the damped TET, Section 4 - to the TET and analysis of attractors in certain classes of forced-damped nonlinear systems.

2 Linear modal analysis of damped systems

It is well-known that generic discrete oscillatory system with viscous damping can be described by the following matrix equation:

$$M\ddot{x} + C\dot{x} + Kx = Q(t) \tag{1}$$

Here M is the mass matrix, C- the damping matrix and K – the stiffness matrix. Vector Q(t) represents external forces acting on the oscillatory system. All these matrices are considered symmetric.

Let us first consider the case without damping, C=0. In this case, as it is well – known [(5)], one can diagonalize both M and K with the help of the same modal transformation:

$$\begin{aligned} &U^T M U = I, \quad U^T K U = D = diag(\omega_1^2, ..., \omega_n^2) \\ &Ux = \xi, \quad \mathrm{U}^T Q = F \end{aligned} \tag{2}$$

Thus, equation (1) with C=0 is decomposed into a set of N independent modal equations, which represent independent forced linear oscillators:

$$\ddot{\xi}_k + \omega_k^2 \xi_k = F_k(t), k = 1..n \tag{3}$$

After trivial solution of these equations, one can easily compute the solution in terms of the initial variables by means of the inverse modal transformation:

$$x = U^T \xi \tag{4}$$

It should be mentioned that the normalization of the modal matrix is uniquely defined by the fact that it makes the mass matrix unit.

If the damping is nonzero, the outlined approach is not applicable straightforwardly, since generically it is not possible to diagonalize *three* matrices simultaneously by means of the same modal matrix. Still, one can outline a particular case when the modal description is still possible – the case of *proportional damping*[(5)]. In this particular case, the damping matrix is proportional to a sum of the mass and the stiffness matrices with some real coefficients:

$$C = \alpha M + \beta K, \quad \alpha, \beta \in R \tag{5}$$

Then, the modal transformation brings the initial system to the following form:

$$\ddot{\xi} + \alpha\dot{\xi} + D(\beta\dot{\xi} + \xi) = F(t) \tag{6}$$

One can see that the modal decomposition succeeds in this case and one obtains a set of N independent forced-damped linear oscillators:

$$\ddot{\xi}_k + 2\zeta_k\omega_k\dot{\xi}_k + \omega_k^2\xi_k = F_k(t), k = 1..n, \qquad \zeta_k = \frac{\alpha + \beta\omega_k^2}{2\omega_k} \tag{7}$$

Here ζ_k is the modal damping coefficient of each mode. It depends on the modal frequency; therefore, In the same system some modes can be underdamed, and the others – overdamped or critically damped.

Generally speaking, one cannot expect that in some particular system the damping matrix will be exactly the linear combination of the stiffness and the damping matrices. Nevertheless, the case of the proportional damping well may have a practical interest – for instance, if the system is highly symmetric.

For example, let us consider a system of two coupled linear oscillators presented in Fig. 1.

Let us consider the symmetric case with the following values of parameters:

$$\begin{gathered} m_1 = m_2 = 1, c_{11} = c_{22} = c_0, \\ \lambda_{11} = \lambda_{22} = \lambda_0 \end{gathered} \tag{8}$$

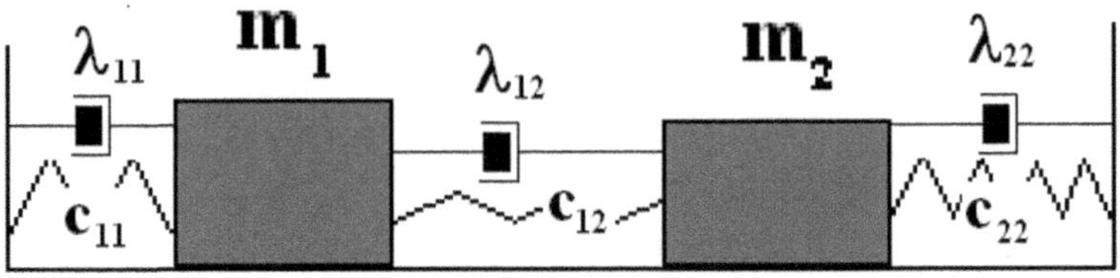

Figure 1. Sketch of the model system

In this case, the system easily can be reduced to the independent modal equations in terms of in-phase and out-of-phase modes:

$$\begin{aligned}
&\ddot{x}_1 + \lambda_0 \dot{x}_1 + c_0 x_1 + c_{12}(x_1 - x_2) + \lambda_{12}(\dot{x}_1 - \dot{x}_2) = 0 \\
&\ddot{x}_2 + \lambda_0 \dot{x}_2 + c_0 x_2 + c_{12}(x_2 - x_1) + \lambda_{12}(\dot{x}_2 - \dot{x}_1) = 0 \\
&V = x_1 + x_2, W = x_1 - x_2 \\
&\ddot{V} + \lambda_0 \dot{V} + c_0 V = 0 \\
&\ddot{W} + (\lambda_0 + 2\lambda_{12})\dot{W} + (c_0 + 2c_{12})W = 0
\end{aligned} \tag{9}$$

Here x_i, i =1,2 are displacements of the individual oscillators, and V and W are displacements of the in-phase and out-of-phase modes respectively. In the case of the proportional damping, the modal matrix is the same as in the undamped case – the **modal shapes do not change**. The damping ratios of two modes can be very different:

$$\zeta_V = \frac{\lambda_0}{2\sqrt{c_0}}, \zeta_W = \frac{\lambda_0 + 2\lambda_{12}}{2\sqrt{c_0 + 2c_{12}}} \tag{10}$$

Let us adopt the following numeric values of parameters;

$$\begin{aligned}
&\lambda_0 = 0.1, c_0 = 1, \lambda_{12} = 1.5, c_{12} = 0.1 \Rightarrow \zeta_V = 0.05, \zeta_W = 1.41 \\
&\dot{x}_1(0) = 1, x_1(0) = 0, x_2(0) = 0, \dot{x}_2(0) = 0
\end{aligned} \tag{11}$$

Time series for the oscillator displacements are presented in Fig. 2.

One can see that after some brief initial transient both oscillators have almost the same displacements – in other terms, only the in-phase mode is observed. Needless to say, it is related to the fact that for given example the out-of-phase mode has much higher modal damping coefficient. In fact, the out-of-phase mode is overdamped in this case.

For generic symmetric damping matrix, $U^T CU$ is not diagonal; therefore standard modal analysis is not applicable. Still, the equations of motion can be diagonalized in the state space [(5)]. The state vector is defined as:

$$y = (x^T \dot{x}^T)^T \tag{12}$$

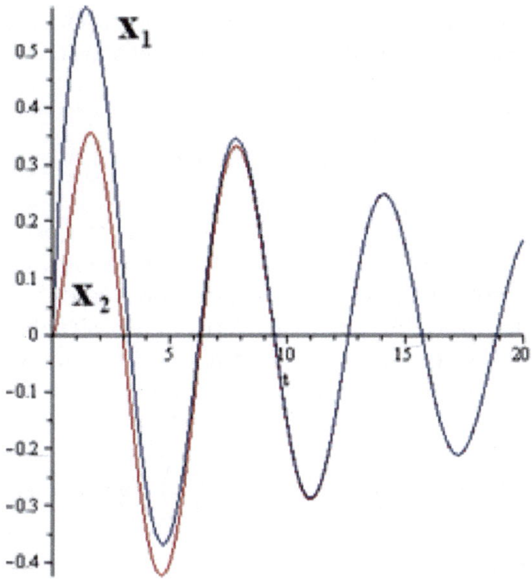

Figure 2. Time series - the case of proportional damping

The equations of motion can be reshaped as:

$$\ddot{x} = -M^{-1}C\dot{x} - M^{-1}Kx + M^{-1}Q \tag{13}$$

In terms of the state space introduced above, the equations of motion can be presented as:

$$\dot{y} = Ay + BQ, A = \begin{pmatrix} 0 & I \\ -M^{-1}K & -M^{-1}C \end{pmatrix}, \quad \mathrm{B}=\begin{pmatrix} 0 \\ M^{-1} \end{pmatrix} \tag{14}$$

Let us consider the unforced system: **Q**=0. In this case the solution is governed by eigenvalues and eigenvectors of A. The eigenvalues may be complex and appear in conjugate pairs. The eigenvectors are, generically, *not orthogonal, but biorthogonal*:

$$\begin{array}{l} Aq = \lambda_k q, k = 1, ..2n \\ q_k - \text{system of eigenvectors} \\ p^T A = \lambda_k p \\ p_k - \text{system of "left" eigenvectors} \\ p_j^T A q_k = \lambda_j p_j^T q_k = \lambda_k p_j^T q_k \Rightarrow p_j^T q_k = \delta_{jk} \end{array} \tag{15}$$

"Modal" decomposition in the systems with non-proportional damping is performed with respect to the system of "right" eigenvectors and with the help of the system of "left" eigenvectors:

$$\begin{array}{l} y = \sum\limits_{k=1}^{2n} a_k q_k \Rightarrow a_k = p_k^T y \\ \lambda_k a_k = p_k^T A y \end{array} \tag{16}$$

Thus, in order to accomplish the modal decomposition, the eigenvectors problem should be solved twice – for instance, for A and A^T.

To illustrate the procedure, we'll perform it for arguably the simplest possible example – linear conservative oscillator. Its equation of motion, state space and representation un terms of "complex modes" are presented below:

$$\begin{array}{l} \ddot{x} + \omega^2 x = 0;\ \mathrm{x(0)=x_0},\ \dot{x}(0) = x_1, y = \begin{pmatrix} x \\ \dot{x} \end{pmatrix};\ y(0) = y_0 = \begin{pmatrix} x_0 \\ x_1 \end{pmatrix} \\ \mathrm{A}=\begin{pmatrix} 0 & 1 \\ -\omega^2 & 0 \end{pmatrix} \\ \lambda_1 = i\omega, q_1 = \frac{1}{\sqrt{1+\omega^2}} \begin{pmatrix} 1 \\ i\omega \end{pmatrix}; p_1 = \frac{1}{\sqrt{1+\omega^2}} \begin{pmatrix} i\omega \\ 1 \end{pmatrix} \\ \lambda_2 = -i\omega, q_2 = \frac{1}{\sqrt{1+\omega^2}} \begin{pmatrix} 1 \\ -i\omega \end{pmatrix}; p_2 = \frac{1}{\sqrt{1+\omega^2}} \begin{pmatrix} -i\omega \\ 1 \end{pmatrix} \end{array} \tag{17}$$

The solution of arbitrary initial value problem in terms of the state variables is written as follows:

$$y(t) = p_1^T y_0 \exp(i\omega t) q_1 + p_2^T y_0 \exp(i\omega t) q_2 \tag{18}$$

To grasp the physical sense of this solution, let us multiply both sides by transposed "left" eigenvectors. After easy manipulations, we get:

$$\begin{array}{l} \psi(t) = \dot{x} + i\omega x = (x_1 + i\omega x_0)\exp(i\omega t) = \psi(0)\exp(i\omega t) \\ \psi^*(t) = \dot{x} - i\omega x = (x_1 - i\omega x_0)\exp(-i\omega t) = \psi^*(0)\exp(-i\omega t) \end{array} \tag{19}$$

Thus, we have obtained interesting simplification of the simplest possible linear oscillator – complexification. Physically, it corresponds to consideration of rotations instead of oscillations. This idea is, of course, not so necessary for the linear oscillator, but allows nice simplifications for the problems related to quasilinear and even strongly nonlinear oscillations. Some details on use of this procedure will be given in the following Sections.

Let us reconsider the 2DOF damped system depicted in Fig. 1 - this time with non-proportional damping. The set of parameters in current example is selected as:

$$m_1 = m_2 = 1, c_{11} = c_{22} = c_0, \\ \lambda_{22} = \lambda_0 = 0, \lambda_{11} \neq 0 \tag{20}$$

The matrix in the state space looks like

$$A = \begin{pmatrix} 0 & 0 & 1 & 0 \\ 0 & 0 & 0 & 1 \\ -(c_0 + c_{12}) & c_{12} & -\lambda_{11} & 0 \\ c_{12} & -(c_0 + c_{12}) & 0 & 0 \end{pmatrix} \tag{21}$$

It is not easy to write down the explicit expressions for eigenvalues and eigenvectors. To assess this problem, we consider the following numeric example:

$$c_0 = 1, \lambda_{11} = 1.5, c_{12} = 0.1 \tag{22}$$

The eigenvalues and eigenvectors are expressed as:

$$\begin{aligned} \lambda_{1,2} = & -0.747 \pm 0.73i, q_{1,2} = \begin{pmatrix} 0.49 \pm 0.48i \\ 0.0012 \pm 0.044i \\ -0.721 \pm 0i \\ -0.033 \mp 0.032i \end{pmatrix} \\ \lambda_{3,4} = & -0.003 \pm 1.05i, q_{3,4} = \begin{pmatrix} 0.0439 \mp 0.00026i \\ 0.002 \pm 0.689i \\ 0.000135 \pm 0.046i \\ -0.722 \pm 0i \end{pmatrix} \end{aligned} \tag{23}$$

With the help of these data, it is easy to write down the closed – form solution; it will not be very instructive. So, one can see that even in a system of linear damped oscillators it is possible to perform the "modal" decomposition. However, generically these "modes" do not allow deeper insight into the system dynamics, and, so, they are used relatively rarely. However, the idea of "complexification" can be put to good use in the nonlinear analysis. Typical example of this sort will be presented in the next Section.

3 NNMs and their bifurcations in damped nonlinear systems.

Let us consider a paradigmatic system which exhibits bifurcation of the NNMs – two Duffing oscillators with linear coupling:

$$\ddot{x}_1 + \varepsilon\lambda\dot{x}_1 + x_1 + \tfrac{4}{3}\varepsilon x_1^3 + \varepsilon\alpha(x_1 - x_2) = 0 \\ \ddot{x}_2 + \varepsilon\lambda\dot{x}_2 + x_2 + \tfrac{4}{3}\varepsilon x_2^3 + \varepsilon\alpha(x_2 - x_1) = 0 \tag{24}$$

Without damping, the antisymmetric mode in the system can lose stability, giving rise to two localized stable NNMs. It is clearly seen at the Poincare section in Fig.3.

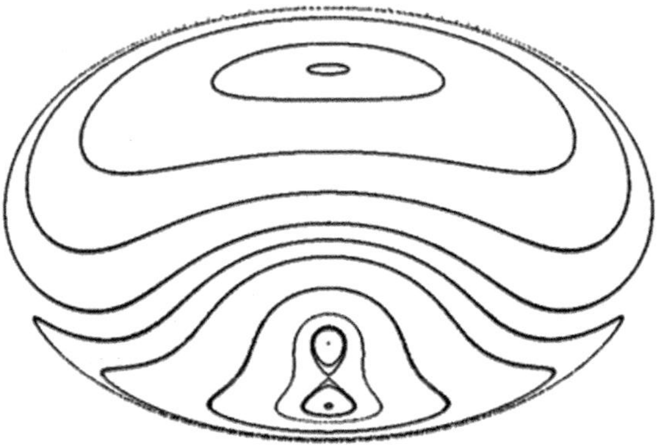

Figure 3. Poincare section for Hamiltonian system (24)

However, here we are interested in the response with nonzero damping. Note that the damping, the nonlinearity and the coupling between the oscillators are adopted to be small. If the initial conditions are set at the localized NNM, we obtain the time series presented in Fig. 4.

It is clear that the dynamics does not converge to the antisymmetic mode after passing the bifurcation point. Instead, we observe a kind of modulated antisymmetric response. To explain this behavior, we'll use the "complexification" technique [(6)].

In this technique, both dependent variables are changed by their "complex" counterparts and then multiple scales expansion is applied:

$$\begin{aligned}&\dot{x}_k + i\omega x_k = \varphi_k \exp(i\omega t), \quad k = 1,2\\&\tau_j = \varepsilon^j t, j = 0,1,...; \ \frac{d}{dt} = \frac{\partial}{\partial\tau_0} + \varepsilon\frac{\partial}{\partial\tau_1} + ...\\&\varphi_k = \varphi_{k0} + \varepsilon\varphi_{k1} + ...\end{aligned} \tag{25}$$

By regular procedure, the conditions for absence of the secular terms are written as:

$$\begin{aligned}&\varphi'_{10} + \frac{\lambda}{2}\varphi_{10} - \frac{i}{2}\left|\varphi_{10}\right|^2\varphi_{10} - \frac{i\alpha}{2}(\varphi_{10} - \varphi_{20}) = 0\\&\varphi'_{20} + \frac{\lambda}{2}\varphi_{20} - \frac{i}{2}\left|\varphi_{20}\right|^2\varphi_{20} - \frac{i\alpha}{2}(\varphi_{20} - \varphi_{10}) = 0\\&{}' \frac{\partial}{\partial\tau_1}\end{aligned} \tag{26}$$

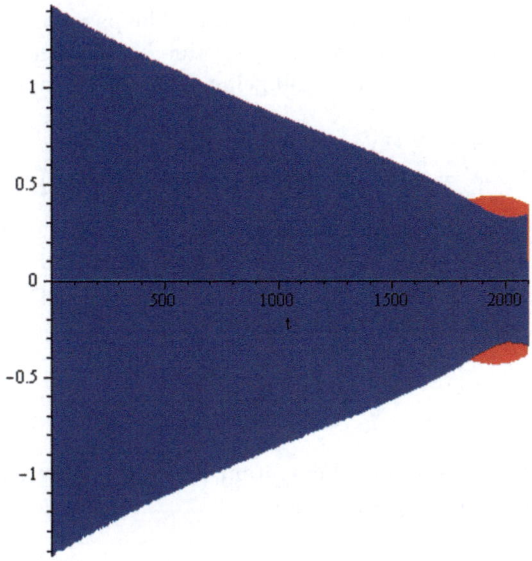

Figure 4. Destruction of the NNM and start of oscillatory response

By additional change of variables, amplitudes and phases of the new dependent variables are figured out:

$$\begin{array}{l} \varphi_{10} = N\cos\theta\exp(i\delta_1), \quad \varphi_{20} = N\sin\theta\exp(i\delta_2), \quad \delta{=}\delta_1\text{-}\delta_2 \\ N' = -\frac{\lambda}{2}N \Rightarrow N = N_0\exp(-\lambda\tau_1/2) \\ \theta' = \frac{\alpha}{2}\sin\delta \\ \delta' = \frac{1}{2}\cos 2\theta\left(N^2 + \frac{2\alpha\cos\delta}{\sin 2\theta}\right) \end{array} \tag{27}$$

In this system, variable N is related to an "occupation number" in the system (energy-like quantity), θ describes the localization of energy of one of the oscillators (θ =0, $\pi/2$ correspond to a complete localization on one of the oscillators, $\theta = \pi/4$ is the delocalized state), δ describes the phase shift.

In the conservative case (λ =0) one obtains:

$$\begin{array}{ll} N = N_0 = \text{const} & \\ \theta' = \frac{\alpha}{2}\sin\delta, & \delta' = \frac{1}{2}\cos 2\theta\left(N_0^2 + \frac{2\alpha\cos\delta}{\sin 2\theta}\right) \end{array} \tag{28}$$

Pure modes correspond to stationary points of (28). They can be summarized as follows:

$$\begin{array}{l} \delta = 0, \quad \cos 2\theta = 0 \Rightarrow \theta = \pm\frac{\pi}{4} - \text{ symmetric/antisymmetric modes} \\ \delta = 0, \quad \theta = -\frac{1}{2}\arcsin\left(\frac{2\alpha}{N_0^2}\right) \text{ - localized mode; } N_0^2 \geq 2\alpha \end{array} \tag{29}$$

Now let us turn to the non-conservative case. The passage through the bifurcation corresponds to the passage of the value N through critical value$\sqrt{2\alpha}$, as follows from (27-29). The following form of System (27) should be considered:

$$\begin{aligned} \theta' &= \tfrac{\alpha}{2}\sin\delta \\ \delta' &= \tfrac{1}{2}\cos 2\theta\left(N_0^2\exp(-\lambda\tau_1) + \tfrac{2\alpha\cos\delta}{\sin 2\theta}\right) \end{aligned} \tag{30}$$

It is easy to see that both the symmetric and the antisymmetric mode are preserved in the damped dynamics. As for the localized mode, one should look at the vicinity of the bifurcation. To be more specific, the bifurcation occurs at time instance τ^*, defined as:

$$\tau^* = -\frac{1}{\lambda}\ln\frac{2\alpha}{N_0^2} \tag{31}$$

In the vicinity of this point the system of slow – flow equations can be simplified by the following expansion:

$$\tau_1 = \tau^* + \eta, \;\; \theta=-\frac{\pi}{4} + \beta(\eta), \;\; |\beta| \ll 1, |\delta| \ll 1 \tag{32}$$

Then, the slow-flow equations in the vicinity of the bifurcation point are reduced to the following Airy equation [(7)]:

$$\beta_{\eta\eta} + 2\alpha^2\lambda\eta\beta = 0 \tag{33}$$

Typical solution for current signs selection is presented in Fig. 5.

One can see that arbitrarily small perturbation "above" the bifurcation point brings about the oscillatory behavior after the passage through the bifurcation – exactly as observed in Fig. 4. So, one can conclude that for weakly nonlinear system, the damping can preserve the NNMs with special symmetry; however, the modes can be destroyed in the vicinity of the bifurcation points.

4 Time scale separation and a problem of targeted energy transfer.

As a preliminary illustrative example of the TET [(3)], we consider a two degree-of-freedom (DOF) dissipative unforced system described by the following equations:

$$\begin{aligned} &\ddot{y}_1 + \lambda_1\dot{y}_1 + y_1 + \lambda_2(\dot{y}_1 - \dot{y}_2) + k(y_1 - y_2)^3 = 0 \\ &\varepsilon\ddot{y}_2 + \lambda_2(\dot{y}_2 - \dot{y}_1) + k(y_2 - y_1)^3 = 0 \end{aligned} \tag{34}$$

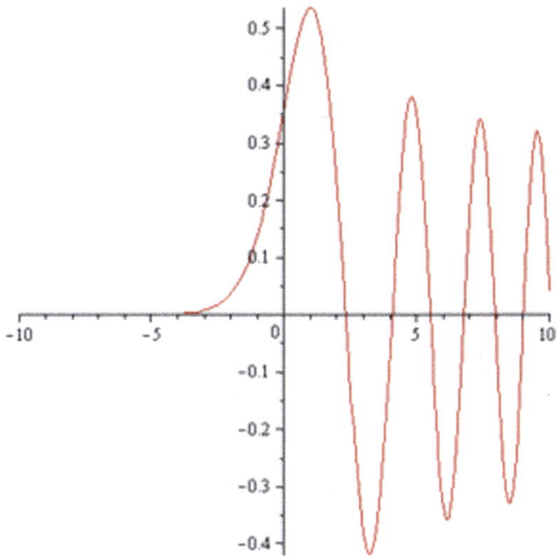

Figure 5. Typical solution for Airy equation

Physically, these equations describe a damped linear oscillator (LO) with mass and natural frequency normalized to unity, and viscous damping coefficient λ_1; and an essentially nonlinear attachment with normalized mass ε, normalized nonlinear stiffness coefficient k, and viscous damping coefficient λ_2. Note that system (34) cannot be regarded as a small perturbation of a linear system due to the strongly nonlinear coupling terms. We simulate numerically system (34) for parameter values $\varepsilon = 0.1$, $k = 0.1$, $\lambda_1 = 0.01$ and $\lambda_2 = 0.01$. The selected initial conditions correspond to an impulse $F = A\,\delta(t)$ imposed to the linear oscillator [where $\delta(t)$ is Dirac's delta function – this impulsive forcing is equivalent to imposing the initial velocity $\dot{y}_1(0+) = A$] with the system being initially at rest, i.e., $y_1(0) = y_2(0) = \dot{y}_2(0) = 0$ and $\dot{y}_1(0+) = A$. Hence, the initial energy is stored only in the LO. The instantaneous transfer of energy from the LO to the nonlinear attachment can be monitored by computing the nondimensional energy ratio κ, which denotes the portion of instantaneous total energy stored in the nonlinear attachment,

$$\kappa = \frac{E_2}{E_1 + E_2}, \quad E_1 = \frac{1}{2}\left(y_1^2 + \dot{y}_1^2\right), \quad E_2 = \frac{\varepsilon}{2}\dot{y}_2^2 + \frac{k}{4}\left(y_1 - y_2\right)^4 \qquad (35)$$

where E_1 and $E_2$are instantaneous energies of the LO and the attachment, respectively. Of course, all quantities in relations (35) are time – dependent.

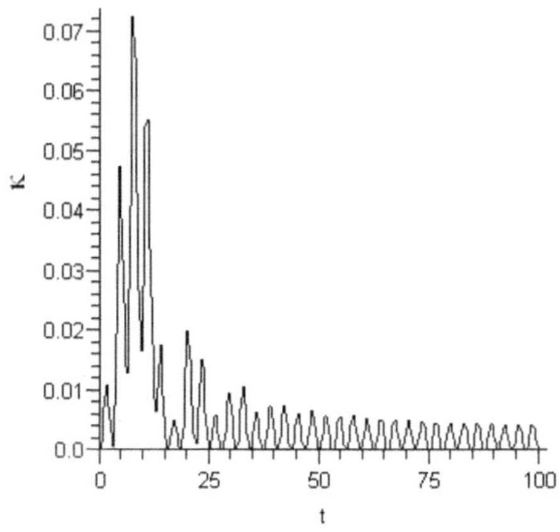

Figure 6. Evolution of the energy ratio κ for impulse strength $A = 0.5$.

In Figures 6 and 7 we depict the evolution of the energy ratio κ for impulse strengths $A = 0.5$ and $A = 0.7$, respectively. From Figure 4.6 it is clear that only a small amount of energy (of the order of 7%) is transferred from the LO to the nonlinear attachment. However, for a slightly higher impulse the energy transferred climbs to almost 95% (cf. Figure 7), and within a rather short time (t ~15, much less than characteristic time of viscous energy dissipation in the LO) almost the entire impulsive energy is passively transferred from the LO to the nonlinear attachment, which acts as nonlinear energy sink. It should be mentioned that the mass of the attachment in this particular example is just 10% of the mass of the LO.

The perturbation approach for analysis of the averaged dynamics is based on an assumption of strong mass asymmetry between the LO and the NES, as described by the small parameter ε in (34); this means that we will focus on linear oscillators with lightweight NESs. This approach does not necessarily assume small damping, and instead relies on perturbation analysis considering the NES mass ε as the small parameter. This approach is considered in this Section, for a system with parameters, $\lambda_1 = 0$, $\lambda_2 = \varepsilon\lambda$, $k = 4\varepsilon/3$. The latter convention does not affect the generality of the analysis, since it may be satisfied by appropriate rescalings of

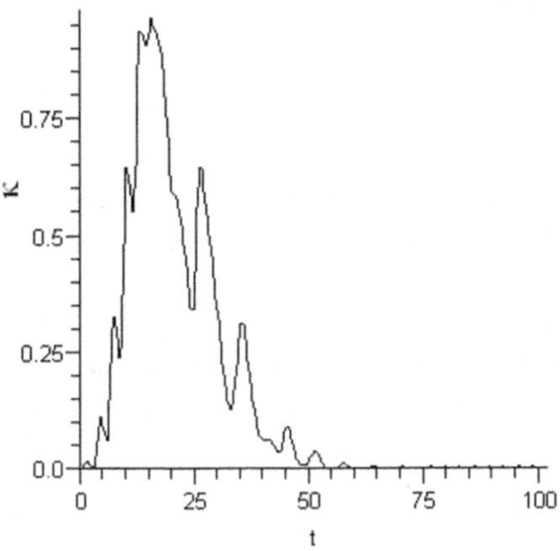

Figure 7. Evolution of the energy ratio κ for impulse strength $A = 0.7$.

the dependent and independent variables of the system.

We start our analysis of fundamental TET by considering the system of averaged (complex modulation) equations. Introducing the following changes of complex variables:

$$\begin{array}{l} \varphi_j e^{it} = \dot{y}_j + iy_j, j = 1,2 \\ \chi_1 = \frac{\varphi_1 + \varepsilon\varphi_2}{1+\varepsilon}, \chi_2 = \varphi_1 - \varphi_2 \end{array} \tag{36}$$

the modulation equations take the form:

$$\begin{array}{l} \dot{\chi}_1 + \frac{i\varepsilon}{2(1+\varepsilon)}(\chi_1 - \chi_2) = 0 \\ \dot{\chi}_2 + \frac{i}{2(1+\varepsilon)}(\chi_2 - \chi_1) + \frac{\lambda(1+\varepsilon)}{2}\chi_2 - \frac{i(1+\varepsilon)}{2}\left|\chi_2\right|^2\chi_2 = 0 \end{array} \tag{37}$$

We recall that the slow flow system (37), was derived under the assumption of 1:1 resonance between the LO and the NES, and so this model is valid only in the neighborhood of the 1:1 resonance manifold of the underlying hamiltonian system. As in (37) the complex coordinates χ_1 and χ_2 describe the oscillations of the center of mass of, and the relative displacement between the LO and the NES, respectively. By successive differentiation and simple algebra, the above averaged system may be reduced to the following single modulation equation governing the slow flow of 1:1 resonance capture in the damped dynamics:

$$\frac{d^2\chi_2}{dt^2}+\frac{d}{dt}\left[\frac{i}{2}\chi_2+\frac{\lambda(1+\varepsilon)}{2}\chi_2-\frac{i(1+\varepsilon)}{2}\left|\chi_2\right|^2\chi_2\right]+\frac{i\varepsilon}{4}\left(\lambda\chi_2-i\left|\chi_2\right|^2\chi_2\right)=0 \tag{38}$$

This equation is integrable for $\lambda = 0$, but here we are interested in the damped case $\lambda > 0$. More precisely, we assume that $\lambda >> \varepsilon$, so we treat λ as an $O(1)$ quantity.

As above, Equation (38) may be analyzed by the multiple scales approach [(8)]. To this end, we introduce the new time scales, $\tau_i = \varepsilon^i t,\ i = 0, 1, ...,$ which are treated as distinct independent variables in the following analysis. Expressing the time derivatives in (38) as,

$$\frac{d}{dt}=\frac{\partial}{\partial\tau_0}+\varepsilon\frac{\partial}{\partial\tau_1}+O\left(\varepsilon^2\right),\quad \frac{d^2}{dt^2}=\frac{\partial^2}{\partial\tau_0^2}+2\varepsilon\frac{\partial^2}{\partial\tau_0\partial\tau_1}+O\left(\varepsilon^2\right) \tag{39}$$

substituting (39) into (38), and retaining only $O(1)$ terms we derive the following first-order modulation equation,

$$\frac{\partial^2\chi_2}{\partial\tau_0^2}+\frac{\partial}{\partial\tau_0}\left[\frac{i}{2}\chi_2+\frac{\lambda}{2}\chi_2-\frac{i}{2}\left|\chi_2\right|^2\chi_2\right]=0 \tag{40}$$

which possesses the following exact first integral of motion:

$$\frac{\partial\chi_2}{\partial\tau_0}+\left[\frac{i}{2}\chi_2+\frac{\lambda}{2}\chi_2-\frac{i}{2}\left|\chi_2\right|^2\chi_2\right]=M(\tau_1,\tau_2,...) \tag{41}$$

In expressing the constant of integration M as function of the slow-scales $\tau_1, \tau_2, ...,$ we recognize that the first integral of motion (41) refers only to the first-order dynamics, i.e., it is only constant at $O(1)$; mathematically the slow variation of the first integral (41) is justified by the fact that the multiple scales of the problem are distinct and independent from each other. Hence, by (41) we allow slow variation of the dynamics, but at higher order (slower) time scales. By the same reasoning, the equilibrium points, $\Phi(\tau_1, \tau_2, ...)$of the first-order system (41) may depend on the higher order slow time scales $\tau_1, \tau_2, ...$ These equilibrium points of the slow flow are computed by solving the following algebraic equation:

$$\frac{i}{2}\Phi+\frac{\lambda}{2}\Phi-\frac{i}{2}\left|\Phi\right|^2\Phi=M(\tau_1,\tau_2,...) \tag{42}$$

Clearly, if an equilibrium is stable it holds that

$$\Phi(\tau_1,\tau_2,...)=\lim_{\tau_0\to+\infty}\chi_2(\tau_0,\tau_1,\tau_2,...)<\infty \tag{43}$$

whereas, it holds that

$$\Phi(\tau_1, \tau_2, ...) = \lim_{\tau_0 \to -\infty} \chi_2(\tau_0, \tau_1, \tau_2, ...) < \infty \tag{44}$$

if that equilibrium is unstable. One can show that the first-order dynamical system (40) does not possess any limit sets besides equilibrium points (for instance by applying Bendixon's criterion [(9)]).

Since we will carry the analysis only up to $O(\varepsilon)$, we omit from here on slow time scales of order higher than one and express the solution of (42) in the following polar form,

$$\Phi(\tau_1) = N(\tau_1)\exp(i\gamma(\tau_1)) \tag{45}$$

Upon substituting into (42) and separating real and imaginary terms, we reduce the computation of the equilibrium points of the slow flow to the following equation,

$$\lambda^2 Z(\tau_1) + Z(\tau_1)\left[1 - Z(\tau_1)\right]^2 = 4\left|M(\tau_1)\right|^2 \tag{46}$$

where $Z(\tau_1) \equiv N^2(\tau_1)$. The number of solutions of equation (46) depends on $|M(\tau_1)|$ and λ. The function on the left-hand side can be either monotonous, or can have a maximum and a minimum. In the former case the change of $|M(\tau_1)|$ has no effect on the number of solutions and equation (46) provides a single positive solution. In the latter case, however, the change of $|M(\tau_1)|$ brings about a pair of saddle – node bifurcations, and hence multiple solutions.

In order to distinguish between the different cases, we check the roots of the derivative with respect to $Z(\tau_1)$ of the left–hand side of (46):

$$1 + \lambda^2 - 4Z + 3Z^2 = 0 \Rightarrow Z_{1,2} = \left[2 \pm \sqrt{1 - 3\lambda^2}\right]/3 \tag{47}$$

It follows that for $\lambda < 1/\sqrt{3}$ there exist two additional real roots and pair of saddle – node bifurcations, whereas at the critical damping value $\lambda = 1/\sqrt{3}$ the two saddle – node bifurcation points coalesce forming the typical structure of a cusp. Extending these results to equation (46), if a single equilibrium exists, this equilibrium is stable with respect to the time scale τ_0. If three equilibrium points exist, two of them are stable focuses or nodes, and the other is an unstable saddle with respect to the time scale τ_0. Therefore, the $O(1)$ dynamics is attracted always to a stable branch.

The characteristic rate of attraction of the dynamics near a node may be evaluated by linearizing equation (41), and considering the following perturbation of the dynamics near an equilibrium point:

$$\chi_2(\tau_0, \tau_1) = \Phi(\tau_1) + \delta(\tau_0), \quad |\delta| << |\Phi| \tag{48}$$

Upon substitution of (48) into (41) yields the following linearized equation,

$$\frac{\partial \delta}{\partial \tau_0} + \left[\frac{i}{2}\delta + \frac{\lambda}{2}\delta - i\left|\Phi\right|^2 \delta - \frac{i}{2}\Phi^2\delta^*\right] = 0 \tag{49}$$

where asterisk denotes complex conjugate. Rewriting equation (49) as,

$$\left(\frac{\partial}{\partial \tau_0} + \frac{i}{2} + \frac{\lambda}{2} - i\left|\Phi\right|^2\right)\delta = \frac{i}{2}\Phi^2\delta^* \tag{50}$$

taking its complex conjugate and combining the two equations, we derive an expression that explicitly computes the evolution of the perturbation $\delta(\tau_0)$ (note that Φ depends only on τ_1 and not on the time scale τ_0),

$$\left[\frac{\partial^2}{\partial \tau_0^2} + \lambda\frac{\partial}{\partial \tau_0} + \frac{1}{4}(1 + \lambda^2 - 4Z + 3Z^2)\right]\delta = 0 \;\Rightarrow\; \delta = \delta_0 \exp\left[(-\lambda \pm i\omega)\,t/2\right] \tag{51}$$

where, $\omega = \sqrt{3Z^2 - 4Z + 1}$. Solution (51) reveals that the linearized dynamics in the vicinity of the equilibrium points depends on λ and Z.

The following possible alternatives are now described. For relatively large values of damping above the critical value, $\lambda > 1/\sqrt{3}$, there exists a single stable node in the $O(1)$ dynamics. For $Z > 1$ or $Z < 1/3$ the attraction of the dynamics to that node is through oscillations [i.e., ω is real – underdamped cases], whereas for $1 > Z > 1/3$ the attraction is through a decaying motion (i.e., ω is imaginary – the overdamped case).

For relatively small damping values, $\lambda < 1/\sqrt{3}$, the situation is more complex, since there exist two additional real equilibrium points given by (47). For $Z > 1$ or $Z < 1/3$ the attraction of the dynamics to the stable node is oscillatory (underdamped cases), whereas, for $1 > Z > Z_1$ or $Z_2 > Z > 1/3$ the attraction is through a decaying motion (overdamped cases). For $Z_1 > Z > Z_2$ we obtain an unstable equilibrium, and the linearized model predicts exponential growth in the dynamics.

In summary, as Z slowly decreases due to its dependency on the slow-time scale τ_1, and depending on the damping value λ, the $O(1)$ dynamics undergoes qualitative changes (bifurcations). In particular, if $\lambda > 1/\sqrt{3}$ we anticipate the dynamics to remain always stable, since in that case there exists a single slowly-varying attracting manifold of the $O(1)$ averaged flow. However, if $\lambda < 1/\sqrt{3}$ the dynamics becomes unstable, in which case we expect that the $O(1)$ averaged flow will make a sudden transition from one attracting manifold to another for slowly decreasing Z. In order to study this complicated damped transition, one should investigate the slow evolution of the equilibrium of the $O(1)$ averaged flow $\Phi(\tau_1)$.

To this end, we consider the $O(\varepsilon)$ terms in the multiple – scale expansion:

$$2\frac{\partial^2 \chi_2}{\partial \tau_0 \partial \tau_1} + \frac{\partial}{\partial \tau_1}\left[\frac{i}{2}\chi_2 + \frac{\lambda}{2}\chi_2 - \frac{i}{2}|\chi_2|^2\chi_2\right] +$$
$$+\frac{\partial}{\partial \tau_0}\left[\frac{\lambda}{2}\chi_2 - \frac{i}{2}|\chi_2|^2\chi_2\right] + \frac{i}{4}\left[\lambda\chi_2 - j|\chi_2|^2\chi_2\right] = 0 \qquad (52)$$

We are interested in the behavior of the solution of the $O(\varepsilon)$ averaged flow in the neighborhood of a stable equilibrium point, or equivalently, in the neighborhood of the damped NNM invariant manifold $\Phi(\tau_1) = \lim_{\tau_0 \to +\infty} \chi_2(\tau_0, \tau_1)$. Therefore, by taking the limit $\tau_0 \to +\infty$ in equation (52) we obtain the following equation which describes the evolution of the dynamics at the slower time scale τ_1:

$$\frac{\partial}{\partial \tau_1}\left(\frac{i}{2}\Phi + \frac{\lambda}{2}\Phi - \frac{i}{2}|\Phi|^2\Phi\right) + \frac{i}{4}(\lambda\Phi - i|\Phi|^2\Phi) = 0 \qquad (53)$$

In deriving this equation we take into account that on the slowly-varying, stable invariant manifold there is no dependence of the dynamics on τ_0, since $\Phi(\tau_1)$ was defined previously as the equilibrium point of the $O(1)$ averaged flow (4.39-4.40). Hence, the differential equation (53) describes the slow evolution of the stable equilibrium points of equation (38) (these are equilibrium points with respect to the fast time scale τ_0, but not with respect to the slow time scale τ_1 and to slow time scales of higher orders, which, however are omitted from the present analysis). The slowly varying equilibrium $\Phi(\tau_1)$ provides an $O(\varepsilon)$ approximation to the *damped NNM manifold* of the dynamics of the system ; this is an *invariant manifold* of the damped dynamics and can be regarded as the analytical continuation for weak damping of the corresponding NNM of the underlying hamiltonian system [(10),(11)].

Rearranging equation (53) in the form,

$$\left(\frac{i}{2} + \frac{\lambda}{2} - i|\Phi|^2\right)\frac{\partial \Phi}{\partial \tau_1} - i\Phi^2\frac{\partial \Phi^*}{\partial \tau_1} = -\frac{i}{4}\left(\lambda\Phi - i|\Phi|^2\Phi\right) \qquad (54)$$

and adding to it its complex conjugate, we obtain the following explicit expression for the slowly varying derivative of the equilibrium point of the $O(1)$ slow flow:

$$\frac{\partial \Phi}{\partial \tau_1} = \frac{-\lambda\Phi + i\left(|\Phi|^2\Phi - 3|\Phi|^4\Phi - \lambda^2\Phi\right)}{2\left(1 + \lambda^2 - 4|\Phi|^2 + 3|\Phi|^4\right)} \qquad (55)$$

Using the polar representation, $\Phi(\tau_1) = N(\tau_1)\exp(i\theta(\tau_1))$, and separating real and imaginary parts, equation (55) yields the following set of real differential equations governing the slow evolution of the magnitude and phase of the stable equilibrium points of the $O(1)$ averaged flow (i.e., of the stable damped NNM manifolds),

$$\begin{aligned}\frac{\partial N}{\partial \tau_1} &= \frac{-\lambda N}{2(1+\lambda^2-4Z+3Z^2)} \\ \frac{\partial \theta}{\partial \tau_1} &= \frac{(Z-3Z^2-\lambda^2)}{2(1+\lambda^2-4Z+3Z^2)}\end{aligned} \tag{56}$$

where $Z(\tau_1) \equiv N^2(\tau_1)$. The first of equations (56) can be integrated exactly by quadratures to yield,

$$(1+\lambda^2)\ln Z(\tau_1) - 4Z(\tau_1) + (3/2)Z^2(\tau_1) = K - \lambda\tau_1 \tag{57}$$

where K is a constant of integration [it actually depends on the higher order time scales $\tau_2, \tau_3, ...$, but these are not considered here as the analysis is restricted to $O(\varepsilon)$].

Expression (57) implicitly determines the evolution of $Z(\tau_1)$ and, consequently, of $N(\tau_1)$. The slow evolution of the phase $\gamma(\tau_1)$ is described by the second of equations (56), and may be computed by direct integration once $Z(\tau_1)$ is known; due to the implicit form of (57), however, this task cannot be performed analytically and requires a numerical solution.

Essential information concerning the qualitative behavior of the solution may be extracted from relation (56) even without explicitly solving it. Indeed, for sufficiently strong damping, $\lambda > 1/\sqrt{3}$, the denominator on the right-hand-side terms is always positive, and the first equation describes a monotonous decrease of $Z(\tau_1)$ towards zero with increasing τ_1. In other words, we conjecture that the slowly varying dynamics remains always on the in-phase damped NNM manifold. By contrast, for relatively weak damping, $\lambda < 1/\sqrt{3}$, the velocity $\partial Z/\partial\tau_1$ is a negative quantity for $Z > Z_1$, but becomes divergent as the limit $Z \to Z_1$ is approached from above.

We cannot proceed to any statement regarding the sign of the velocity when the amplitude is in the range $Z_2 > Z > Z_1$, as the equilibrium point is unstable there; therefore, we infer that as Z decreases below the critical amplitude Z_1 the damped dynamics should be attracted to a NNM damped manifold distinct from the in-phase one. This other manifold is a weakly nonlinear (linearized) branch of the damped out-of-phase NNM invariant manifold. Of course, this conclusion is valid only for the averaged system (37 - 38), which was derived under the condition of 1:1 resonance capture.

The previous analytical findings are illustrated by performing numerical simulations of the averaged system (37) for parameters $\varepsilon = 0.05, \lambda = 0.2$, and initial conditions $\chi_1(0) = 0.7 + 0i, \chi_2(0) = 0.7 + 0i$. The time evolution of the square of the modulation of the envelope of the NES response, $|\chi_2|^2$, is depicted in Figure 8. Clearly, both the magnitude and frequency of the envelope modulation of the NES response tend to zero as the trajectory approaches the critical value Z_1 =0.979. In the vicinity of this value, the trajectory jumps to the alternative stable attractor. This point may be further illustrated using the three-dimensional plot depicted in Figure 9, where the real and imaginary parts of the complex envelope modulation of the NES, χ_2, are plotted in a parametric plot for increasing time. The damped trajectory of the envelope modulation of the NES starts from zero, gets attracted initially by the stable damped in-phase NNM manifold, before making a transition (jump) to the weakly nonlinear, low energy stable NNM manifold.

In order to check the validity of the asymptotic approximations, we performed direct simulations of the original set (34) (i.e., of the exact system before averaging) with the same initial conditions used for the plots of Figures 8 and 9; the result is presented in Figure 10. It is clear from this Figure that the damped dynamics is initially attracted by the damped in-phase NNM manifold , as evidenced by the in-phase 1:1 resonant oscillations of the NES and the LO, with nearly unit frequency. With diminishing amplitude of the NES, the critical amplitude is reached close to $t\tilde{}50$s, and a transition of the damped dynamics to a the out-of-phase linearized low-energy regime takes place, with the motion localizing to the LO. This is in accordance with the predictions of the averaging analysis.

5 Targeted energy transfer in forced systems.

The first attempt of theoretical and experimental analysis of forced system with attached NES was made in paper [(12)], where a capability of vibration suppression was analyzed. It should be mentioned that only pure periodic responses were considered there. The other complication – possibility of co-existence of periodic and quasiperiodic regimes in forced systems with the NES – has been revealed in paper [(13)]. Analytic methodology for complete analysis of the periodic and quasiperiodic responses was suggested in [(14),(15)].

It will be demonstrated that the steady state response of a primary system under harmonic excitation with an attached NES exhibits not only common steady – state and weakly modulated responses, but also a very special type of responses characterized by very deep modulations of the resulting oscillations. This type of the response is referred to as Strongly Modulated Response *(SMR)*. Moreover, the SMRs are related to relaxation oscillations of the corresponding averaged dynamical flows (the slow-flows of

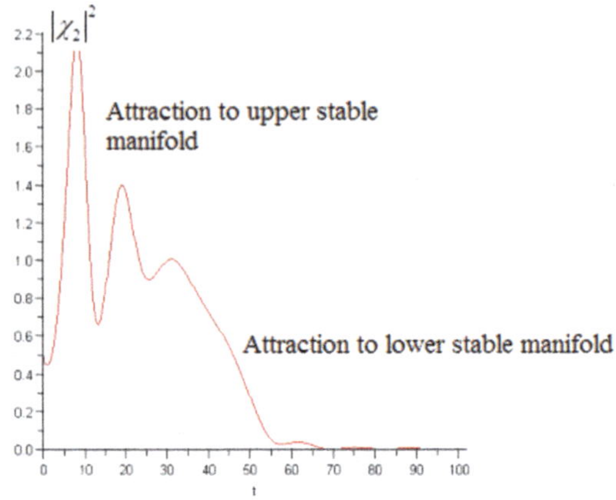

Figure 8. Response of the averaged system (37) in the regime of 1:1 resonance capture, for $\varepsilon = 0.05, \lambda = 0.2$, and initial conditions given by $\chi_1(0) = 0.7 + 0i, \chi_2(0) = 0.7 + 0i$.

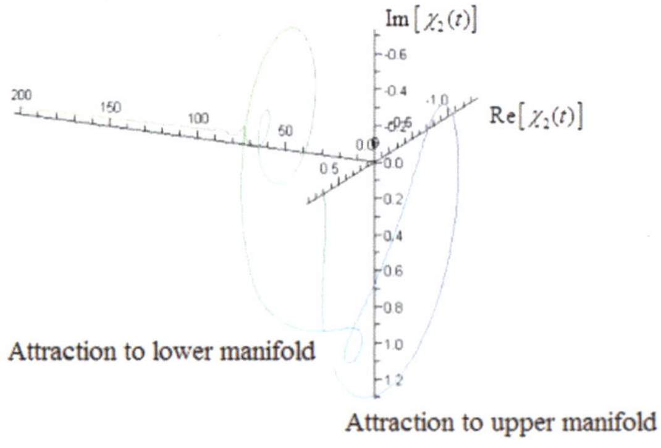

Figure 9. Real and imaginary parts of the complex modulation χ_2 of the NES plotted against time, in the regime of 1:1 resonance capture for $\varepsilon = 0.05, \lambda = 0.2$, and initial conditions $\chi_1(0) = 0.7 + 0i, \chi_2(0) = 0.7 + 0i$.

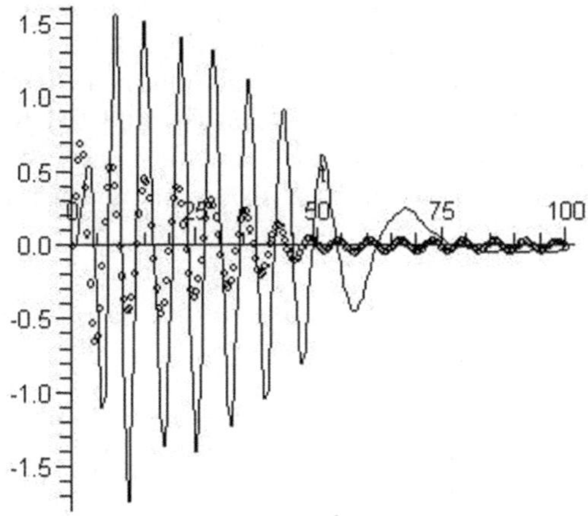

Figure 10. Direct numerical simulation of the damped system (34) for parameters $\varepsilon = 0.05$, $\lambda_1 = 0$, $\lambda_2 = 0.01$, and initial conditions $x(0) = v(0) = \dot{x}(0) = 0$ and $\dot{x}(0) = 0.7$; the dynamics correspond to the analytical results of Figures 8 and 9. The dotted line describes the LO dislacement, the solid corresponds to the NES

the dynamics). The possible application of the NESs as strongly nonlinear vibration absorbers for vibration isolation of harmonically forced single- and multi-DOF primary subsystems is then discussed, and it is shown that under certain conditions, the efficiency of the NESs as vibration isolators can exceed that of properly tuned linear absorbers (or tuned mass dampers – TMDs)

The simplest benchmark model includes a primary SDOF linear oscillator under harmonic external excitation with an ungrounded, lightweight and essentially nonlinear NES attached [(14),(15)]. This system is described by the following set of equations (after reduction to non-dimensional variables):

$$\begin{aligned} &\ddot{y}_1 + \varepsilon\lambda(\dot{y}_1 - \dot{y}_2) + (1 + \varepsilon\sigma)y_1 + (4\varepsilon/3)(y_1 - y_2)^3 = \varepsilon A \cos t \\ &\varepsilon\ddot{y}_2 + \varepsilon\lambda(\dot{y}_2 - \dot{y}_1) + (4\varepsilon/3)(y_2 - y_1)^3 = 0 \end{aligned} \tag{58}$$

where y_1 and $y_2$are the displacements of the linear oscillator and the attachment respectively, $\varepsilon\lambda$ the damping coefficient, εA the amplitude of the external force, and$\varepsilon\sigma$ a frequency detuning parameter. The parameter $\varepsilon << 1$

is the small parameter of the problem which scales the coupling between the two oscillators, the damping forces, the amplitude of the external force, the detuning parameter, and the mass of the NES. The coefficients A, λ, σ are adopted to be of $O(1)$. The coefficient of the nonlinear term can be modified by proper rescaling of the dependent variables and the forcing amplitude; the value $(4\varepsilon/3)$ is chosen for the sake of convenience. The dots denote a differentiation with respect to time. A sketch of this system is presented in Figure 11

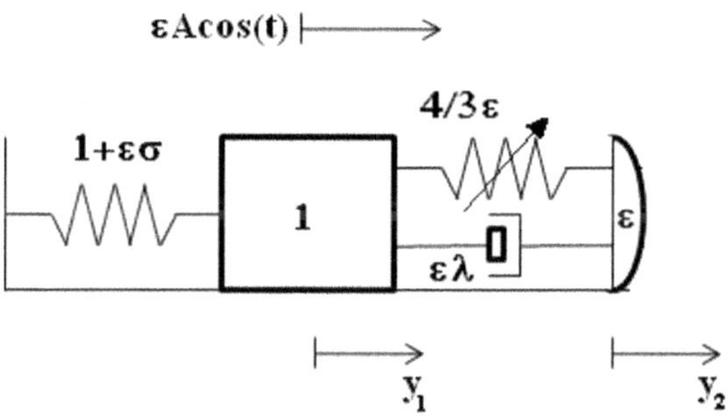

Figure 11. Sketch of the dynamical model described by System (58).

We apply the following coordinate transformations, denoting the center-of-mass and relative displacements of the system,

$$\begin{aligned} v &= y_1 + \varepsilon y_2 \\ w &= y_1 - y_2 \end{aligned} \tag{59}$$

and then switch the analysis to complex variables:

$$\begin{aligned} \varphi_1 \exp(it) &= \dot{v} + iv \\ \varphi_2 \exp(it) &= \dot{w} + iw \end{aligned} \tag{60}$$

It is clear that one should seek periodic solutions of System (1) with dominant frequencies identical to the frequency of the external periodic force, and approximately equal to the eigenfrequency of the linear oscillator (that is, the frequency detuning $\varepsilon\sigma$ provides a slight frequency mismatch). Hence, a *fundamental nonlinear resonances* of System (58) should be studied.

After substitution of Equations (59) and (60) into System (58) and subsequent averaging over the fast oscillations of frequency unity, the following slow–flow (complex modulation) equations are obtained:

$$\begin{aligned}
&\dot{\phi}_1 + \tfrac{i\varepsilon}{2(1+\varepsilon)}\left(\phi_1 - \phi_2\right) - \tfrac{i\varepsilon\sigma(\phi_1+\varepsilon\phi_2)}{2(1+\varepsilon)} = \tfrac{\varepsilon A}{2} \\
&\dot{\phi}_2 + \lambda(1+\varepsilon)\tfrac{\phi_2}{2} + \tfrac{i}{2(1+\varepsilon)}\left(\phi_2 - \phi_1\right) - \tfrac{i\varepsilon\sigma(\phi_1+\varepsilon\phi_2)}{2(1+\varepsilon)} - \tfrac{i(1+\varepsilon)}{2}\left|\phi_2\right|^2\phi_2 = \tfrac{\varepsilon A}{2}
\end{aligned} \tag{61}$$

The system of equations (61) has a complicated structure and cannot be solved analytically. The first step towards analyzing its steady state solutions is to perform local analysis of its equilibrium (fixed points). Such analysis is of significant physical interest, since these points correspond to periodic responses of the system described by System (58). To find the fixed points one equates the time derivatives of System (61) to zero $\left(\dot{\phi}_1 = \dot{\phi}_2 = 0\right)$ thus obtaining the following complex algebraic relations:

$$\begin{aligned}
&\tfrac{i\varepsilon}{2(1+\varepsilon)}\left(\varphi_{10} - \varphi_{20}\right) - \tfrac{i\varepsilon\sigma(\varphi_{10}+\varepsilon\varphi_{20})}{2(1+\varepsilon)} = \tfrac{\varepsilon A}{2} \\
&\lambda(1+\varepsilon)\tfrac{\varphi_{20}}{2} + \tfrac{i}{2(1+\varepsilon)}\left(\varphi_{20} - \varphi_{10}\right) - \tfrac{i\varepsilon\sigma(\varphi_{10}+\varepsilon\varphi_{20})}{2(1+\varepsilon)} - \tfrac{i(1+\varepsilon)}{2}\left|\varphi_{20}\right|^2\varphi_{20} = \tfrac{\varepsilon A}{2}
\end{aligned} \tag{62}$$

The amplitude $N_{20} = |\varphi_{20}|$ provides the first – order approximation for the amplitude of steady state periodic oscillation of the relative response $w = y_1 - y_2$ - i.e., the displacement between the linear oscillator and the NES – cf. System (58). *This amplitude is directly related to the efficiency of steady state TET* in the system considered, since the capacity of the NES to passively absorb and locally dissipate a significant portion of the energy of the linear oscillator is directly tied to the relative response w attaining large amplitudes. Indeed, large amplitudes of the relative response w signifies resonance interaction of the NES with the linear oscillator (which is a prerequisite for the TET), and, in addition, it guarantees that the damping element coupling the NES to the linear oscillator strongly dissipates vibration energy at steady state.

Depending on the parameters, System (61) can have one or three positive (real) solutions. Therefore, due to continuity one expects generically that at certain special critical values of the parameters A, λ and σ two of these solutions will coalesce, yielding bifurcations of steady state periodic solutions; generically, these will be saddle – node (SN) bifurcations. In addition to the SN bifurcations, where a stable steady – state solution of the slow flow simply disappears when it coalesces with an unstable one, there exists one additional generic bifurcation scenario for loss of stability, namely, the realization of Hopf bifurcations [(9)]. It can be studied by linear perturbation of the fixed points corresponding to the solutions of System (58) and corresponds to the case when a pair of complex conjugate roots of the characteristic polynomial passes the imaginary axis. Typical projections of the solutions of System (62) on the two dimensional plane of parameters (A, λ) for $\sigma = 3$ are presented in Figure 12 .

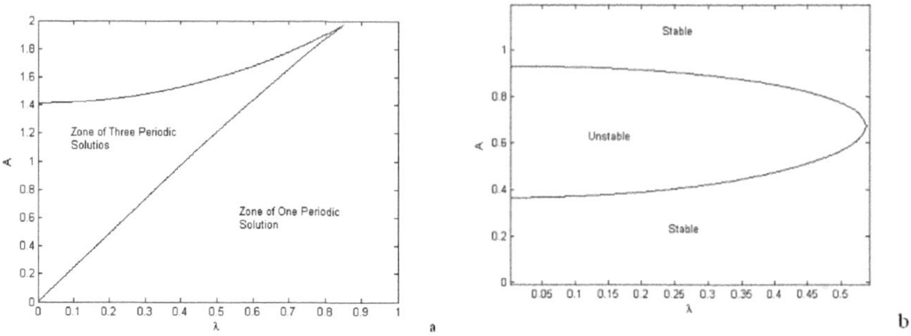

Figure 12. urves of (a) SN bifurcations for $\sigma = 3$; (b) Hopf bifurcations for $\sigma = 0.5$ and $\varepsilon = 0.05$ (no SN bifurcations exist for the parameter values chosen).

Additional important information concerning the local bifurcations of the periodic solutions of the slow flow (61) may be obtained by constructing *frequency response diagrams*; these depict the amplitude N_{20} of the steady state periodic solution as function of the detuning parameterσ, for fixed values of the amplitude of external forcing A, damping λ, and NES mass ε. In Figure 13a representative frequency response diagram is presented; bifurcation points and stability types of branches of solutions are also marked there. Recalling the assumptions of the analysis, the depicted frequency response provides an *approximate fundamental resonance plot* of System (58). Although plots of this type do not convey much new information compared to the previously considered bifurcation diagrams, they are directly applicable to the problem of vibration isolation since they depict the amplitudes of steady state responses in the frequency domain. Regarding the frequency response of Figure 13, one mentions that there exists an upper stable branch of the steady state periodic solutions, corresponding to large-amplitude stable periodic oscillations of the NES relative to the linear oscillator; this branch co-exists with a stable low-amplitude branch of periodic responses corresponding to low-amplitude relative oscillations. Such "detached" branches of the frequency response were recently studied in somewhat similar systems [(16)].

Use of the CX-A approach for the analytical treatment of essentially nonlinear systems assumes that the approximation (which is formally justified only for weakly nonlinear systems), will remain correct in the limit

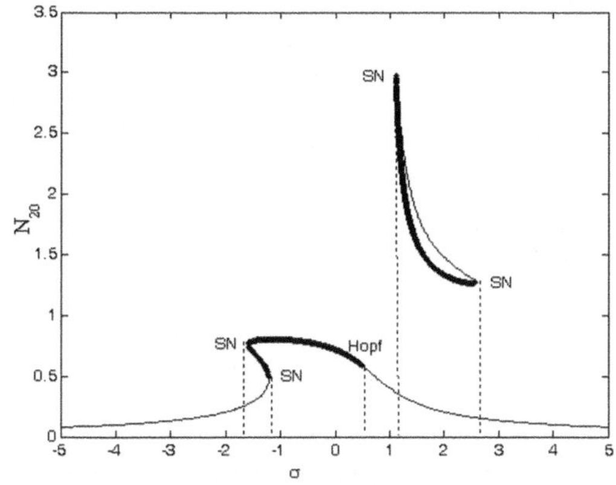

Figure 13. Frequency response diagram (fundamental resonance plot) for $A = 0.4$, $\lambda = 0.2$ and $\varepsilon = 0.01$; bold lines refer to unstable regions of periodic solutions, and thin lines to stable regions (SN and Hopf bifurcations are also noted).

when the small parameter of the problem becomes of order unity. This assumption requires additional verification by direct numerical simulations of the initial System (58), to verify independently the predictions of the analysis. The response regimes for certain sets of parameters are presented in Figures 14a,b.

For the parameter values used for the numerical simulations of Figure 14a, the single real solution of Equation (62) $N_{20} = |\varphi_{20}| = 0.577$. This value is in agreement with the amplitude of the numerical solution depicted in Figure 14a. The parameters for the simulation depicted in Figure 14b were selected in order to study the response in the zone where the analysis predicts that the periodic solution is unstable. Indeed, the numerical solution is in the form of a quasi-periodic oscillation, as evidenced by the slowly modulated fast oscillation of Figure 14b. This type of solutions is referred to as weakly modulated response (WMR).

One can conclude that the analytic approach presented above yields reliable predictions of the behavior of original forced System (58). Moreover, the approach is rather sensitive, since the only difference between the plots of Figures 14a,b is an 8% difference in the forcing amplitude A. Still, this difference brings about qualitatively different responses and the analytic approach succeeds to capture this fact. So, the results of the CX-A technique are reliable and valid, at least in the regime of fundamental nonlinear

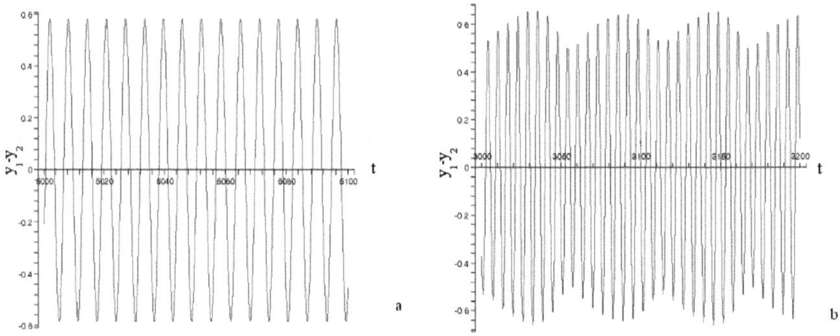

Figure 14. Relative response $y_1(t) - y_2(t)$ of system (58) for (a)$A = 0.225$, $\lambda = 0.2$, $\varepsilon = 0.05$,$\sigma = 0$, $y_1(0) = 0.29$, $\dot{y}_1(0) = 0.25$, $y_2(0) = 0$, and $\dot{y}_2(0) = -0.15$; (b) $A = 0.24$, $\lambda = 0.2$, $\varepsilon = 0.05$, $\sigma = 0$, $y_1(0) = 0.29$, $\dot{y}_1(0) = 0.25$, $y_2(0) = 0$, and $\dot{y}_2(0) = -0.15$.

resonances, and as long as the slow-fast partition of the dynamics in ansatz (60) (which was a basic assumption of the analysis) is justified.

Still, one should pay attention to the special sets of initial conditions used for producing the responses of Figures 4.14a,b. The reason is that the response under consideration should be close enough to the fixed point of the slow flow, since the analysis presented above is only local. It follows that if the initial conditions are not specially tailored, the flow could be attracted to alternative response regimes which do not satisfy the assumptions of the analysis, and hence, are not predicted by it. Indeed, it is a well known feature of (forced or unforced) nonlinear dynamical systems that they may possess qualitatively different co-existing solutions. It is true that in many harmonically excited systems (especially weakly nonlinear ones) steady-state responses such as the ones discussed above (that is, either stable or Hopf – modulated) may be the only types of steady state motions that can be possibly realized by these systems. For the essentially nonlinear System (58), however, this is not the case.

The latter claim is substantiated by performing numerical simulations of the original System (58) for the same parameter values used to generate the transient responses depicted in Figures 14, but now with zero initial conditions. The results are presented in Figures 15 respectively. In both plots one can see a qualitatively new type of response regime involving a *strongly modulated, nearly periodic oscillation.* In the beginning of each cycle the amplitudes of both responses $v = y_1 + \varepsilon y_2$ and $w = y_1 - y_2$ grow slowly. Then, after a certain amplitude threshold is reached, the amplitude of the motion of the center of mass v abruptly decreases, whereas, the relative response between the NES and the linear oscillator, w, is excited with subsequent decay.

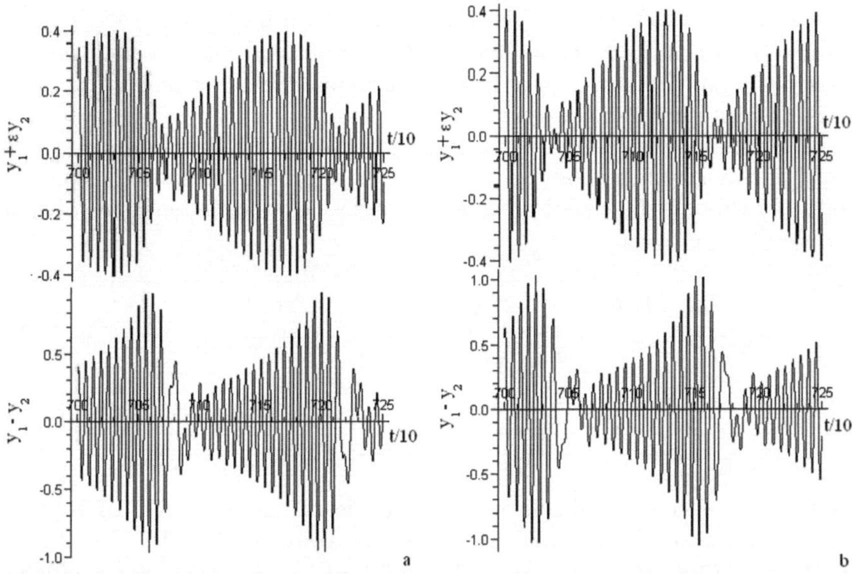

Figure 15. Strongly modulated responses (SMRs) of system (58) for (a) $A = 0.225, \quad \lambda = 0.2, \quad \varepsilon = 0.05, \sigma = 0$ and zero initial conditions; (b) $A = 0.225, \quad \lambda = 0.2, \quad \varepsilon = 0.05, \sigma = 0$ and zero initial conditions.

It is easy to demonstrate that for these values of parameters, the averaged flow has only one fixed point in both cases, and the responses presented in Fig. 14 correspond to exactly these fixed points. The results depicted in Fig.15 are very different indeed. This means that the system can exhibit steady state response regimes (e.g., the presented strongly modulated ones) which in principle *cannot be captured by local analysis of the fixed points of the averaged flow.*

In order to distinguish this type of steady state response from those derived by the local analysis of the previous Sections and related to the fixed (equilibrium) points of the slow flow (61), this response is denoted as a

Strongly Modulated Response (SMR). The width of the amplitude modulation is equal to the response amplitude, and the analytical treatment of such responses poses distinct challenges. Indeed, as discussed above in order to analyze SMRs local analysis of the slow flow equations (61) is insufficient, and, rather *global analysis* of the dynamics is required. In general, such a challenging analytical problem is hardly solvable, since the slow flow (61) is essentially nonlinear and evolves in four dimensional phase space. Still, assuming that the mass ε (which can also be regarded as a mass ratio in the normalized System (58)) is sufficiently small, it may be used as a small parameter for performing singular perturbation analysis. It should be mentioned that in the local analysis of the previous Sections the smallness of ε was not required and not assumed.

The analysis of SMRs starts from combining the two first-order equations of the slow flow from System (61) through simple manipulations, and reducing the slow flow to the following single second order complex ordinary differential equation,

$$\begin{gathered}\frac{d^2\varphi_2}{dt^2}+\frac{d}{dt}\left[\alpha\varphi_2-\frac{i(1+\varepsilon)}{2}\left|\varphi_2\right|^2\varphi_2+\frac{i\varepsilon(1-\sigma)}{2(1+\varepsilon)}\varphi_2\right]+\\ \frac{i\varepsilon(1-\sigma)}{2(1+\varepsilon)}\left[\alpha\varphi_2-\frac{i(1+\varepsilon)}{2}\left|\varphi_2\right|^2\varphi_2-\frac{\varepsilon A}{2}\right]-\frac{i\varepsilon\beta(1+\varepsilon\sigma)}{2(1+\varepsilon)}\varphi_2=\frac{\varepsilon A\beta}{2}\end{gathered}\tag{63}$$

where,

$$\alpha=\frac{\lambda(1+\varepsilon)^2+j\left(1-\varepsilon^2\sigma\right)}{2(1+\varepsilon)},\quad \beta=\frac{j(1+\varepsilon\sigma)}{2(1+\varepsilon)}$$

In the sequence, a multiple scales analysis of the solutions of Equation (63) is performed by the multiple-scales analysis (4.26, 4.40). Substituting expansions (39) into Equation (63) and setting equal to zero the coefficients of powers of ε, one derives the following hierarchy of problems at successive orders of approximation:

$$O\left(\varepsilon^0\right):\ \frac{\partial^2\varphi_2}{\partial\tau_0^2}+\frac{\partial}{\partial\tau_0}\left[\frac{\lambda\varphi_2}{2}+\frac{i\varphi_2}{2}-\frac{i}{2}\left|\varphi_2\right|^2\varphi_2\right]=0\tag{64}$$

$$\begin{gathered}O\left(\varepsilon^1\right):\ 2\frac{\partial^2\varphi_2}{\partial\tau_0\partial\tau_1}+\frac{\partial}{\partial\tau_1}\left[\frac{\lambda\varphi_2}{2}+\frac{i\varphi_2}{2}-\frac{i}{2}\left|\varphi_2\right|^2\varphi_2\right]+\frac{(1-\sigma)}{4}\left|\varphi_2\right|^2\varphi_2+\\ +\frac{\partial}{\partial\tau_0}\left[\frac{\lambda\varphi_2}{2}+\frac{i(1-\sigma)\varphi_2}{2}-\frac{i}{2}\left|\varphi_2\right|^2\varphi_2\right]+\left[\frac{\sigma}{4}+\frac{i\lambda(1-\sigma)}{4}\right]\varphi_2-\frac{iA}{4}=0\end{gathered}\tag{65}$$

$\cdots$

Equation (64) describes the leading order approximation of the evolution of the slow flow (averaged) dynamics. Similarly to (40), this equation can be trivially integrated,

$$\frac{\partial}{\partial\tau_0}\phi_2+\left(\frac{j}{2}\phi_2+\frac{\lambda}{2}\phi_2-\frac{j}{2}\left|\phi_2\right|^2\phi_2\right)=C(\tau_1,\tau_2,\ldots)\tag{66}$$

where $C(\tau_1, \tau_2, ...)$ is an arbitrary function of higher order time scales. Higher order time scales are not considered in the current analysis, since as shown below the dynamical phenomena in question are captured by the leading order approximations. Then, one obtain the slow invariant manifold (SIM) described by (46-refeq46). The phase $\theta(\tau_1)$ of the fixed point is evaluated by:

$$\theta(\tau_1) = \arg C(\tau_1) - \tan^{-1}\left[\frac{1 - Z(\tau_1)}{\lambda}\right] \tag{67}$$

In the case $\lambda < 1/\sqrt{3}$ the fold lines $L_{1,2} = \{N(\tau_1) = \sqrt{Z_{1,2}},\ \theta(\tau_1) \in [0, 2\pi)\}$ divide the stable and unstable branches of the SIM. In Figure 16 the projection of the two-dimensional SIM on the plane $(N, 4|C|^2)$is depicted; the fold lines correspond to the local maximum and minimum points of the SIM (Figure 16).

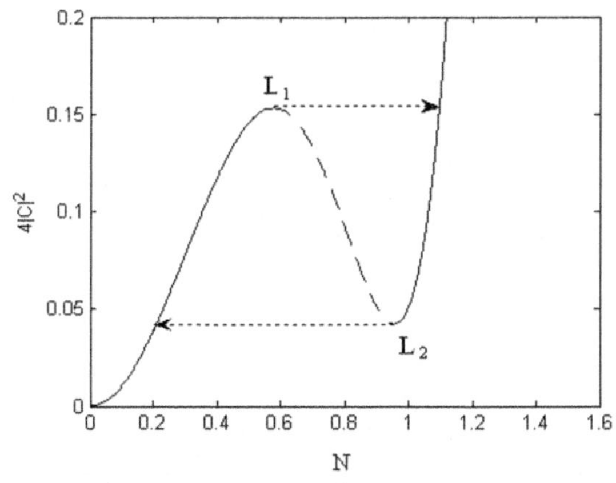

Figure 16. Projection of the slow invariant manifold of the system (SIM) for $\lambda = 0.2$; the unstable branch is denoted by dashed line, and arrows denote hypothetic transitions (jumps) in the regime of relaxation oscillations.

It is well-known [(17), (18)] that such a folding structure of the SIM may give rise to *relaxation-type oscillations*, characterized by sudden transitions (jumps) of the response during each cycle (the hypothetic sudden transitions between the two stable branches are denoted by arrows at Figure 16). *The conjecture is that such relaxation oscillations occur in the SMRs described above.* Still, such motions may be possible only if the dynamical flow can reach the fold lines $L_{1,2}$, while following the two branches of the SIM with respect to the slow time scale τ_1.

In order to assess this possibility one should investigate the behavior of the flow on the SIM given by $\Phi(\tau_1)$. To this end, the $O(\varepsilon)$ subproblem (65) derived by the multiple scales expansion. Considering equation (65) in the limit $\tau_0 \to +\infty$, and taking into account the asymptotic stability of points on the stable branches with respect to time scale τ_0, one derives the following equation for motion *on* the SIM governed by the slow time scale τ_1:

$$\frac{\partial \Phi}{\partial \tau_1} = \frac{2\left[\left(\lambda - i + 2i\left|\Phi\right|^2\right)G + i\Phi^2 G^*\right]}{\lambda^2 + 1 - 4\left|\Phi\right|^2 + 3\left|\Phi\right|^4} \tag{68}$$

where

$$G = -\frac{1-\sigma}{4}\left|\Phi\right|^2\Phi - \left[\frac{\sigma}{4} + \frac{i\lambda(1-\sigma)}{4}\right]\Phi + \frac{iA}{4}$$

Expressing this complex relation in terms of its modulus and phase through the polar transformation $\Phi(\tau_1) = N(\tau_1)\exp\left[i\theta(\tau_1)\right]$, one obtains the dynamical system on the cylinder $(N, \theta) \in \left(R^+ \times S^1\right)$ governing the slow evolution on the SIM at time scale τ_1.

For particular case A=0 the phase portrait of system (68) is presented in Figure 17.

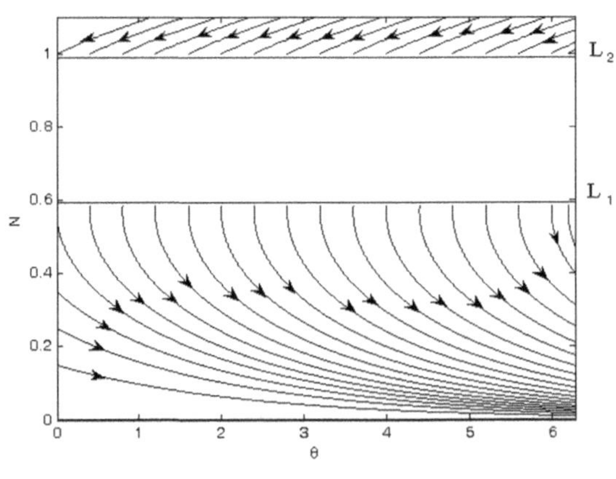

Figure 17. Phase portrait of the slow evolution of the SIM for the case of no external harmonic excitation, $A = 0$.

The phase trajectories on the upper stable branch slowly evolve directed towards the fold line L_2, whereas the trajectories on the lower stable branch are not able to reach the fold line L_1. This means that although the dynamics can make a sudden transition (jump) from the upper stable branch of the SIM to the lower one, it cannot make a similar transition back. Actually, this ends up as a rather trivial observation since in the absence of external harmonic forcing the dynamics cannot reach a nontrivial steady state regime, as it is damped out by dissipation towards the state of trivial (zero) equilibrium.

In order to allow for jumps from the lower stable branch of the SIM back to the upper one (and, therefore, to provide the necessary condition for the occurrence of relaxation oscillations) the slow flow in the vicinity of L_1 should undergo bifurcation. That is, at some subset of L_1 the orbits of the slow flow lines on the lower branch of the SIM should become tangent to L_1. Such points correspond to fixed points of a *desingularized* slow flow [(18)], where the numerator of equation (68) vanishes. In order to investigate these special points, one should compute the fixed points of the slow flow equation (68) for arbitrary amplitudes of the external harmonic excitation A. The appropriate condition reads,

$$\left(\lambda - i + 2i\left|\Phi\right|^2\right)G + i\Phi^2 G^* = 0 \tag{69}$$

and possesses two sets of solutions (fixed points). The first set is trivial and is computed by setting $G = 0$; this solution corresponds to fixed points of the initial equation (62), i.e., to fixed points of the global flow that (quite naturally) lie on the SIM.

The other set of solutions satisfies the following conditions:

$$\begin{gathered} 3\left|\Phi\right|^4 - 4\left|\Phi\right|^2 + 1 + \lambda^2 = 0 \\ \exp\left(2i \arg G\right) = -i\Phi^2 / \left[\lambda + i\left(2\left|\Phi\right|^2 - 1\right)\right] \end{gathered} \tag{70}$$

The first equation in (70) coincides with the equation for the fold lines; therefore, as expected, the solutions of this type describe the folded singularities of the slow flow. Bifurcations of such singularities correspond to violations of the transversality condition and, therefore, yield qualitative changes of the flow in the vicinity of the fold line. Specifically, these bifurcations result in a switch of the directions of the flow lines and therefore provide the necessary conditions for relaxation oscillations. What is even more interesting is that equations (70) may be solved in closed form. Indeed, introducing again the polar transformation $\Phi(\tau_1) = N(\tau_1)\exp\left[i\theta(\tau_1)\right]$, (70) yields the following solutions for the positions of the singularities on the fold lines L_1 and L_2:

$$L_1: \quad N_1 = Z_1^{1/2}, \quad \theta = \Theta_{1,2} \equiv \gamma_{01} \pm \cos^{-1}\left[\frac{\lambda N_1}{A\sqrt{\left(1 - N_1^2\right)^2 + \lambda^2}}\right]$$

$$L_2: \quad N_2 = Z_2^{1/2}, \quad \theta = \Theta_{3,4} \equiv \gamma_{02} \pm \cos^{-1}\left[\frac{\lambda N_2}{A\sqrt{\left(1-N_2^2\right)^2+\lambda^2}}\right]$$

$$N_{1,2} = (4/3) \pm \left[(4/3)^2 - 4\left(1+\lambda^2\right)/3\right]^{1/2} \tag{71}$$

$$\gamma_{0k} = \sin^{-1}\left[\frac{\lambda}{\sqrt{\left(1-N_k^2\right)^2+\lambda^2}}\right], \quad k = 1,2$$

Hence, for sufficiently weak external harmonic excitations, that is, for amplitudes of the harmonic excitation below the first critical threshold,

$$A < A_{1\text{crit}} = \frac{\lambda N_1}{\sqrt{\left(1-N_1^2\right)^2+\lambda^2}} \tag{72}$$

no bifurcation close to the lower fold line L_1 can occur. Then, the slow flow in the vicinity of both fold lines of the SIM remains qualitatively similar to that depicted in Fig. 17, providing no possibility for the occurrence of relaxation oscillations (and thus of SMRs) in the slow flow (61). As the forcing amplitude approaches the value $A \to A_{1\text{crit}}$ from below, a SN bifurcation occurs at L_1, as $\theta \to \gamma_{01}$, and a pair of singularities is formed; in the interval between these points, the flow in the vicinity of L_1 reverses direction. The representative phase portrait describing the evolution of the slow flow on the SIM for the amplitude of the external harmonic force in the range $A_{1crit} < A$ is presented in Figure 18.

It follows from the plot of Figure 8 that after the occurrence of the SN bifurcations close to the fold lines L_1 and L_2, there exists a subset of orbits on the SIM that carry the flow to L_1, thus providing the possibility for a jump to the upper stable branch of the SIM, and, hence, to relaxation oscillations. Indeed, the flow can reach the fold L_2 and then jump down again to L_1, thus closing the loop of the relaxation oscillation and giving rise to the SMR.

Still, the condition $A > A_{1crit}$ is *necessary*, but by no means *sufficient* for the occurrence of SMRs in the slow flow (61). In other terms, if this condition is valid then sudden transitions (jumps) between the stable branches of the SIM *may occur*, but there is no guarantee that the series of these transitions will accumulate to stable attractors of the slow dynamics in the form of SMRs. In order to obtain the missing sufficient conditions for the occurrence of SMRs one should investigate more delicate aspects of the slow flow dynamics (61). This is performed next.

Studying carefully the phase portrait depicted in Figure 18, one observes that there exists an interval of θ, namely, $\Theta_1 < \theta < \Theta_2$, where all orbits on the SIM arrive to the fold line L_1 and then depart from it. In the regime of

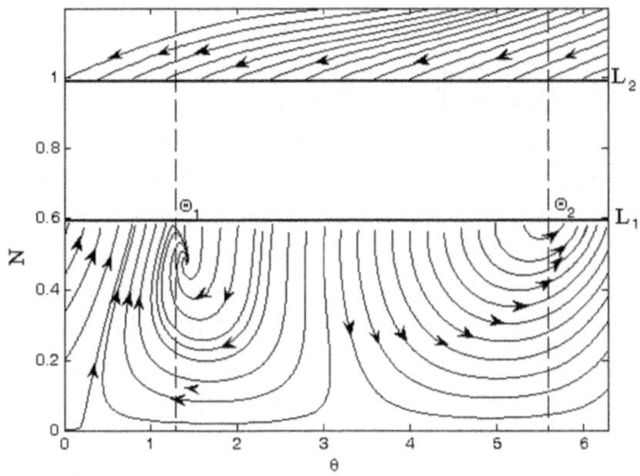

Figure 18. Phase portrait of the slow evolution of the SIM for the case when $A_{1crit} < A$ and $\sigma = 0.5$; only the stable branches of the SIM are depicted.

relaxation oscillations, an orbit in the previously constructed phase cylinder $(N, \theta) \in (R^+ \times S^1)$ initially jumps from a point of this interval on L_1 to the upper branch of the SIM; then it slowly evolves following an orbit of the slow flow towards the upper fold line L_2, before jumping back to the lower stable branch of the SIM; following an orbit of the slow flow it moves towards the lower fold line L_1 reaches it in one of the points of the interval $\theta \in [\Theta_1, \Theta_2]$; following this the orbit jumps up to the upper branch of the SIM and the cycle of the relaxation oscillation (SMR) continues indefinitely.

Therefore, it is natural to consider this relaxation regime in terms of a one-dimensional map P of the interval $[\Theta_1, \Theta_2]$ of the fold line L_1 into itself:

$$P : [\Theta_1, \Theta_2] \xrightarrow{P} [\Theta_1, \Theta_2], \;\; \theta \to P(\theta)$$

In the regime of relaxation oscillations (SMRs) this map takes a point on the fold line L_1 of the cylinder $(N, \theta) \in (R^+ \times S^1)$ and maps it into L_1 under the action of the super - slow flow (68). Clearly, a stable SMR will correspond to an attractor of this map (for example, a period$-k$ fixed point), so the conditions for existence of this attractor will provide the necessary and sufficient conditions for existence of the corresponding SMR in the slow flow (61), and, hence, also in the original dynamical system (1) for NES massε sufficiently small.

In order to construct the one-dimensional map P, one considers separately its 'slow' and 'super-slow' components during a cycle of the relaxation oscillation. As far as the super-slow components of the map are concerned, these correspond to the parts of the relaxation cycle spent on the lower and the upper stable branches of the SIM. Hence, we may use equations (68) and directly connect the 'exit' and 'landing' points on the fold lines L_1 and L_2. Due to complexity of the associated expressions, in the following developments the 'slow' components of the map P are evaluated numerically.

As for the 'slow' components of the map, it is clear that the function ϕ_2 should be continuous at the points of transition between the 'super-slow' and the 'slow' components. Therefore, to model the jumps which provide the 'fast' components of the map we should define appropriately the complex invariant $C(\tau_1)$. If the value of $C(\tau_1)$ is known at the point of start of the jump (the 'exit' point) between the two fold lines, it is possible to compute the amplitude N and phase θ corresponding to the point of 'landing' of the jump unambiguously, and thus to complete the definition of the map P. The procedure of numerical integration should be performed twice, however, one for each of the two stable (upper and lower) branches of the SIM; hence, two values for the invariant $C(\tau_1)$ should be computed for each of the two 'slow' components of the map in order to determine the 'landing' points of the jumps in L_1 and L_2.

These two slow components correspond to the two jumps between fold lines during each cycle of the relaxation oscillation. Fortunately, these fast elements of the mapping cycle can be written down in closed form. For example, if one knows the values of N and θ at the 'exit' point of the jump on the fold line L_1, say (N_1, θ_{01}), and denotes the 'landing' point on the upper stable branch of the SIM by (N_u, θ_u), one may compute the value of N_u by exploiting the invariance of $C(\tau_1)$ on the fast component of the jump. Similarly, the jump from an 'exit' point (N_2, θ_{02}) on the upper fold line $L_2$to the point (N_d, θ_d) on the lower stable branch of the SIM is described by the discrete map. These maps are explicitly expressed as follows:

$$\begin{aligned} N_1 \to N_u &= \sqrt{\tfrac{2}{3}\left(1+\sqrt{1-3\lambda^2}\right)}, \theta_{01} \to \theta_u = \theta_{01} + \tan^{-1}\left[\frac{9\lambda\sqrt{1-3\lambda^2}}{-1+15\lambda^2-\sqrt{1-3\lambda^2}}\right] \\ N_2 \to N_d &= \sqrt{\tfrac{2}{3}\left(1-\sqrt{1-3\lambda^2}\right)}, \theta_{02} \to \theta_d = \theta_{02} - \tan^{-1}\left[\frac{9\lambda\sqrt{1-3\lambda^2}}{-1+15\lambda^2+\sqrt{1-3\lambda^2}}\right] \end{aligned} \quad (73)$$

It should be stressed that for each point of the interval $[\Theta_1, \Theta_2]$ only one computation is required for a single cycle of the map. The outlined construction of the one-dimensional Poincaré map P is somewhat similar to the procedure developed in [34] for analyzing chaotic attractors in regimes of relaxation oscillations occurring in low-dimensional phase spaces. Clearly, not every orbit which starts from the lower fold line L_1 of the SIM will land within the interval $[\Theta_1, \Theta_2]$, since it may be attracted to alternative attractors lying either on the upper or lower stable branches of the SIM. Of course, only those points which are mapped into this interval can carry sustained relaxation oscillations and yield SMRs.

Representative examples of return maps are illustrated in Figure 19. The map is defined for all points of the interval $\theta \in [\Theta_1, \Theta_2]$, since all of these points are mapped into the same interval under the action of the map, which is clearly contracting. Therefore applying the contracting map theorem one proves the existence of a stable attractor of the map in the interval $[\Theta_1, \Theta_2]$, which corresponds to a sustained regime of relaxation oscillations and thus to an SMR of the slow flow (61). In this case, the attractor is the stable period-one fixed point $\theta_e \approx 0.51$.

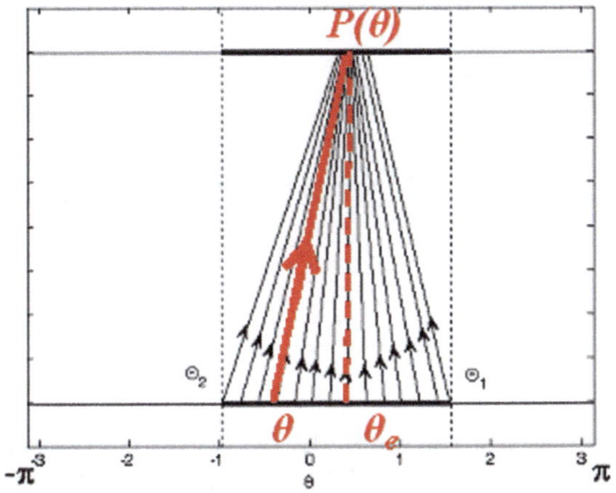

Figure 19. ne-dimensional map P for $A = 0.6$, $\lambda = 0.2$ and $\sigma = 1$; the stable attractor of the map is denoted by dashed line.

By varying the detuning parameter and by carefully studying the structure of the map, we may determine the value of σ for which the period-one attractor of the map disappears, and thus investigate the dynamical mechanism responsible for its appearance. Hence, one obtains an analytic tool for determining the frequency region of existence of SMRs. For the system considered with $A = 0.6$ and $\lambda = 0.2$, the boundaries of the detuning parameter within which the SMR exists are determined as, $\sigma_R = 2.69 > \sigma > \sigma_L = -2.0546$. Considering the transformations relating the slow flow to the exact equations of motion, one concludes that the SMRs in system (68) exists in an $O(\varepsilon)$ neighborhood of the exact resonance.

The previous global analysis of the dynamics identified the mechanism of creation and annihilation of the stable and unstable periodic orbits (limit cycle oscillations, LCOs) of the slow flow in the neighborhood of the upper boundary $\sigma = \sigma_R$ of the frequency detuning range of existence of SMRs. A projection of a representative stable LCO (SMR) on the phase cylinder $(N, \theta) \in (R^+ \times S^1)$ is presented in Figure 20. This orbit clearly depicts the slow evolution of the dynamics on the upper and lower stable branches of the SIM (denoted by solid lines), and the fast transitions (jumps – denoted by dashed lines) when the orbit reaches the fold lines L_1 and L_2.

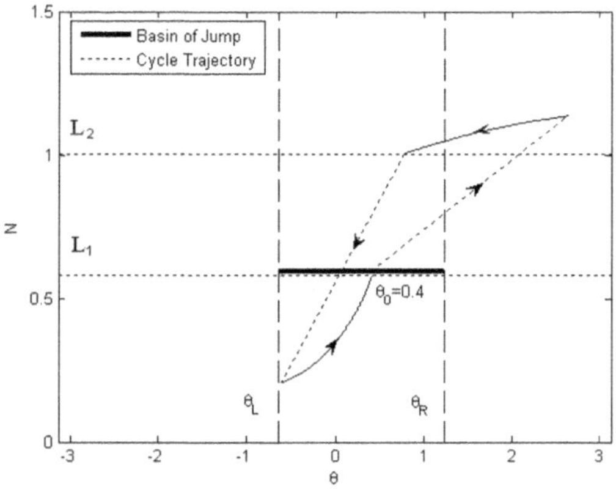

Figure 20. Projection of the stable LCO corresponding to a SMR for $A = 0.3$, $\lambda = 0.2$ and $\sigma = 0.5$: dashed lines refer to 'slow' jumps between the two stable branches of the SIM; solid lines refer to 'super - slow' evolutions on the stable branches of the SIM.

Once the dynamics is reduced to the one-dimensional map P, one expects that the system will exhibit generic bifurcations that occur in general classes of this type of dynamical systems.

It should be mentioned that the analytical approach developed in this Section is valid only in the limit $\varepsilon \to 0$, i.e., only for the case of lightweight NESs. In [(15)] comparisons between the previous analytical results and direct numerical simulations of system (68) were performed that validated the previous asymptotic analysis.

Suppression of the LCO in the self-excited Van der Pol (VdP) oscillator with the NES attached was considered in [(19)] (see also [(3)], v. II). The approach involved extensive numeric simulations and local analysis of the averaged flow equations. Main result of these studies is that the NES is capable of suppressing the LCO in the primary structure with the self-excitations.

- The first mechanism is is characterized by a recurrent series of suppressed burst-outs of the heave and pitch modes of the wing, leading eventually to complete suppression of the aeroelastic instabilities;
- The second mechanism is characterized by intermediate or partial suppression of LCOs. In other terms, the LCOs persist but have relatively small amplitude;
- The third mechanism is the most effective for suppressing the aeroelastic instabilities, as it results in complete and permanent elimination of LCOs.

In all works mentioned in the previous paragraph, efficient suppression of the LCOs with the help of the nonlinear energy sink was observed, either numerically or experimentally. In the following sections, the theoretical explanation of these results will be presented [(20), (4)].

As the simplest possible model of the self – excited system with the NES, let us consider common Van der Pol oscillator with attached purely cubic NES II (ungrounded) with linear damping. The initial system of equations describing this system can be written as follows:

$$\begin{aligned}&m_1\tfrac{d^2}{dt^2}x_1 + c\tfrac{d}{dt}x_1(x_1^2 - A^2) + qx_1 + \gamma(\tfrac{d}{dt}x_1 - \tfrac{d}{dt}x_2) + K(x_1 - x_2)^3 = 0\\ &m_2\tfrac{d^2}{dt^2}x_2 + \gamma(\tfrac{d}{dt}x_2 - \tfrac{d}{dt}x_1) + K(x_2 - x_1)^3 = 0\end{aligned} \tag{74}$$

In this system, m_i, i=1,2 are the masses of the primary oscillator and the NES respectively, q is the linear stiffness of the VDP oscillator, coefficients c and A characterize the combination of positive and negative damping in the VDP model, γ and K are the damping coefficient and the nonlinear stiffness of the NES respectively. After rescaling, system (74) is reduced to a non-dimensional form with four independent parameters:

$$\begin{aligned}&\ddot{u}_1 + \varepsilon\alpha\dot{u}_1(u_1^2 - 1) + u_1 + \varepsilon\lambda(\dot{u}_1 - \dot{u}_2) + \tfrac{4}{3}\varepsilon k(u_1 - u_2)^3 = 0\\ &\ddot{u}_2 + \lambda(\dot{u}_2 - \dot{u}_1) + \tfrac{4}{3}k(u_2 - u_1)^3 = 0\end{aligned} \tag{75}$$

where

$$\begin{aligned}&\tau = \omega t,\ \ \omega{=}\sqrt{q/m_1},\ u_i = x_i/A,\ \ i = 1,2,\\ &\varepsilon{=}m_2/m_1,\ \ \alpha{=}cA^2/m_2\omega^2,\ \ \lambda = \gamma/m_2\omega,\ \ \mathrm{k}{=}3KA^2/4m_2\omega^2\end{aligned},$$

the dot denotes the differentiation with respect to τ.

Let us introduce new coordinate, representing a relative displacement in the NES:

$$w = u_1 - u_2 \tag{76}$$

With account of (46), system (45) is rewritten as follows:

$$\begin{array}{l} \ddot{u}_1 + \varepsilon\alpha\dot{u}_1(u_1^2 - 1) + u_1 + \varepsilon\lambda\dot{w} + \frac{4}{3}\varepsilon k w^3 = 0 \\ \ddot{w} + (1+\varepsilon)\lambda\dot{w} + \frac{4}{3}(1+\varepsilon)k w^3 + \varepsilon\alpha\dot{u}_1(u_1^2 - 1) + u_1 = 0 \end{array} \tag{77}$$

System (77) is the basis for further analysis. As in the sections above, the motion of the system is studied in the vicinity of 1:1 resonance manifold, where all variables oscillate with frequency close to the natural frequency of the VDP oscillator. Change of variables

$$\varphi = (\dot{u}_1 + i u_1)\exp(-it), \;\; \xi = (w + iw)\exp(-it)$$

and subsequent averaging over the "fast" unit frequency yields the following "slow – flow" equations:

$$\begin{array}{l} \dot{\varphi} = -\frac{\varepsilon\alpha}{2}\varphi\left(\frac{|\varphi|^2}{4} - 1\right) - \frac{\varepsilon\lambda}{2}\xi + \frac{i\varepsilon k}{2}\left|\xi\right|^2\xi \\ \dot{\xi} = \frac{i}{2}(\varphi - \xi) - \frac{(1+\varepsilon)\lambda}{2}\xi + \frac{i(1+\varepsilon)k}{2}\left|\xi\right|^2\xi - \frac{\varepsilon\alpha}{2}\varphi\left(\frac{|\varphi|^2}{4} - 1\right) \end{array} \tag{78}$$

Further reduction of System (78) is performed by splitting the complex variables φ and ξ into modulus and argument parts:

$$\begin{array}{l} \varphi = R\exp(i\delta_1), \;\; \xi{=}P\exp(i\delta_2), \;\; \delta{=}\delta_1 - \delta_2 \\ \dot{R} = -\frac{\varepsilon\alpha}{2}R\left(\frac{R^2}{4} - 1\right) - \frac{\varepsilon\lambda}{2}P\cos\delta + \frac{\varepsilon k}{2}P^3\sin\delta \\ \dot{P} = -\frac{R}{2}\sin\delta - \frac{(1+\varepsilon)\lambda}{2}P - \frac{\varepsilon\alpha}{2}R\left(\frac{R^2}{4} - 1\right)\cos\delta \\ \dot{\delta} = \frac{1}{2} - \frac{(1+\varepsilon)k}{2}P^2 - \frac{R}{2P}\cos\delta + \frac{\varepsilon}{2}\left(\frac{\lambda P}{R}\sin\delta + \frac{kP^3}{R}\cos\delta + \frac{\alpha R}{P}\left(\frac{R^2}{4} - 1\right)\sin\delta\right) \end{array} \tag{79}$$

Fixed points of system (79) can be described by single algebraic equation [(20)], but the problem is that the system under consideration possesses dynamic regimes which are not related to fixed points of the slow flow (78). Numerous examples of such behavior will be presented below. Treatment of these regimes requires global analysis of system (79); the latter seems to be impossible without further simplifying assumptions.

It is natural to suggest that the NES has small weight compared to the mass of the principal oscillator. Besides, the averaging procedure performed above has a chance to be valid if the frequency of vibrations is close to the natural frequency of the primary linear oscillator. In other terms, one can adopt the following assumptions concerning the order of magnitude of coefficients in system (79):

$$\varepsilon << 1, \alpha, k, \lambda \sim O(1) \tag{80}$$

In this case, just similarly to the treatments in Section 4.3, system (79) can be considered as problem of singular perturbation with two "slow" and one "super-slow" variable. The term "slow" is related to the evolution of the averaged flow (80), whereas the term "fast" is reserved for the oscillations which were averaged out. So, the analysis of the initial problem requires consideration of three time scales.

In order to perform the analysis, we introduce the time scales as: $\tau_0 = \tau, \tau_1 = \varepsilon\tau$. At "slow" time scale, system (79) is reduced to the following form:

$$\begin{aligned} \frac{\partial R}{\partial \tau_0} &= 0 \\ \frac{\partial P}{\partial \tau_0} &= -\frac{R}{2}\sin\delta - \frac{\lambda}{2}P \\ \frac{\partial \delta}{\partial \tau_0} &= \frac{1}{2} - \frac{k}{2}P^2 - \frac{R}{2P}\cos\delta \end{aligned} \tag{81}$$

From (81) it follows that $R = R(\tau_1)$ and two last equations can be considered as planar system. This planar system has no limit cycles and all trajectories are attracted to fixed points, which, in turn, depend on the "super – slow" time τ_1:

$$N(\tau_1) = \lim_{\tau_0 \to \infty} P(\tau_0, \tau_1), \;\; \Delta(\tau_1) = \lim_{\tau_0 \to \infty} \delta(\tau_0, \tau_1)$$

From (81) one obtains:

$$\begin{aligned} &\sin\Delta(\tau_1) = -\frac{\lambda N(\tau_1)}{R(\tau_1)}, \quad \cos\Delta(\tau_1) = \frac{N(\tau_1)(1-kN^2(\tau_1))}{R(\tau_1)} \\ &R^2(\tau_1) = N^2(\tau_1)(\lambda^2 + (1-kN^2(\tau_1))^2) \end{aligned} \tag{82}$$

In the following we will omit for brevity the explicit dependence on τ_1. Besides, we use the notations $Y = R^2, \;\; Z = N^2$. In these notations, the last equation of (82) is rewritten as

$$Y = Z(\lambda^2 + (1-kZ)^2) \tag{83}$$

Quite naturally, it is a slight modification of the SIM (46), since the same NES is used. For the analysis of the super-slow flow, it is enough to take only the first equation of system (79). In the limit $\tau_0 \to \infty$ it yields:

$$\frac{\partial R}{\partial \tau_1} = -\frac{\alpha}{2}R\left(\frac{R^2}{4} - 1\right) - \frac{\lambda}{2}N\cos\Delta + \frac{k}{2}N^3\sin\Delta \tag{84}$$

Substituting equations for $\Delta(\tau_1)$ from (82) and multiplying by R, one obtains:

$$\frac{\partial Y}{\partial \tau_1} = \alpha Y(1 - Y/4) - \lambda Z \tag{85}$$

Substituting the expression for the SIM (83) into (85), one obtains the following equation for Z(τ_1):

$$\frac{\partial Z}{\partial \tau_1} = \frac{Z\left[\alpha(\lambda^2 + (1-kZ)^2)(1 - \frac{Z}{4}(\lambda^2 + (1-kZ)^2)) - \lambda\right]}{1 + \lambda^2 - 4kZ + 3k^2 Z^2} \tag{86}$$

Equation (86) can be solved in quadratures, but it is rather awkward task. In order to analyze the dynamics of the system qualitatively, we need only know how the fixed points of (86) are situated at the SIM. For this sake, one has to find zeros of the numerator in (86). It involves solution of fifth – power algebraic equation, and is rather difficult in parametric form (one has to use elliptic theta – functions etc.). Instead, in the next section we are going to present these fixed points graphically, as intersection of the SIM (83) and parabola defined by the right – hand side of (85) equal to zero:

$$\alpha Y(1 - Y/4) - \lambda Z = 0 \tag{87}$$

In this review, only the most interesting response regimes are presented. More complete analysis can be found in [(20)].

Let us consider the following set of parameters:

$$\alpha = 0.15, \lambda = 0.15, k = 0.36, \varepsilon = 0.05 \tag{88}$$

In Figure 21 we present two curves: the SIM (83), denoted by the thick line, and parabola (87) – by the thin line. They intersect only for Z=0 and it is the only stable fixed point of the flow. It corresponds to complete elimination of the LCO.

Let us check this prediction numerically. In Fig. 22 numeric simulation of the initial system (45) is presented. Fig. 22a describes the response of the primary VDP oscillator and 22b – that of the NES (relative displacement$w(t) = u_1(t) - u_2(t)$). Initial conditions for the simulation are specified in the figure caption.

As one can see, the LCO in this case is very efficiently suppressed. More prolonged simulation yields for u_1 at t=5000 the maximum displacement close to 0.0007. Of course, the actual displacement will never be exactly zero, but still this case can qualify for complete elimination of the LCO.

It is possible to demonstrate that this regime exists only provided that [(20)]:

$$\frac{\lambda}{\alpha} > 1 + \lambda^2 \Rightarrow \alpha < \frac{\lambda}{1+\lambda^2} = \frac{1}{\lambda + 1/\lambda} \tag{89}$$

Quite obviously, the latter condition can be fulfilled only for α<0.5. So, if α is not too large, one can choose the NES parameters in order to achieve

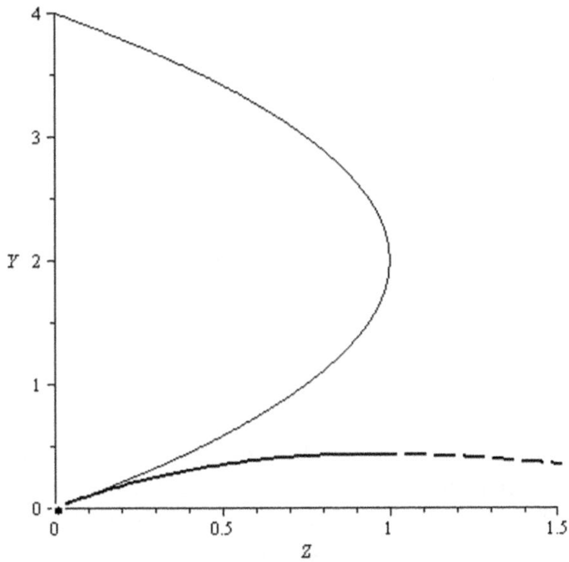

Figure 21. The case of complete LCO elimination. Thick line corresponds to the SIM (83), thin line – to parabola (87). The small circle at Z=0 denotes stable fixed point

the complete elimination of the LCO. If it is not the case, one can try to use higher value of the relative mass ε, thus achieving lower effective value of α. If by some reason it is not feasible, on should rely on alternative (but less efficient) mechanisms of the LCO suppression.

Let us consider the next set of parameters, where the LCO is not eliminated completely:

$$\alpha = 0.6, \lambda = 0.3, k = 0.4, \varepsilon = 0.05 \tag{90}$$

The diagram of the super-slow flow on the SIM is presented in Fig. 23. One can see that the only stable fixed point corresponds to small – amplitude steady – state oscillations at the lower branch of the SIM. In this case the LCO does not disappear completely but is suppressed partially.

Numeric verification of this regime is presented in Figs. 24a,b. One can see that the regime of partial LCO suppression is indeed observed. Moreover, the amplitudes both of the primary oscillator and the NES are successfully predicted in Fig. 24 (square roots of the ordinate and the abscissa of the stable fixed point respectively).

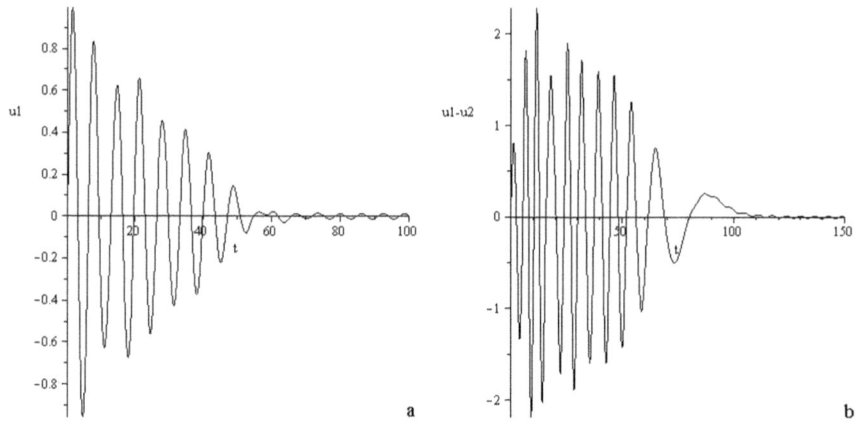

Figure 22. Numeric simulation for the case of the complete LCO suppression (set of parameters (88). Initial conditions: $u_1(0) = 0, \dot{u}_1(0) = 0.36, u_2(0) = 0, \dot{u}_2(0) = 0$. a) Time series for primary VDP oscillator u_1(t); b)Time series for relative NES displacement w(t)= u_1(t)- u_2(t).

The last example of possible response regimes describes a global bifurcation – a birth of saddle – node pair on the upper branch of the SIM (Figure 25). This Figure is produced for the set of parameters

$$\alpha = 1.3, \lambda = 0.3, k = 0.4, \varepsilon = 0.05 \qquad (91)$$

Figure 25 suggests co-existence of two stable response regimes – high – amplitude LCO oscillations (node at the upper branch) and the SMR. These regimes have well- defined basins of attraction. Namely, the trajectories with all initial conditions above the dashed horizontal line passing through the upper saddle will be attracted to the stable node, and below this line – to the SMR. In order to verify this prediction we simulate the response for two sets of initial conditions, denoted as points 1 and 2 in Fig. 25. The results are presented in Figs. 26 a,b and are in complete agreement with the above prediction.

If α grows even further, the SMR disappears. It happens when the upper saddle crosses the "landing point" of the upper SIM branch Z_u. One of possible sets of parameters for this critical situation is

$$\alpha = 2.318, \lambda = 0.3, k = 0.4 \qquad (92)$$

Corresponding diagram is presented in Fig. 27.

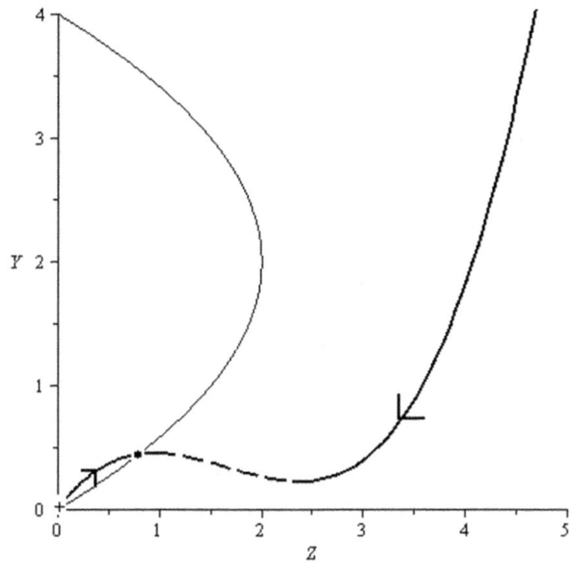

Figure 23. The case of partial LCO elimination. The small cross at Z=0 denotes unstable fixed point, the arrows at the SIM branches denote the directions of the super – slow flow.

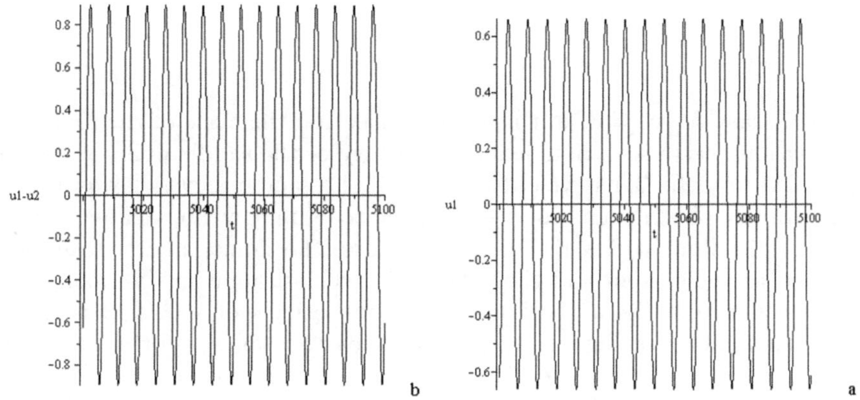

Figure 24. Numeric simulation for the case of the partial LCO suppression – set of parameters (90) Initial conditions: $u_1(0) = 0, \dot{u}_1(0) = 1, u_2(0) = 0, \dot{u}_2(0) = 0$. a) Time series for primary VDP oscillator $u_1(t)$; b)Time series for relative NES displacement $w(t) = u_1(t) - u_2(t)$

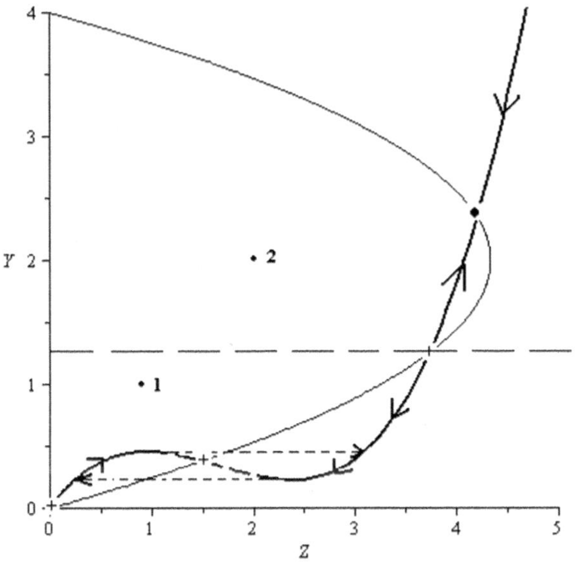

Figure 25. Co-existence of stable SMR and stable LCO regime (set of parameters (91)). The super – slow flow allows the SMR (denoted by dashed arrows and arrows at the SIM). Horizontal dashed line divides between basins of attractions of two stable regimes. Point 1 has coordinates (1,1) Point 2 – (2,2).

It is easy to see that in this critical situation the cycle of the relaxation oscillations becomes a homoclinic trajectory of the saddle point at the upper SIM branch. This cycle disappears if the saddle moves downwards on the branch. Such scenario is equivalent to Shilnikov homoclinic bifurcation of the limit cycles [(17)]. In this case, the homoclinic connection is formed in the system and then the limit cycle disappears. Normally, it is not easy to prove that the homoclinic connection exists in a system with dimensionality 3 or higher. In our case, however, for $\varepsilon \to 0$ this trajectory can be easily demonstrated. Existence of such connection is preserved under small variations of parameters [(17)]; therefore we can state that this bifurcation scenario should exist also for ε small, but finite. When the SMR approaches the homoclinic connection, one should expect significant growth of the modulation period – at the connection it will be "exactly infinite". Such elongation, observed numerically [(20)], confirms the suggested scenario of the bifurcation.

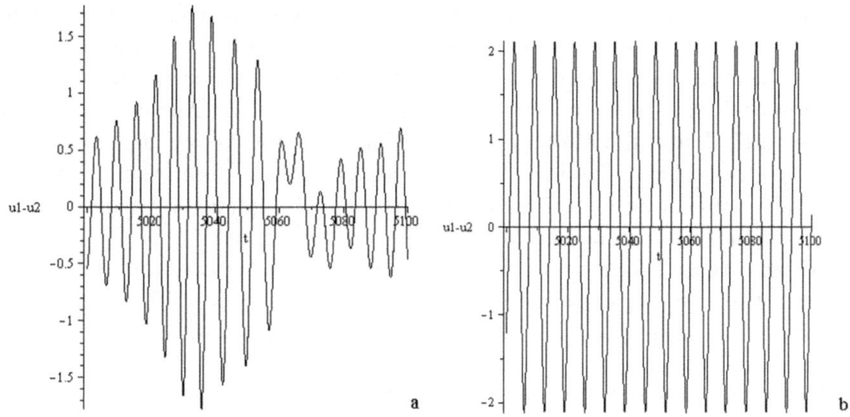

Figure 26. Relative NES displacement for the case of co-existence, set of parameters (91). Initial conditions correspond to points 1 and 2: (a) $u_1(0) = 0, \dot{u}_1(0) = 1, u_2(0) = 0, \dot{u}_2(0) = 0$, (b) $u_1(0) = 0, \dot{u}_1(0) = 1.4, u_2(0) = 0, \dot{u}_2(0) = 0$

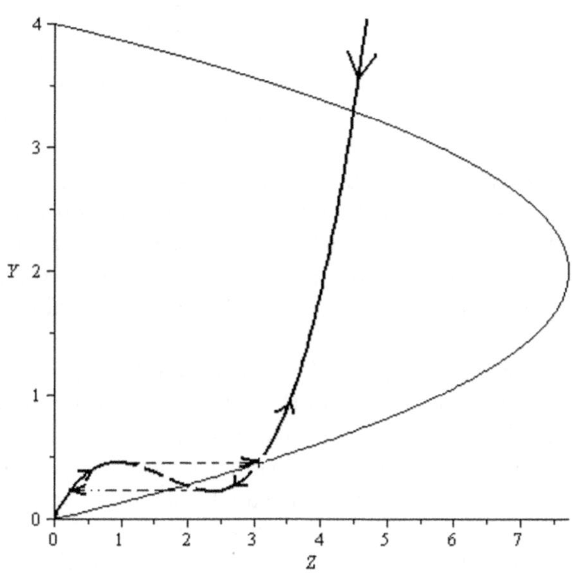

Figure 27. Homoclinic connection leading to the Shil'nikov bifurcation

The methodology presented in this Section paves a way for global asymptotic analysis of the self – excited systems with the NES. Periodic responses, global bifurcations of different types, as well as co-existence and basins of attraction of various self – excitation regimes are revealed. All that is possible due to radical simplifications related to "slow – super-slow" decomposition of the averaged flow. Qualitatively, the coincidence between the theory and numeric simulations is complete. Quantitatively, some deviations in the critical values of parameters (up to 10%) were observed.

At this step, it is instructive to analyze common features and differences between the cases of external forcing and self – excitation in systems with the NES. The first obvious common feature is the relevance of three different time scales. The first one is related to "fast" oscillations of the primary system with its eigenfrequency, and to the frequency of the external force. Averaging over this fast frequency yields the slow – flow equations, like (37), (61), (79) and others. It should be mentioned here, that in all examples described in this review, the nonlinearity is strong and therefore the averaging is performed beyond formal framework of the averaging theory. Still, good numeric coincidences confirm that in the context of the problems considered the procedure can be used, although with some caution.

The slow – flow equations, in turn, also involve two different time scales. At this stage, the separation of "slow" and "super-slow" time scales is justified by use of small parameter ε related to the mass ratio between the NES and the primary mass, as well as to the characteristic scales of the damping and the forcing terms. The dimensionality of the "slow" flow both for the external and self – excitation is equal to 2; more exactly, the "slow" flow exists in the space $R^+ \times S^1$ in both cases. The "super-slow" flow, however, is very different: for the systems with external excitation it is two – dimensional ($R^+ \times S^1$), whereas for the systems with self – excitation it turns out to be one – dimensional (R^+). These similarities of the slow flow and difference with respect to the super-slow flow have crucial consequences for the dynamics of both systems.

First of all, similar structure of the slow flow brings about similarity of possible dynamical responses. In both systems one observes responses with constant amplitude (single or co-existing); these steady-state responses can loose stability via Hopf bifurcation, leading to weakly modulated response (WMR). The latter responses were revealed both for externally excited and self – excited systems. In addition, both systems exhibit the strongly modulated responses (SMRs), related to relaxation oscillations of the slow flow on the slow invariant manifold (SIM); the latter is similar for both types of systems. In the same time, the difference in the dimensionality of the super – slow flow brings about qualitatively different generic bifurcations of the steady-state responses as well as the SMRs. For the systems with external excitations, the two – dimensional super – slow flow causes the "jump" domain at the fold line to be one – dimensional; then, the discrete map which governs the behavior of the SMRs, is one – dimensional ($R^1 \rightarrow R^1$). Therefore, the generic bifurcations of the SMRs are the same as the generic bifurcations of the one – dimensional discrete maps (saddle – node, period doubling). Both these bifurcations of the SMRs were observed in the complete system [(14), (15)], where they are equivalent to the saddle – node bifurcations of tori and to global change of the structure of the attractor, respectively. For the systems with self – excitation the super – slow flow is one – dimensional and the "jump domain" is a single point. Of course, the above bifurcations are not possible for this "map", however, some other generic possibilities arise. The bifurcations of SMRs in such a system occur due to passage of the fixed points of the super – slow flow through the fold points (Hopf bifurcation and canard explosion, [(17))] or through the "landing" points at the SIM (Shilnikov bifurcation). Another possibility is related to the saddle – node bifurcation of the fixed points at the SIM (heteroclynic bifurcation). All these bifurcations are possible, in principle, also for the system with higher dimensionality of the super – slow flow, but they will be not generic and will have co-dimension two. So, it is extremely difficult to find the values of parameters to observe these bifurcations, taking into account the approximate nature of the whole approach.

The difference in dimensionalities of the super –slow flow between the externally excited and the self – excited systems reveals itself also in different relationship between the SMRs and other types of the response. In the systems with self – excitation, the full – scale SMR appears as a result of "canard explosion" [(20)]. To illustrate this point, let us consider the system of VdP oscillator with the NES with the following set of parameters:

$$\alpha = 1, \lambda = 0.3, k = 0.4, \varepsilon = 0.05 \tag{93}$$

one obtains the following diagram (intersection of the SIM (83) and parabola (85), Figure 28):

Both fixed points at this diagram are unstable and the only possible response regime is the stable SMR. It is interesting to investigate how the transition from the steady – state low – amplitude response to the SMR occurs. If one compares Figure 21 and Figure 28, it is clear that the transition occurs when the stable fixed point crosses the fold to the unstable SIM branch. In the lowest – order approximation used here, the fully developed SMR immediately substitutes the steady – state response However, initial averaged system (79) is smooth and such singular behavior is impossible in it. Namely, the transition described here is regular Hopf bifurcation and

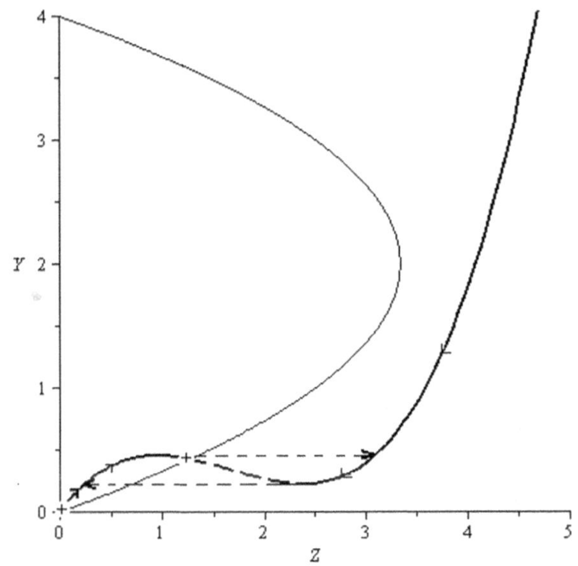

Figure 28. The case of stable SMR. Both fixed points are unstable, the super slow flow allows the SMR (denoted by dashed arrows and arrows at the SIM)

the limit cycle born should have small amplitude. This contradiction arises since the lowest – order approximation is insufficient in the vicinity of the fold. Typical behavior in such situation is exactly the *canard explosion* [(17)]: the amplitude of the limit cycle grows to full – scale relaxation oscillations with exponentially small variation of the governing parameter. The analysis presented here treats only the shape of the modulation envelope; so, it is instructive to verify whether such mechanism will be observed for the modulation amplitude in complete dynamic flow (74). In order to see sharp transition, lower value of ε =0.005 is used. Results of numeric simulation are presented in Fig. 29, the values of parameters are mentioned in figure captions.

One can see that the Hopf bifurcation of the modulation envelope occurs at about $\alpha \approx 0.706$ and the amplitude of the limit cycle in the modulation amplitude is small, as required. This regime can be identified as the weakly modulated response (WMR), mentioned above. This amplitude steadily grows up to $\alpha \approx 0.716$ and then "explodes" to full – scale SMR at $\alpha \approx 0.717$. This picture completely coincides with the canard explosion scenario men-

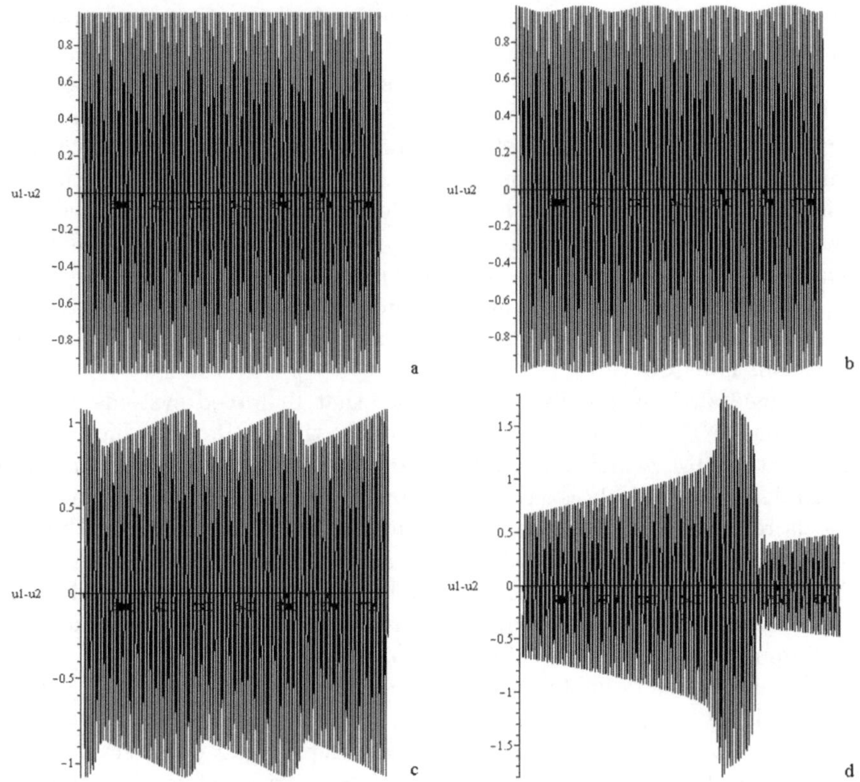

Figure 29. Canard explosion of the modulation in the complete flow. NES displacement versus time is presented. The parameters are $\varepsilon = 0.005$, $\lambda = 0.3$, $k = 0.4$. Parameter α is varied: a) $\alpha = 0.705$; b) $\alpha = 0.708$; c) $\alpha = 0.716$, d) $\alpha = 0.717$

tioned above. Analytic estimation yields $\alpha \approx 0.716$, in excellent agreement with the numeric results.

So, in the case of the self-excited systems the relationship between the WMR and the SMR is straightforward – the SMR develops from the WMR by smooth variation of the system parameters. For the systems with external excitation, the situation is completely different – the SMR generically arises as a result of saddle – node bifurcation of limit cycles. It should be mentioned, however, that also in this case the WMR and the SMR can be related, for instance, through a mechanism of subcritical Hopf bifurcation [(21)]; however, to establish this fact, more refined analysis taking into account high – order terms in approximate solution of singularly perturbed equations is required.

In general, it is possible to conclude that in forced systems with the small-mass NES the necessary condition for the SMRs is S-shaped structure of the SIM, which is completely determined by the *slow flow*. In all examples presented above, this slow flow is described by two scalar ODEs and is related to particular structure and coupling of the NES. However, the sufficient conditions for the SMR, as well as the generic bifurcations, are governed by the *super-slow* flow and the recurrence map of the jump domain into itself. More exactly, the dimensionality of the recurrence maps determines possible generic bifurcations.

To clarify this point further, let us consider the example with 2D slow flow and 3D super – slow flow. Such behavior is realized if two linear oscillators with close frequencies are considered; one of these oscillators is capable of self – excitation and another is attached to the NES. The equations of motion for this model look as follows:

$$\begin{aligned} &\ddot{u}_1 + \varepsilon\alpha\dot{u}_1(u_1^2 - 1) + u_1 + \varepsilon p(u_1 - u_2) = 0 \\ &\ddot{u}_2 + u_2 + \varepsilon p(u_2 - u_1) + \varepsilon\lambda(\dot{u}_2 - \dot{v}) + \tfrac{4}{3}\varepsilon k(u_2 - v)^3 = 0 \\ &\ddot{v} + \lambda(\dot{v} - \dot{u}_2) + \tfrac{4}{3}k(v - u_2)^3 = 0 \end{aligned} \tag{94}$$

Here $u_{1,2}$ are displacements of the linear oscillators, v – that of the NES. Coefficient εp denotes the weak linear coupling between two linear oscillators. For this system, one predicts that the averaging over the unit frequency will bring about 5D average flow equations; the slow flow will be 2D, as in the previous examples, whereas the super – slow flow will be 3D (2 variables related to the amplitudes of the linear oscillators and the third one related to the phase difference between them). Therefore, the SIM will have qualitatively the same structure as in all previous examples; however, the "jump" recurrence map will have the dimensionality$R^2 \to R^2$. One of immediate consequences of this "complication" is that one can expect that this nonlinear map will have considerable regions in the space of parameter, corresponding to chaotic attractors. In the initial state space of System (94), such chaotic attractors will correspond to chaotic modulations of the SMRs. Such behavior was observed neither in externally forced nor in self – excited systems with lower dimensions. However, for system described by equations (94) such SMR with chaotic-like modulations turns out to be generic. An example of such dynamic response is presented in Figure 30.

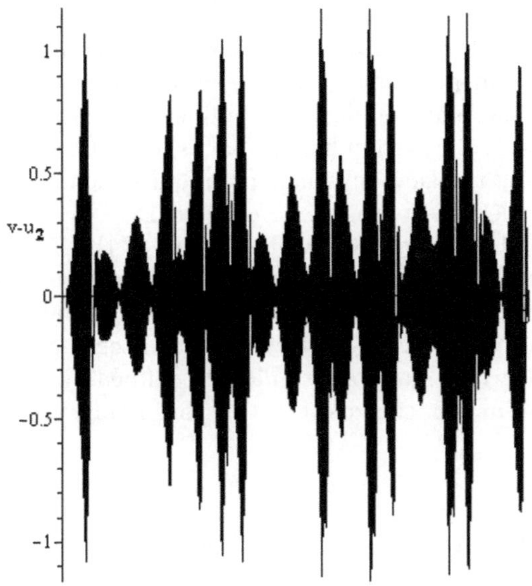

Figure 30. Relative displacement of the NES, System (94). The parameters are ε =0.05, α =0.5, p=1, k=1, λ =0.2, initial conditions are zero besides$\dot{u}_1(0) = 0.1$

This example demonstrates that one can predict generic behavior of the dynamic systems with the NES based on dimensionalities of the slow and super – slow flows in averaged equations of motion.

6 Concluding remarks

Usually one perceives the nonlinear normal modes as particular exact solutions of the nonlinear dynamical system which are "synchronous", or, more generally, belong to a certain low-dimensional invariant manifold. In rare instances these exact solutions can be found analytically. More often one can assess them with the help of approximate procedures, such as an averaging or a harmonic balance; these approximate predictions may be checked with the help of appropriately designed numeric procedures.

From the results presented above one can conclude that the direct extension of this concept of the NNM for the damped case is hardly possible. For analytic continuation one should use "complex" modes of the linear damped systems. Moreover, loosely defined NNMs in the damped system do not survive passage through the bifurcation - in other terms, they may be fundamentally unstable, even if their "conservative" counterparts are perfectly stable. Moreover, it is not clear why such direct extension would be useful and what problems it could help to solve.

However, some "shift of paradigm" allows obtaining useful notion of the damped NNMs. If the system has special asymptotic structure, the damped NNM appears as the slow invariant manifold which attracts the dynamics at fast time scale. As it was demonstrated in Sections 3 and 4, such notion of the NNM allows predictive treatment of the TET problem, both in the damped and forced-damped cases. This case is very different from the "regular" invariant manifolds that appear in conservative problems. In the latter, the choice of "master" variables is in a sense arbitrary; in the TET problem the invariant manifold is uniquely defined by the asymptotic structure of the system. So, the NNM in this case is provably *special solution.*

Bibliography

[1] A.F. Vakakis, L.I. Manevitch, Y.V. Mikhlin, V.N. Pilipchuk and A.A. Zevin *Normal Modes and Localization in Nonlinear Systems* Wiley Interscience, New York, 1996.

[2] L.I. Manevitch, Y.V. Mikhlin and V.N. Pilipchuk *Method of Normal Vibrations for Essentially Nonlinear Systems* Nauka, Moscow, 1989 (in Russian).

[3] A.F. Vakakis, O.V. Gendelman, G. Kerschen, L.A. Bergman, M.D. McFarland and Y.S. Lee *Nonlinear Targeted Energy Transfer in Mechanical and Structural Systems* v.I and v.II, Springer, 2009.

[4] O.V. Gendelman *Targeted Energy Transfer in Systems with External and Self Excitation* Proceedings of the Institution of Mechanical Engineers, Part C, Journal of Mechanical Engineering Science, invited review, 225, 2007-2043, 2011.

[5] L. Meirovitch *Fundamentals of Vibrations* McGraw-Hill, 2001.

[6] L.I. Manevitch and O.V. Gendelman *Tractable models in Solid Mechanics* Springer, 2011.

[7] F.W.J. Olver, D.W. Lozier, R.F. Boisvert, C.W. Clark *NIST Handbook of Mathematical Functions* NIST and Cambridge University Press, 2010.

[8] A.H. Nayfeh and D.T. Mook *Nonlinear Oscillations* Wiley and sons, 1995.

[9] J. Guckenheimer and P. Holmes *Nonlinear Oscillations, Dynamical System, and Bifurcation of Vector Fields* Springer Verlag, New York, 1983.

[10] S.W. Shaw, C. Pierre *Nonlinear normal modes and invariant manifolds* Journal of Sound and Vibration, 150, 170-173, 1991.

[11] S.W. Shaw, C. Pierre *Normal modes for nonlinear vibratory systems* Journal of Sound and Vibration, 164, 85-124, 1993.

[12] X. Jiang, D.M. McFarland, L.A. Bergman and A.F. Vakakis *Steady State Passive Nonlinear Energy Pumping in Coupled Oscillators: Theoretical and Experimental Results* Nonlinear Dynamics 33, 7-102, 2003.

[13] O.V. Gendelman, E. Gourdon and C.-H. Lamarque *Quasiperiodic Energy Pumping in Coupled Oscillators under Periodic Forcing* Journal of Sound and Vibration 294, 651-662, 2006.

[14] Y. Starosvetsky and O.V. Gendelman *Strongly modulated response in forced 2DOF oscillatory system with essential mass and potential asymmetry* Physica D 237, 1719-1733, 2008.

[15] Y. Starosvetsky and O.V. Gendelman *Response Regimes of Linear Oscillator Coupled to Nonlinear Energy Sink with Harmonic Forcing and Frequency Detuning* Journal of Sound and Vibration 315, 746-765, 2008.

[16] G. Gatti, I. Kovacic and M.J. Brennan *On the response of a harmonically excited two degree-of-freedom system consisting of a linear and a nonlinear quasi-zero stiffness oscillator* Journal of Sound and Vibration 329, 1823-1835, 2010.

[17] V.I. Arnold, V.S. Afrajmovich, Y.S. Ilyashenko and L.P. Shilnikov *Dynamical Systems V. Encyclopedia of Mathematical Sciences* Springer Verlag, Berlin and New York, 1994.

[18] J. Guckenheimer, M. Wechselberger and L.S. Young *Chaotic attractors of relaxation oscillators* Nonlinearity 19,701-720, 2006

[19] Y.S. Lee, A.F. Vakakis, L.A. Bergman and D.M. McFarland *Suppression of limit cycle oscillations in the Van der Pol oscillator by means of passive nonlinear energy sinks (NESs)* Structural Control and Health Monitoring 13, 41-75, 2006.

[20] O.V. Gendelman and T. Bar *Bifurcations of Self - Excitation Regimes in Van der Pol Oscillator with Nonlinear Energy Sink* Physica D 239: 220-229, 2010.

[21] Y. Starosvetsky and O.V. Gendelman *Bifurcations of attractors in forced system with nonlinear energy sink: effect of mass asymmetry* Nonlinear Dynamics 59, 711-731, 2010.

Computation of Nonlinear Normal Modes through Shooting and Pseudo-Arclength Computation

G. Kerschen
Department of Aerospace and Mechanical Engineering
University of Liège
Liège, Belgium

1 Introduction

Most existing constructive techniques for computing NNMs are based on asymptotic approaches and rely on fairly involved mathematical developments. In this context, algorithms for the numerical continuation of periodic solutions are really quite sophisticated and advanced (see, e.g., (1; 2), and the AUTO and MATCONT softwares). These algorithms have been extensively used for computing the forced response and limit cycles of nonlinear dynamical systems. Interestingly, there have been very few attempts to compute the periodic solutions of conservative mechanical structures (i.e., NNM motions) using numerical continuation techniques. One of the first approaches was proposed by Slater in (3) who combined a shooting method with sequential continuation to solve the nonlinear boundary value problem that defines a family of NNM motions. Similar approaches were considered in Lee et al. (4) and Bajaj et al. (5). A more sophisticated continuation method is the so-called asymptotic-numerical method. It is a semi-analytical technique that is based on a power series expansion of the unknowns parameterized by a control parameter. It is described in the next chapter. In this study, a shooting procedure is combined with the so-called pseudo-arclength continuation method for the computation of NNM motions. We show that the NNM computation is possible with limited implementation effort, which holds promise for a practical and accurate method for determining the NNMs of nonlinear vibrating structures.

This chapter is organized as follows. In the next section, the proposed algorithm for NNM computation is presented. Its theoretical background is first recalled, and the numerical implementation is then described. Improvements are also presented for the reduction of the computational bur-

G. Kerschen (Ed.), *Modal Analysis of Nonlinear Mechanical Systems*, CISM International Centre for Mechanical Sciences DOI 10.1007/ 978-3-7091-1791-0_5 © CISM Udine 2014

den. The proposed algorithm is then demonstrated using three different nonlinear vibrating systems in Section 3.

2 Numerical Computation of NNMs

The numerical method proposed here for the NNM computation relies on two main techniques, namely a shooting technique and the pseudo-arclength continuation method. It is summarized in Figure 1.

2.1 Shooting Method

The equations of motion are

$$\mathbf{M}\,\ddot{\mathbf{x}}(t) + \mathbf{K}\,\mathbf{x}(t) + \mathbf{f}_{\rm nl}\left\{\mathbf{x}(t), \dot{\mathbf{x}}(t)\right\} = 0 \tag{1}$$

where $\mathbf{M}$ is the mass matrix; $\mathbf{K}$ is the stiffness matrix; $\mathbf{x}$, $\dot{\mathbf{x}}$ and $\ddot{\mathbf{x}}$ are the displacement, velocity and acceleration vectors, respectively; $\mathbf{f}_{\rm nl}$ is the nonlinear restoring force vector, assumed to be regular. In principle, systems with nonsmooth nonlinearities can be studied with the proposed method, but they require a special treatment. The equations of motion can be recast into state space form

$$\dot{\mathbf{z}} = \mathbf{g}(\mathbf{z}) \tag{2}$$

where $\mathbf{z} = [\mathbf{x}^* \quad \dot{\mathbf{x}}^*]^*$ is the $2n$-dimensional state vector, and star denotes the transpose operation, and

$$\mathbf{g}(\mathbf{z}) = \begin{pmatrix} \dot{\mathbf{x}} \\ -\mathbf{M}^{-1}\left[\mathbf{K}\mathbf{x} + \mathbf{f}_{\rm nl}(\mathbf{x}, \dot{\mathbf{x}})\right] \end{pmatrix} \tag{3}$$

is the vector field. It is assumed that the mass matrix is invertible. The solution of this dynamical system for initial conditions $\mathbf{z}(0) = \mathbf{z}_0 = [\mathbf{x}_0^* \quad \dot{\mathbf{x}}_0^*]^*$ is written as $\mathbf{z}(t) = \mathbf{z}\,(t, \mathbf{z}_0)$ in order to exhibit the dependence on the initial conditions, $\mathbf{z}\,(0, \mathbf{z}_0) = \mathbf{z}_0$. A solution $\mathbf{z}_p(t, \mathbf{z}_{p0})$ is a periodic solution of the autonomous system (2) if $\mathbf{z}_p(t, \mathbf{z}_{p0}) = \mathbf{z}_p(t+T, \mathbf{z}_{p0})$, where T is the minimal period.

The NNM computation is carried out by finding the periodic solutions of the governing nonlinear equations of motion (2). In this context, the *shooting method* is probably the most popular numerical technique (1; 2; 6). It solves numerically the two-point boundary-value problem defined by the periodicity condition

$$\mathbf{H}(\mathbf{z}_{p0}, T) \equiv \mathbf{z}_p(T, \mathbf{z}_{p0}) - \mathbf{z}_{p0} = \mathbf{0} \tag{4}$$

$\mathbf{H}(\mathbf{z}_0, T) = \mathbf{z}(T, \mathbf{z}_0) - \mathbf{z}_0$ is called the *shooting function* and represents the difference between the initial conditions and the system response at time

T. Unlike forced motion, the period T of the free response is not known a priori.

The shooting method consists in finding, in an iterative way, the initial conditions $\mathbf{z}_{p0}$ and the period T that realized a periodic motion. To this end, the method relies on direct numerical time integration and on the Newton-Raphson algorithm.

Starting from some assumed initial conditions $\mathbf{z}_{p0}^{(0)}$, the motion $\mathbf{z}_p^{(0)}(t, \mathbf{z}_{p0}^{(0)})$ at the assumed period $T^{(0)}$ can be obtained by numerical time integration methods (e.g., Runge-Kutta or Newmark schemes). In general, the initial guess $(\mathbf{z}_{p0}^{(0)}, T^{(0)})$ does not satisfy the periodicity condition (4). This is illustrated in Figure 2 for a Duffing oscillator

$$\ddot{x} + x + 0.5x^3 = 0 \tag{5}$$

Two pairs of initial conditions, $[x(0)\ \dot{x}(0)] = 0.9 \times [4.9009\ 0]$ and $[x(0)\ \dot{x}(0)] = 1.1 \times [4.9009\ 0]$, are two approximations to the actual solution, $[x(0)\ \dot{x}(0)] = [4.9009\ 0]$, for the current period $T = 2.0215\,\text{s}$. The former (latter) approximation yields a motion with a too large (small) period.

A Newton-Raphson iteration scheme is therefore to be used to correct an initial guess and to converge to the actual solution. The corrections $\Delta\mathbf{z}_{p0}^{(0)}$ and $\Delta T^{(0)}$ are found by expanding the nonlinear function

$$\mathbf{H}\left(\mathbf{z}_{p0}^{(0)} + \Delta\mathbf{z}_{p0}^{(0)}, T^{(0)} + \Delta T^{(0)}\right) = 0 \tag{6}$$

in Taylor series

$$\mathbf{H}\left(\mathbf{z}_{p0}^{(0)}, T^{(0)}\right) + \left.\frac{\partial \mathbf{H}}{\partial \mathbf{z}_{p0}}\right|_{(\mathbf{z}_{p0}^{(0)}, T^{(0)})} \Delta\mathbf{z}_{p0}^{(0)} + \left.\frac{\partial \mathbf{H}}{\partial T}\right|_{(\mathbf{z}_{p0}^{(0)}, T^{(0)})} \Delta T^{(0)} + \text{H.O.T.} = 0 \tag{7}$$

and neglecting higher-order terms (H.O.T.).

The initial conditions $\mathbf{z}_{p0}$ and the period T characterizing the periodic solution are computed through the iterative procedure

$$\mathbf{z}_{p0}^{(k+1)} = \mathbf{z}_{p0}^{(k)} + \Delta\mathbf{z}_{p0}^{(k)} \text{ and } T^{(k+1)} = T^{(k)} + \Delta T^{(k)} \tag{8}$$

with

$$\left.\frac{\partial \mathbf{H}}{\partial \mathbf{z}_{p0}}\right|_{(\mathbf{z}_{p0}^{(k)}, T^{(k)})} \Delta\mathbf{z}_{p0}^{(k)} + \left.\frac{\partial \mathbf{H}}{\partial T}\right|_{(\mathbf{z}_{p0}^{(k)}, T^{(k)})} \Delta T^{(k)} = -\mathbf{H}\left(\mathbf{z}_{p0}^{(k)}, T^{(k)}\right) \tag{9}$$

where k is the shooting iteration index. Convergence is achieved when $\mathbf{H}(\mathbf{z}_{p0}, T) \approx 0$ to the desired accuracy. In the neighborhood of the solution, the convergence is fast (i.e., quadratic convergence for an exact evaluation of the Jacobian matrix). However, it should be kept in mind that the Newton-Raphson method is a local algorithm; the convergence is guaranteed only when the initial guess is sufficiently close to the solution.

Each shooting iteration involves the time integration of the equations of motion to evaluate the current shooting residue

$$\mathbf{H}\left(\mathbf{z}_{p0}^{(k)}, T^{(k)}\right) = \mathbf{z}_p^{(k)}(T^{(k)}, \mathbf{z}_{p0}^{(k)}) - \mathbf{z}_{p0}^{(k)}.$$

As evidenced by equation (9), the shooting method also requires the evaluation of the partial derivatives of $\mathbf{H}(\mathbf{z}_0, T) = \mathbf{z}(T, \mathbf{z}_0) - \mathbf{z}_0$. The $2n \times 1$ vector $\partial\mathbf{H}/\partial T$ is given by

$$\begin{aligned} \frac{\partial \mathbf{H}}{\partial T}(\mathbf{z}_0, T) &= \left.\frac{\partial \mathbf{z}(t, \mathbf{z}_0)}{\partial t}\right|_{t=T} \\ &= \mathbf{g}(\mathbf{z}(T, \mathbf{z}_0)) \end{aligned} \qquad (10)$$

The $2n \times 2n$ Jacobian matrix $\partial\mathbf{H}/\partial\mathbf{z}_0$ is provided by

$$\frac{\partial \mathbf{H}}{\partial \mathbf{z}_0}(\mathbf{z}_0, T) = \left.\frac{\partial \mathbf{z}(t, \mathbf{z}_0)}{\partial \mathbf{z}_0}\right|_{t=T} - \mathbf{I} \qquad (11)$$

where $\mathbf{I}$ is the $2n \times 2n$ identity matrix. There are basically two means of computing the Jacobian matrix $\partial\mathbf{z}(t, \mathbf{z}_0)/\partial\mathbf{z}_0$

1. This matrix represents the variation of the solution $\mathbf{z}(t, \mathbf{z}_0)$ at time t when the initial conditions $\mathbf{z}_0$ are perturbed. It can therefore be evaluated through a numerical finite-difference analysis by perturbing successively each of the initial conditions and integrating the equations of motion (2).
2. An alternative computation is obtained by differentiating the equations of motion (2) with respect to the initial conditions $\mathbf{z}_0$

$$\frac{\partial}{\partial \mathbf{z}_0}\left[\dot{\mathbf{z}}(t, \mathbf{z}_0)\right] = \frac{\partial}{\partial \mathbf{z}_0}\left[\mathbf{g}(\mathbf{z}(t, \mathbf{z}_0))\right] \qquad (12)$$

It follows

$$\frac{d}{dt}\left[\frac{\partial \mathbf{z}(t, \mathbf{z}_0)}{\partial \mathbf{z}_0}\right] = \left.\frac{\partial \mathbf{g}(\mathbf{z})}{\partial \mathbf{z}}\right|_{\mathbf{z}(t, \mathbf{z}_0)} \left[\frac{\partial \mathbf{z}(t, \mathbf{z}_0)}{\partial \mathbf{z}_0}\right] \qquad (13)$$

with

$$\frac{\partial \mathbf{z}(0, \mathbf{z}_0)}{\partial \mathbf{z}_0} = \mathbf{I} \qquad (14)$$

since $\mathbf{z}(0, \mathbf{z}_0) = \mathbf{z}_0$. Hence, the matrix $\partial \mathbf{z}(t, \mathbf{z}_0)/\partial \mathbf{z}_0$ at $t = T$ can be obtained by numerically integrating over T the initial-value problem defined by the ordinary differential equations (ODEs) (13) with the initial conditions (14).

In addition to the integration of the current solution $\mathbf{z}(t, \mathbf{x}_0)$ of (2), these two methods for computing $\partial \mathbf{z}(t, \mathbf{z}_0)/\partial \mathbf{z}_0$ require $2n$ numerical integrations of $2n$-dimensional dynamical systems, which may be computationally intensive for large systems. However, equations (13) are linear ODEs and their numerical integration is thus less expensive. The numerical cost can be further reduced if the solution of equations (13) is computed together with the solution of the nonlinear equations of motion in a single simulation. We note that the finite-difference procedure is required when $\mathbf{g}$ is nondifferentiable, i.e., when the nonlinearities are nonsmooth.

In the present case, the phase of the periodic solutions is not fixed. If $\mathbf{z}(t)$ is a solution of the autonomous system (2), then $\mathbf{z}(t + \Delta t)$ is geometrically the same solution in state space for any Δt. The initial conditions $\mathbf{z}_{p0}$ can be arbitrarily chosen anywhere on the periodic solution. This is illustrated in Figure 2 for the Duffing oscillator (5) where different initial conditions corresponding to the same periodic solution are shown. Hence, an additional condition has to be specified in order to remove the arbitrariness of the initial conditions. Mathematically, the system (9) of $2n$ equations with $2n + 1$ unknowns needs a supplementary equation, termed the *phase condition.*

Different phase conditions have been proposed in the literature (1; 2). For instance, the simplest one consists in setting one component of the initial conditions vector to zero, as in (7). This is illustrated in Figure 2 where the depicted periodic solution of the Duffing oscillator is characterized by a zero initial velocity. A phase condition particularly suitable for the NNM computation is utilized in the present study and is discussed in Section 2.3.

In summary, the NNM computation is carried out by solving the augmented two-point boundary-value problem defined by

$$\mathbf{F}(\mathbf{z}_{p0}, T) \equiv \left\{ \begin{array}{ccc} \mathbf{H}(\mathbf{z}_{p0}, T) & = & 0 \\ h(\mathbf{z}_{p0}) & = & 0 \end{array} \right. \tag{15}$$

where $h(\mathbf{z}_{p0}) = 0$ is the phase condition.

An important characteristic of NNMs is that they can be stable or unstable, which is in contrast to linear theory where all modes are neutrally stable. In this context, instability means that small perturbations of the initial conditions that generate the NNM motion lead to the elimination of the mode oscillation. Nonetheless, the unstable NNMs can be computed using the shooting procedure.

The stability analysis can be performed when an NNM motion has been computed by the shooting algorithm. The *monodromy matrix* Φ_T of a periodic orbit $\mathbf{z}_p(t, \mathbf{z}_{p0})$ of period T is defined by its $2n \times 2n$ Jacobian matrix evaluated at $t = T$

$$\Phi_T(\mathbf{z}_{p0}) = \left. \frac{\partial \mathbf{z}_p(t, \mathbf{z}_{p0})}{\partial \mathbf{z}_{p0}} \right|_{t=T} \tag{16}$$

Perturbing the initial conditions with the vector $\Delta\mathbf{z}_0$ and expanding the perturbed solution $\mathbf{z}(T, \mathbf{z}_{p0} + \Delta\mathbf{z}_0)$ in Taylor series yields

$$\Delta\mathbf{z}(T) = \Phi_T(\mathbf{z}_{p0})\Delta\mathbf{z}_0 + \mathcal{O}(\|\Delta\mathbf{z}_0\|^2) \tag{17}$$

where $\Delta\mathbf{z}(T) = \mathbf{z}(T, \mathbf{z}_{p0} + \Delta\mathbf{z}_0) - \mathbf{z}_p(T, \mathbf{z}_{p0})$.

Equations (17) shows that the monodromy matrix provides the first-order variation of the periodic solution after one period. After m periods, one obtains

$$\Delta\mathbf{z}(mT) = [\Phi_T(\mathbf{z}_{p0})]^m \Delta\mathbf{z}_0 + \mathcal{O}(\|\Delta\mathbf{z}_0\|^2) \tag{18}$$

The linear stability of the periodic solution calculated by the shooting algorithm is studied by computing the eigenvalues of its monodromy matrix Φ_T. The $2n$ eigenvalues, termed *Floquet multipliers*, provide the exponential variations of the perturbations along the eigendirections of the monodromy matrix. If a Floquet multiplier has a magnitude larger than one, then the periodic solution is unstable; otherwise, it is stable in the linear sense.

2.2 Continuation of Periodic Solutions

As discussed previously, the conservative system (2) comprises at least n different families of periodic orbits (i.e., NNMs), which can be regarded as nonlinear extensions of the LNMs of the underlying linear system. Due to the frequency-energy dependence, the modal parameters of an NNM vary with the total energy. An NNM family, governed by equations (15), therefore traces a curve, termed an NNM branch, in the $(2n+1)$-dimensional space of initial conditions and period $(\mathbf{z}_{p0}, T)$. As stated before, there may also exist additional NNMs (i.e., bifurcating NNMs) that are essentially nonlinear with no linear counterparts.

In this study, the NNMs are determined using methods for the *numerical continuation* of periodic motions (also called *path-following methods*) (1; 2). Starting from the corresponding LNM at low energy, the computation is carried out by finding successive points $(\mathbf{z}_{p0}, T)$ of the NNM branch. The space $(\mathbf{z}_{p0}, T)$ is termed the continuation space.

Different methods for numerical continuation have been proposed in the literature. The so-called pseudo-arclength continuation method is used herein.

Sequential Continuation

The simplest and most intuitive continuation technique is the *sequential continuation* method. This procedure is first explained due to its straightforward implementation. Moreover, it provides the fundamental concepts of continuation methods.

The sequential continuation of the periodic solutions governed by (15) is carried out in three steps:

1. A periodic solution $(\mathbf{z}_{p0,(1)}, T_{(1)})$ at sufficiently low energy (i.e., in the neighborhood of one LNM) is first computed using the shooting method. The period and initial conditions of the selected LNM are chosen as an initial guess.
2. The period is incremented, $T_{(j+1)} = T_{(j)} + \Delta T$.
3. From the current solution $(\mathbf{z}_{p0,(j)}, T_{(j)})$, the next solution $(\mathbf{z}_{p0,(j+1)}, T_{(j+1)})$ is determined by solving (15) using the shooting method with the period fixed:

$$\mathbf{z}_{p0,(j+1)}^{(k+1)} = \mathbf{z}_{p0,(j+1)}^{(k)} + \Delta\mathbf{z}_{p0,(j+1)}^{(k)} \tag{19}$$

where

$$\left.\frac{\partial \mathbf{F}}{\partial \mathbf{z}_{p0}}\right|_{(\mathbf{z}_{p0,(j+1)}^{(k)}, T_{(j+1)})} \Delta\mathbf{z}_{p0,(j+1)}^{(k)} = -\mathbf{F}(\mathbf{z}_{p0,(j+1)}^{(k)}, T_{(j+1)}) \tag{20}$$

The initial conditions of the previous periodic solution are used as a prediction $\mathbf{z}_{p0,(j+1)}^{(0)} = \mathbf{z}_{p0,(j)}$. Superscript k is the iteration index of the shooting procedure, whereas subscript j is the index along the NNM branch.

Eventually, one NNM branch is computed.

Pseudo-Arclength Continuation

The sequential continuation method parameterizes an NNM branch using the period T. It has two main drawbacks:

1. Because the convergence of the Newton-Raphson procedure depends critically on the closeness of the initial guess to the actual solution, the sequential continuation requires fairly small increments ΔT.
2. Because the value of the period is fixed during the Newton-Raphson corrections, it is unable as such to deal with turning points. This is illustrated in Figure 3 where no solution exists for a period larger than the period at the turning point.

For better performance, a continuation algorithm uses a better prediction than the last computed solution. In addition, corrections of the period are also considered during the shooting process. The pseudo-arclength continuation method relies on these two improvements in order to optimize the path following of the branch.

Starting from a known solution $(\mathbf{z}_{p0,(j)}, T_{(j)})$, the next periodic solution $(\mathbf{z}_{p0,(j+1)}, T_{(j+1)})$ on the branch is computed using a *predictor step* and a *corrector step.*

At step j, a prediction $(\tilde{\mathbf{z}}_{p0,(j+1)}, \tilde{T}_{(j+1)})$ of the next solution $(\mathbf{z}_{p0,(j+1)}, T_{(j+1)})$ is generated along the tangent vector to the branch at the current point $\mathbf{z}_{p0,(j)}$

$$\left[\begin{array}{c} \tilde{\mathbf{z}}_{p0,(j+1)} \\ \tilde{T}_{(j+1)} \end{array}\right] = \left[\begin{array}{c} \mathbf{z}_{p0,(j)} \\ T_{(j)} \end{array}\right] + s_{(j)} \left[\begin{array}{c} \mathbf{p}_{z,(j)} \\ p_{T,(j)} \end{array}\right] \tag{21}$$

where $s_{(j)}$ is the predictor stepsize. The tangent vector $\mathbf{p}_{(j)} = [\mathbf{p}^*_{z,(j)} \; p_{T,(j)}]^*$ to the branch defined by (15) is solution of the system

$$\left[\begin{array}{cc} \left.\dfrac{\partial \mathbf{H}}{\partial \mathbf{z}_{p0}}\right|_{(\mathbf{z}_{p0,(j)}, T_{(j)})} & \left.\dfrac{\partial \mathbf{H}}{\partial T}\right|_{(\mathbf{z}_{p0,(j)}, T_{(j)})} \\ \left.\dfrac{\partial h}{\partial \mathbf{z}_{p0}}^*\right|_{(\mathbf{z}_{p0,(j)})} & 0 \end{array}\right] \left[\begin{array}{c} \mathbf{p}_{z,(j)} \\ p_{T,(j)} \end{array}\right] = \left[\begin{array}{c} \mathbf{0} \\ 0 \end{array}\right] \tag{22}$$

with the condition $\left\|\mathbf{p}_{(j)}\right\| = 1$. The star denotes the transpose operator. This normalization can be taken into account by fixing one component of the tangent vector and solving the resulting overdetermined system using the Moore-Penrose matrix inverse; the tangent vector is then normalized to 1. For illustration, the predictor step is shown schematically in Figure 4.

The prediction is corrected by a shooting procedure in order to solve (15) in which the variations of the initial conditions and the period are forced to be orthogonal to the predictor step. At iteration k, the corrections

$$\begin{aligned} \mathbf{z}^{(k+1)}_{p0,(j+1)} &= \mathbf{z}^{(k)}_{p0,(j+1)} + \Delta \mathbf{z}^{(k)}_{p0,(j+1)} \\ T^{(k+1)}_{(j+1)} &= T^{(k)}_{(j+1)} + \Delta T^{(k)}_{(j+1)} \end{aligned} \tag{23}$$

are computed by solving the overdetermined linear system using the Moore-

Penrose matrix inverse

$$\begin{bmatrix} \left.\frac{\partial \mathbf{H}}{\partial \mathbf{z}_{p0}}\right|_{(\mathbf{z}^{(k)}_{p0,(j+1)},T^{(k)}_{(j+1)})} & \left.\frac{\partial \mathbf{H}}{\partial T}\right|_{(\mathbf{z}^{(k)}_{p0,(j+1)},T^{(k)}_{(j+1)})} \\ \left.\frac{\partial h}{\partial \mathbf{z}_{p0}}^{*}\right|_{(\mathbf{z}^{(k)}_{p0,(j+1)})} & 0 \\ \mathbf{p}^{*}_{z,(j)} & p_{T,(j)} \end{bmatrix} \begin{bmatrix} \Delta \mathbf{z}^{(k)}_{p0,(j+1)} \\ \Delta T^{(k)}_{(j+1)} \end{bmatrix} = \begin{bmatrix} -\mathbf{H}(\mathbf{z}^{(k)}_{p0,(j+1)}, T^{(k)}_{(j+1)}) \\ -h(\mathbf{z}^{(k)}_{p0,(j+1)}) \\ 0 \end{bmatrix} \quad (24)$$

where the prediction is used as initial guess, i.e, $\mathbf{z}^{(0)}_{p0,(j+1)} = \tilde{\mathbf{z}}_{p0,(j+1)}$ and $T^{(0)}_{(j+1)} = \tilde{T}_{(j+1)}$.

The last equation in (24) corresponds to the orthogonality condition for the corrector step. We note that the partial derivatives in (24) are evaluated numerically, as explained previously.

This iterative process is carried out until convergence is achieved. The convergence test is based on the relative error of the periodicity condition:

$$\frac{\|\mathbf{H}(\mathbf{z}_{p0}, T)\|}{\|\mathbf{z}_{p0}\|} = \frac{\|\mathbf{z}_p(T, \mathbf{z}_{p0}) - \mathbf{z}_{p0}\|}{\|\mathbf{z}_{p0}\|} < \epsilon \quad (25)$$

where ϵ is the prescribed relative precision.

For illustration, the corrector step is shown schematically in Figure 4.

2.3 An Integrated Approach for the NNM Computation

The algorithm proposed for the computation of NNM motions is a combination of shooting and pseudo-arclength continuation methods, as shown in Figure 1. Starting from the LNM motion at low energy, there are two steps within the algorithm:

1. The predictor step is *global* and goes from one NNM motion at a specific energy level to another NNM motion at a somewhat different energy level. For an efficient and robust NNM continuation, the stepsize $s_{(j)}$ is to be carefully controlled. A small stepsize leads to a small number of corrector iterations, but it requires a large number of continuation steps to follow an NNM branch. For a large stepsize, the number of corrector iterations is high, and the convergence is slow. The Newton-Raphson procedure may even break down if the prediction is not close enough to the actual solution. Continuation may therefore be computationally intensive in both cases. The step-

size has to be adjusted, possibly in an automatic and flexible manner. Various adaptive stepsize control procedures are discussed in (1).

2. The corrector step is *local* and refines the prediction to obtain the actual solution at a specific energy level. The size of the corrections during the corrector step is determined by the solutions of the overdetermined system (24).

So far, the NNMs have been considered as branches in the continuation space $(\mathbf{z}_{p0}, T)$. An appropriate graphical depiction of the NNMs is to represent them in a FEP. This FEP can be computed in a straightforward manner: (i) the conserved total energy is computed from the initial conditions realizing the NNM motion; and (ii) the frequency of the NNM motion is calculated directly from the period.

A widely-used method for solving first-order equations such as (2) is the Runge-Kutta scheme. In structural dynamics where second-order systems are encountered, Newmark's family of methods is probably the most widespread technique for solving linear and nonlinear large-scale stiff structural systems. This family of numerical time integration methods is considered in this study.

The precision of the integration scheme, which is chosen by the end-user, directly influences the accuracy of the NNM computation. In fact, the computed solution can be regarded as an *exact* solution if the sampling frequency used to integrate the equations is sufficiently high. This is practically the only approximation in the proposed algorithm.

Unlike sequential continuation, the evolution path of this predictor-corrector method is parameterized by the distance $s_{(j)}$ along the tangent predictor, also referred to as arclength continuation parameter in the literature. As mentioned previously, the stepsize has to be carefully controlled for a robust and efficient NNM computation.

The stepsize control used herein relies on the evaluation of the convergence quality by the number of iterations of the corrector step. The stepsize is controlled so that the corrector step requires on average the desirable number of iterations $N^\star$. At each step, the stepsize is updated according to the ratio between the desirable number N^* and the previous number $N_{(j-1)}$ of iterations:

$$s_{(j)} = \left(\frac{N^*}{N_{(j-1)}} \right) s_{(j-1)} \tag{26}$$

In practice, the ratio $r = N^*/N_{(j-1)}$ is often bounded to make the adaptation stepsize more robust and to prevent the continuation from jumping between different branches [1]. The stepsize can also be bounded ($s_{(j)} < s_{max}$)

[1] Another possibility to avoid branch jumping phenomena is to limit the angle between consecutive predictor steps.

to obtain enough discretized points on the branch during the continuation. In case of no convergence (i.e., when the residue increases or when the process requires more correction iterations than the prescribed maximum N_{max}), the stepsize is halved until convergence is achieved.

As a final remark, we note that the sign of the stepsize is chosen in order to follow the branch in the same direction, i.e.,

$$\left[s_{(j)}\mathbf{p}_{(j)}\right]^{*}\left[s_{(j-1)}\mathbf{p}_{(j-1)}\right] > 0 \tag{27}$$

According to the previous predictor step and the current tangent vector, the sign of $s_{(j)}$ is therefore given by

$$\text{sign}\left(s_{(j)}\right) = \text{sign}\left(s_{(j-1)}\mathbf{p}_{(j)}^{*}\mathbf{p}_{(j-1)}\right) \tag{28}$$

Reduction of the Computational Burden

The algorithm described so far may become computationally intensive when dealing with large-scale systems (i.e., systems with many DOFs). Because a practical and computationally tractable calculation of the NNM motions is targeted, two properties of some NNM families can be exploited to speed up the computation:

1. All symmetric NNM branches Snm obey the symmetry condition

$$\mathbf{z}_p\left(\frac{T}{2}, \mathbf{z}_{p0}\right) + \mathbf{z}_{p0} = 0 \tag{29}$$

   For these branches, the shooting procedure can be performed over the half period $T/2$ by searching the initial conditions and the period T that solve this modified periodicity condition. Because the time integrations represent the main computational cost of the algorithm, this modified periodicity condition reduces the computational burden by a factor close to 2.
2. For branches of NNMs represented by an open loop in the configuration space, a suitable phase condition is to set all the velocities to zero. The initial velocities are eliminated from the unknowns of the linear systems to solve at each Newton-Raphson iteration. These systems have therefore $2n+1$ equations with $n+1$ unknowns $\mathbf{x}_{p0}$ and T.

One advantage is that these modifications can be very naturally integrated in the basic algorithm in Figure 1.

In summary, two variants of the basic algorithm described in the previous section have been developed:

1. The *general strategy* consists in (i) carrying out the shooting over the period T and (ii) setting only one velocity to zero. This strategy can compute all possible NNM motions.
2. The *practical strategy* exploits the modified periodicity and phase conditions. The fundamental NNM motions of nonlinear structures often obey these conditions (e.g., the fundamental NNMs of nonlinear systems with odd nonlinearities are necessarily symmetric, because the loss of symmetry requires the so-called symmetry-breaking bifurcation).

Targeting a reduction of the computational burden, but without lack of generality, an approach that integrates the two variants is used. Starting from the LNMs, the fundamental NNM motions are first computed using the practical strategy; a subset of tongues can also be computed using this methodology. We note that the application of this strategy should often suffice for most engineering structures.

When a detailed analysis of the unsymmetric NNMs and those represented by a closed curve in the configuration space is required, the general strategy can then be utilized for computing these NNMs.

3 Numerical Experiments

In what follows, the proposed NNM computation method is demonstrated using three different nonlinear vibrating systems, namely a weakly nonlinear 2DOF system, a discrete model of a nonlinear bladed disk and a nonlinear cantilever beam discretized by the finite element method.

3.1 Weakly Nonlinear 2DOF System

The governing equations of motion are

$$\begin{aligned} \ddot{x}_1 + (2x_1 - x_2) + 0.5\, x_1^3 &= 0 \\ \ddot{x}_2 + (2x_2 - x_1) &= 0 \end{aligned} \tag{30}$$

The two LNMs of the underlying linear system are in-phase and out-of-phase modes for which the two DOFs vibrate with the same amplitude. The natural eigenfrequencies are $f_1 = 1/2\pi \simeq 0.159$ Hz and $f_2 = \sqrt{3}/2\pi \simeq 0.276$ Hz.

The integrated approach described in Section 2.3 is applied to this system. The computation of the fundamental NNMs is first performed using the modified phase and periodicity conditions. The in-phase backbone $S11+$ is depicted in Figure 5(a), whereas the out-of-phase backbone $S11-$ is given in Figure 6. Though a large energy range is investigated, these figures show that the continuation method discretizes the two branches using very few points. Large stepsizes are therefore employed, and only a few seconds are required to computed each branch for 100 integration time steps per half period using a 2GHz processor. This is an important feature when targeting a computationally tractable calculation of the NNMs. The two backbones are depicted together in the FEP in Figure 7. The family of in-phase NNM motions is also represented in a three-dimensional projection of the phase space in Figure 5(b) and in the configuration space in Figure 5(c).

The NNM continuation can now be carried out at higher energy levels. The FEP for the in-phase mode is depicted in Figure 8. It can be observed that a recurrent series of tongues of internally resonant NNMs (i.e., $S31$, $S51$, $S71$, etc.) continue the backbone branch $S11+$ through turning points (fold bifurcations). Due to these turning points, smaller stepsizes are necessary, which renders the tongue calculation more demanding computationally. By contrast, at higher energy on $S11-$, the 1:1 out-of-phase motion persists, and $S11-$ extends to infinity. The complete FEP calculated using the practical strategy is shown in Figure 9.

We now move to the general strategy for the computation of unsymmetric NNMs and NNMs represented by a closed curve in the configuration space. These NNMs are generally generated through bifurcations (e.g., symmetry-breaking bifurcations for unsymmetrical NNMs). Because the tangent is not uniquely defined at the bifurcation point, they require a branching strategy to be effectively computed (1). In this study, a perturbation technique is used to carry out branch switching, once the bifurcation point is located using the Floquet multipliers. The resulting FEP is displayed in Figure 10 and shows two unsymmetrical tongues ($U21$ and $U41$).

The NNM stability is also briefly discussed. Because the monodromy matrix is computed during the continuation procedure, its eigenvalues, the Floquet multipliers, are obtained as a by-product. The stability of the fundamental in-phase and internally resonant NNMs is presented in Figure 11. Clearly, the bifurcation points, which include fold and symmetry-breaking bifurcations, are characterized by a change of stability. The evolution of the Floquet multipliers along $S31$ is shown in the complex plane in Figure 12. This figure shows the mechanism of loss of stability; a pair of Floquet multipliers leaves the unit circle through 1.

3.2 Nonlinear Bladed Disk System

The NNM motions of a more complex structure, a nonlinear bladed disk, are now investigated. The bladed disk is composed of 20 sectors assembled with cyclic periodicity; a single sector is represented in Figure 13. Each sector is modeled using disk (M) and blade (m) lumped masses, coupled by linear (k) and cubic (k_{nl}) springs. The nonlinear springs can be representative of geometrically nonlinear effects in the blades. The disk masses are connected together by linear springs K. The equations of motion of this 40-DOF system are

$$\begin{aligned} m\,\ddot{x}_i + k(x_i - X_i) + k_{nl}(x_i - X_i)^3 &= 0 \\ M\,\ddot{X}_i + K(X_i - X_{i+1}) + K(X_i - X_{i-1}) + & \\ k(X_i - x_i) + k_{nl}(X_i - x_i)^3 &= 0 \end{aligned} \qquad (31)$$

for $i = 1, \ldots, 20$; $X_{21} = X_1$, $X_0 = X_{20}$ (conditions of cyclic periodicity) and $M = 1$, $m = 0.3$, $K = 1$, $k_l = 1$, $k_nl = 0.5$. X_i and x_i are the displacements of the disk and blade masses of the ith sector, respectively.

As the structure is not fixed, the first mode is a rigid-body mode, which is obviously not affected by the nonlinearities. Due to periodicity of the structure, several elastic modes of the underlying linear system appear in pair at the same frequencies, with only a spatial shift between them. It is the case of the first two LNMs that possess one nodal diameter. Their nonlinear extensions are represented in the FEP of Figure 14. The nonlinearities have a slight, but identical (i.e., the backbones cannot be distinguished in the FEP), influence on these two NNMs. Therefore, the continuation can be performed by means of large stepsizes, and only a few seconds are necessary to obtain the backbones for 100 time steps over the half period.

Nonlinear systems with spatial symmetries can possess similar NNMs. This is illustrated in Figures 15 and 16 for the 21st NNM of the structure. While its frequency is altered by the nonlinearities in the system, the distribution of the modal shape remains unchanged, which characterizes similar modes.

Finally, a higher mode (i.e., the 38th) for which the corresponding LNM possesses one nodal circle and nine nodal diameters is investigated (see Figures 17 and 18). Due to the presence of a nodal circle, the blade and disk masses vibrate in an out-of-phase fashion, which enhances the nonlinear effects. As a result, the frequency and the shape of the NNM are strongly affected by the nonlinearities. Furthermore, for increasing energy, there is a clear spatial confinement of vibrational energy to two sectors of the system. Interestingly, this occurs in a perfectly symmetric bladed disk assembly, which has no counterpart in linear theory. We note that the computation of this branch needs approximately one minute for 100 time steps over the half period.

3.3 Nonlinear Cantilever Beam

As a final example, a planar cantilever beam discretized by 20 finite elements and with a cubic spring at the free end is considered. The geometrical and mechanical properties of the beam are listed in Table 1. This models a real nonlinear beam that was used as a benchmark for nonlinear system identification during the European Action COST F3 (8).

The FEP of the first mode and the related NNM motions are plotted in Figures 19 and 20. The frequency of the NNM motions on the backbone increases with increasing energy levels, which highlights the hardening characteristic of the cubic nonlinearity. The FEP also highlights the presence of one tongue, revealing the existence of a 5:1 internal resonance between the first two modes. When the energy gradually increases along the tongue, a smooth transition from the first to the second mode occurs (see (e) and (f) in Figure 20). The computation of the backbone branch up to the tongue needs 5 minutes with 200 time steps over the half period. Due to the presence of turning points, the computation of the tongue is more expensive and demands 10 minutes.

The second NNM is plotted in the FEP of Figure 21. Besides the NNM backbone, three tongues are present. The first tongue corresponds to a 5:1 internal resonance between the second and fourth nonlinear modes of the beam. Similarly, a 7:1 internal resonance between the second and fifth modes, and a 9:1 internal resonance between the second and sixth modes are observed.

Similar dynamics were observed for the higher modes and are not further described herein.

4 Conclusion

In this chapter, a numerical method for the computation of nonlinear normal modes (NNMs) of nonlinear mechanical structures is introduced. The approach targets the computation of the undamped modes of structures discretized by finite elements and relies on the continuation of periodic solutions. The proposed procedure was demonstrated using different nonlinear structures, and the NNMs were computed accurately in a fairly automatic manner. Complicated NNM motions were also observed, including a countable infinity of internal resonances and strong motion localization.

This method represents a first step toward a practical NNM computation with limited implementation effort. For instance, the NNMs of a real-life aircraft are computed in (9) using the proposed method.

Bibliography

[1] R. Seydel, *Practical Bifurcation and Stability Analysis, From Equilibrium to Chaos*, Springer, Berlin, 1994.

[2] A.H. Nayfeh, B. Balachandran, *Applied Nonlinear Dynamics: Analytical, Computational and Experimental Method*, John Wiley and Sons, New York, 1995.

[3] J.C. Slater, A numerical method for determining nonlinear normal modes, *Nonlinear Dynamics* 10 (1996), 19-30.

[4] Y.S. Lee, G. Kerschen, A.F. Vakakis, P.N. Panagopoulos, L.A. Bergman, D.M. McFarland, Complicated dynamics of a linear oscillator with a light, essentially nonlinear attachment, *Physica D* 204 (2005), 41-69.

[5] F.X. Wang, A.K. Bajaj, Nonlinear normal modes in multi-mode models of an inertially coupled elastic structure, *Nonlinear Dynamics* 47 (2007), 25-47.

[6] H.B. Keller, *Numerical Methods in Bifurcation Problems*, Springer, Berlin, 1987.

[7] R. Arquier, S. Bellizzi, R. Bouc, B. Cochelin, Two methods for the computation of nonlinear modes of vibrating systems at large amplitudes, *Computers & Structures* 84 (2006), 1565-1576.

[8] F. Thouverez, Presentation of the ECL benchmark, *Mechanical Systems and Signal Processing* 17 (2003), 195-202.

[9] G. Kerschen, M. Peeters, J.C. Golinval, C. Stephan, Nonlinear modal analysis of a full-scale aircraft, *AIAA Journal of Aircraft* 50 (2013), 1409-1419.

Table 1. Geometrical and mechanical properties of the planar cantilever beam

Length (m)	Width (m)	Thickness (m)	Young's mod. (N/m^2)	Density (kg/m^3)	Nonlin. coeff. (N/m^3)
0.7	0.014	0.014	2.05e11	7800	$6\,10^9$

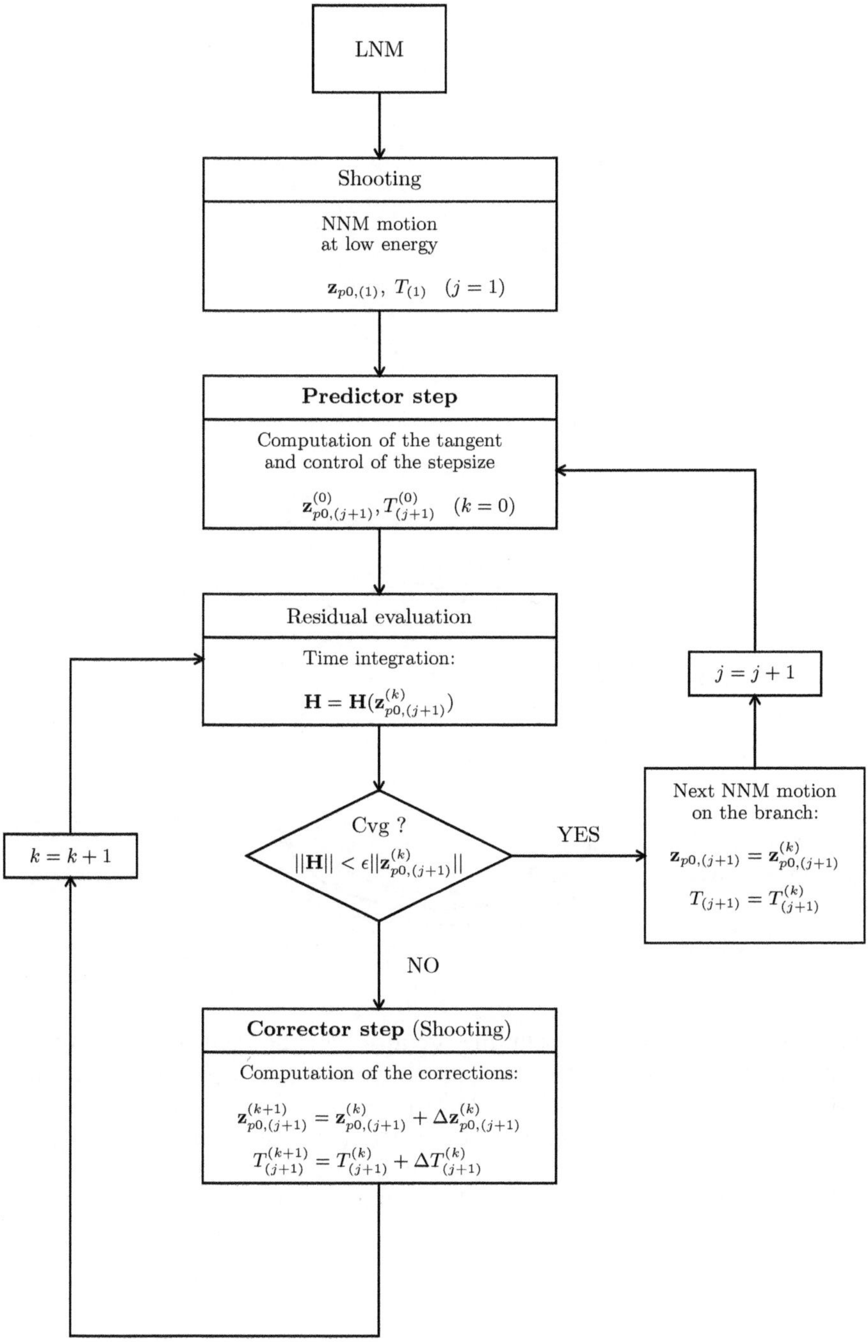

Figure 1. Algorithm for NNM computation.

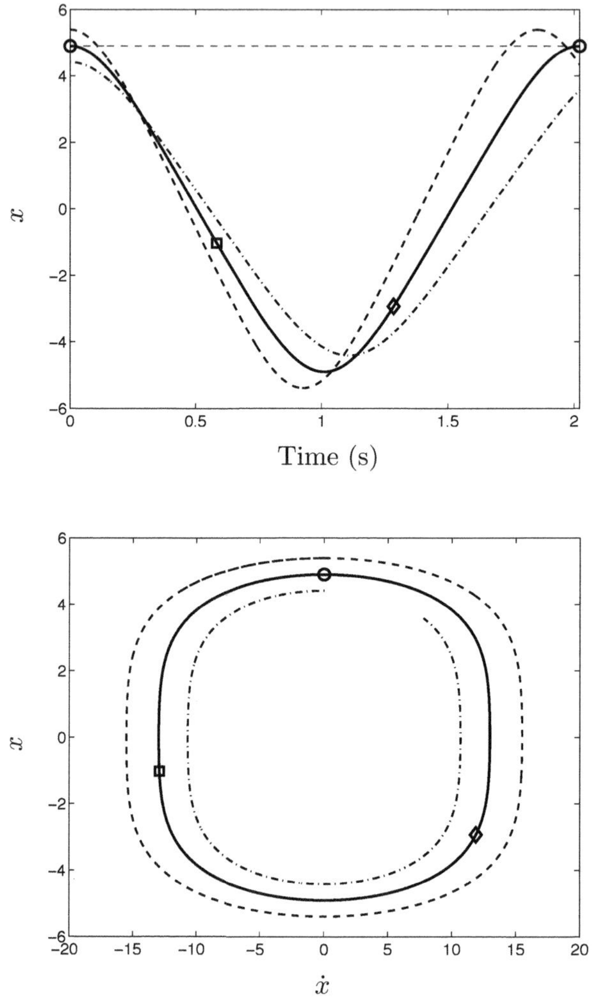

Figure 2. Solutions of the Duffing oscillator for different initial conditions. Top plot: time series; bottom plot: phase space. —— : periodic solution for $[x(0)\ \dot{x}(0)] = [4.9009\ 0]$ and $T = 2.0215$s; − − − : solution for $[x(0)\ \dot{x}(0)] = 1.1 \times [4.9009\ 0]$; −·−·− : solution for $[x(0)\ \dot{x}(0)] = 0.9 \times [4.9009\ 0]$. Markers represent different initial conditions of the periodic solution; ∘: $[x\ \dot{x}] = [4.9009\ 0]$; □: $[x\ \dot{x}] = [-1.0313\ -12.9188]$; ◊: $[x\ \dot{x}] = [-2.9259\ 11.8894]$.

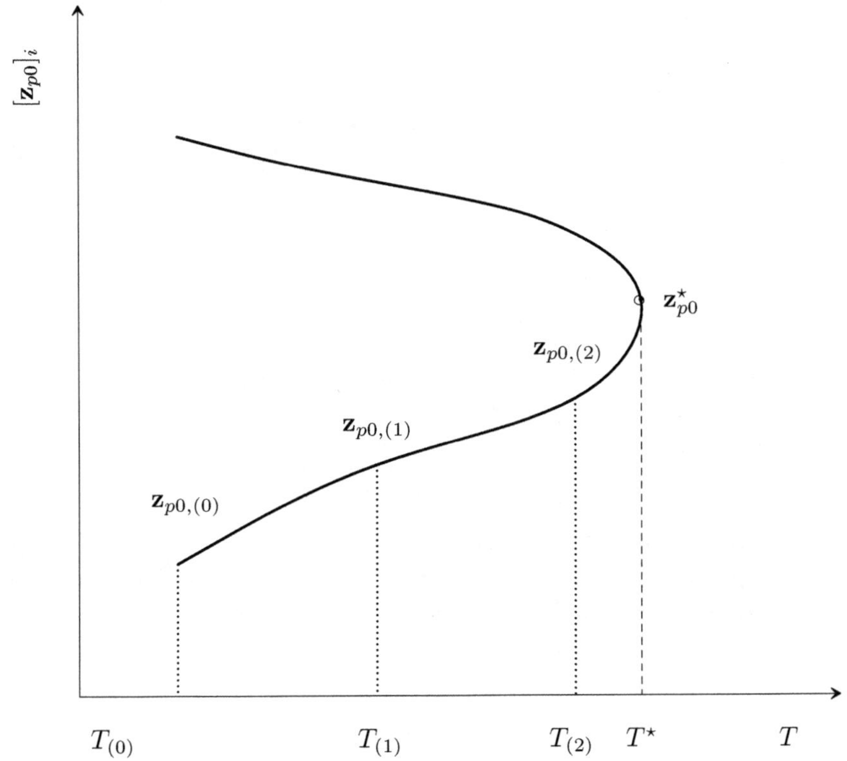

Figure 3. Turning point $(T^{*}, \mathbf{z}^{\star}_{p0})$ in the continuation space. Failure of the sequential continuation for $T \geq T^{\star}$.

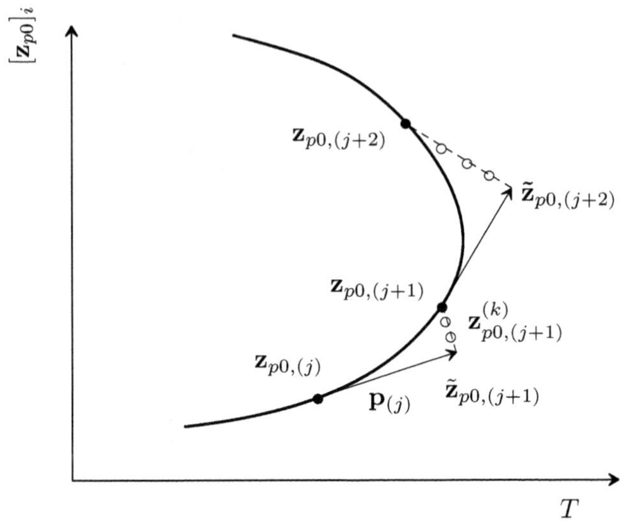

Figure 4. Pseudo-arclength continuation method: branch (——) with a turning point; predictor step (→) tangent to the branch; corrector steps (○ ○ ○) perpendicular to the predictor step.

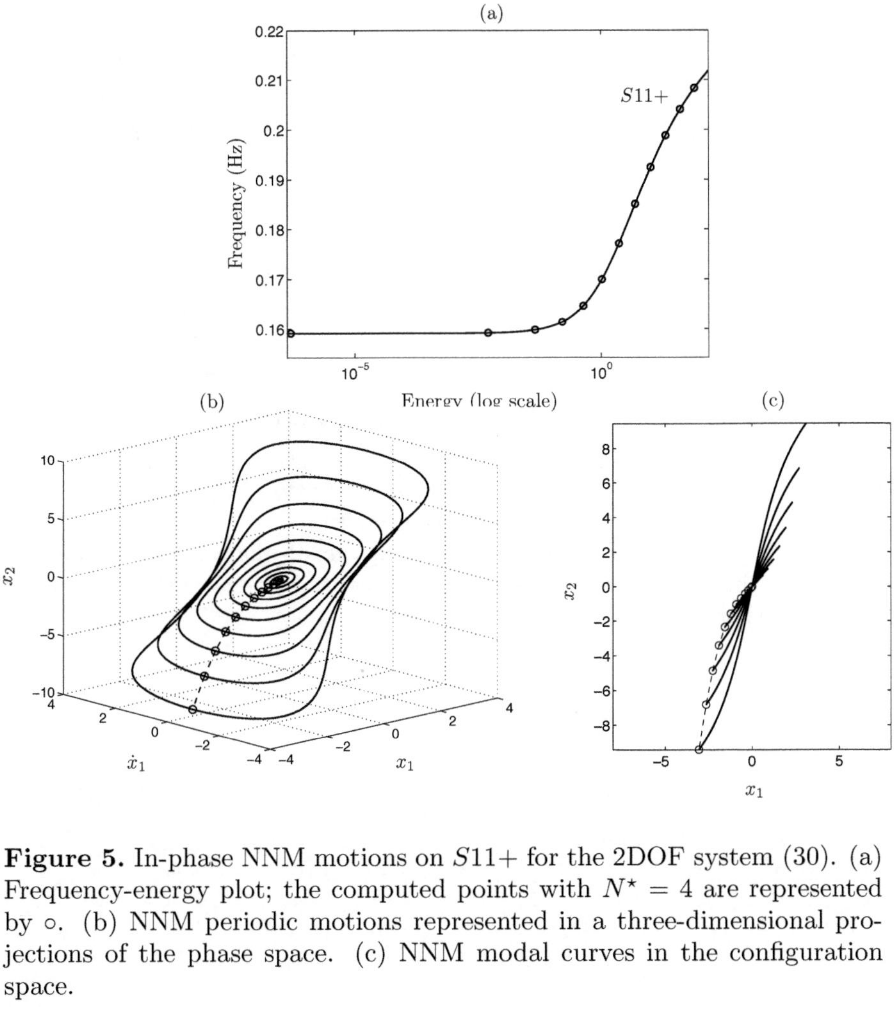

Figure 5. In-phase NNM motions on $S11+$ for the 2DOF system (30). (a) Frequency-energy plot; the computed points with $N^{\star} = 4$ are represented by $\circ$. (b) NNM periodic motions represented in a three-dimensional projections of the phase space. (c) NNM modal curves in the configuration space.

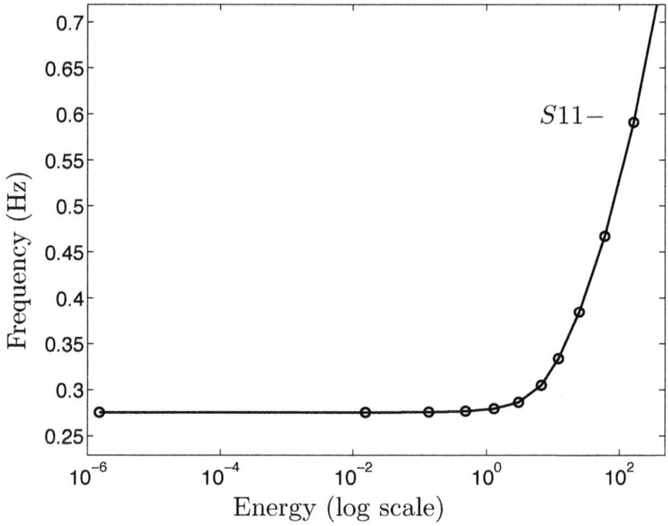

Figure 6. Frequency-energy plot gathering out-of-phase NNM motions on $S11-$ for the 2DOF system (30). The computed points with $N^{\star} = 4$ are represented by $\circ$.

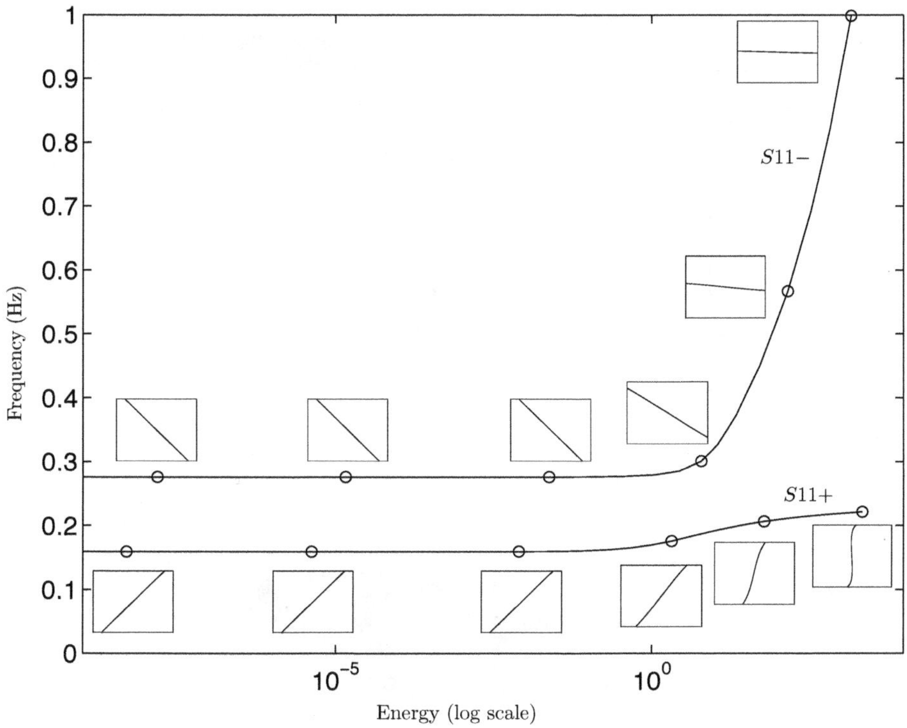

Figure 7. Frequency-energy plot of the 2DOF system computed with the proposed numerical method. NNM motions depicted in the configuration space are inset. The horizontal and vertical axes in these plots are the displacements of the first and second DOFs, respectively; the aspect ratio is set so that increments on the horizontal and vertical axes are equal in size to indicate whether or not the motion is localized to a particular DOF.

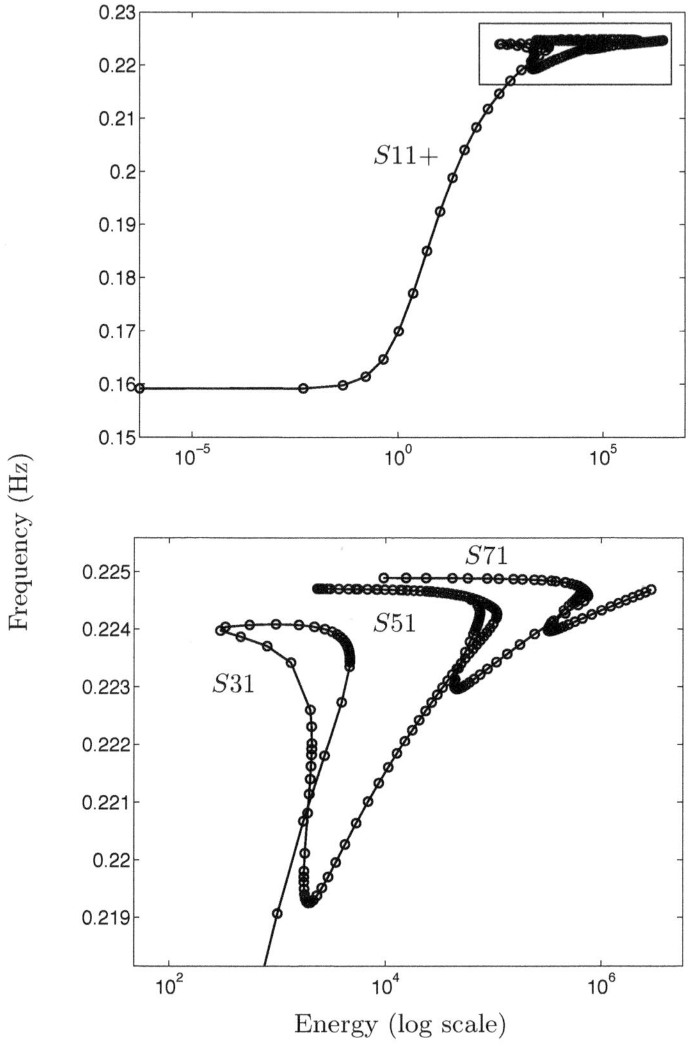

Figure 8. NNMs at high energy for the 2DOF system (30). Top plot: in-phase backbone $S11+$ and tongues branches (internally resonant NNMs). Bottom plot: close-up of the recurrent series of tongues ($S31$, $S51$ and $S71$) at high energy. The computed points are represented by ◦

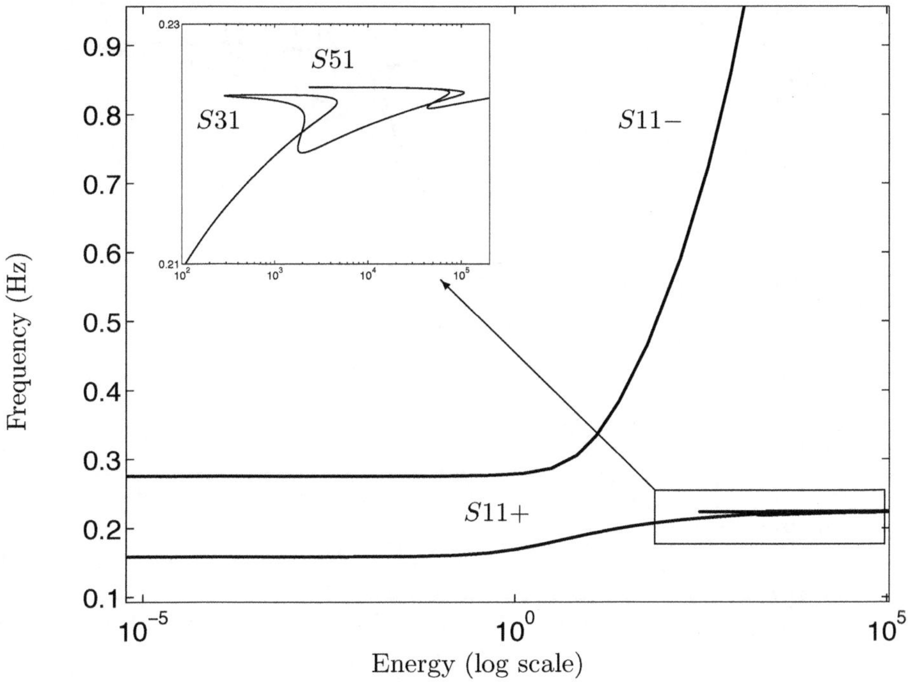

Figure 9. Frequency-energy plot of the 2DOF system

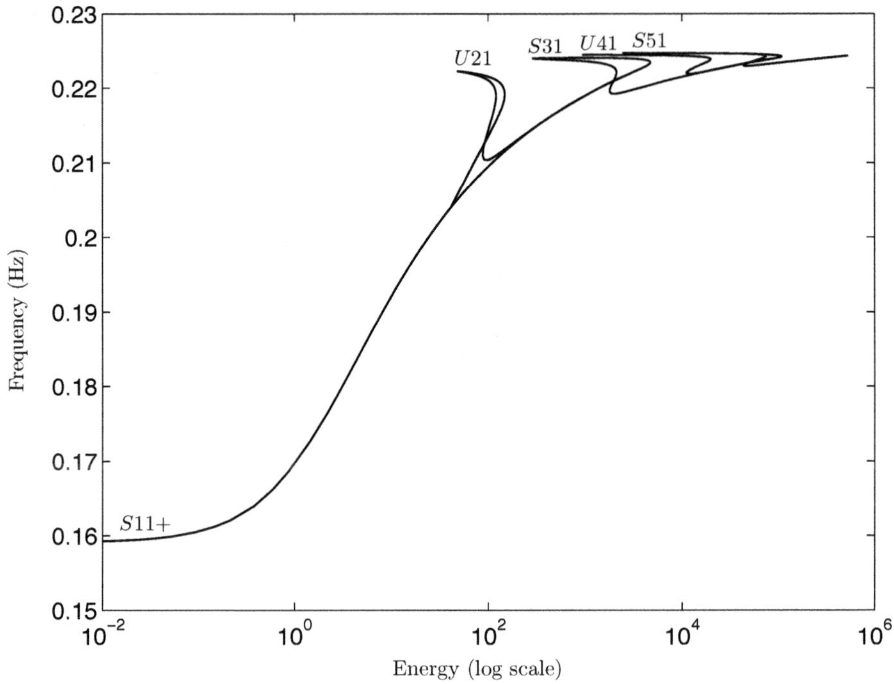

Figure 10. $S11+$ at higher energy levels and internally resonant NNMs ($U21$, $S31$, $U41$, $S51$).

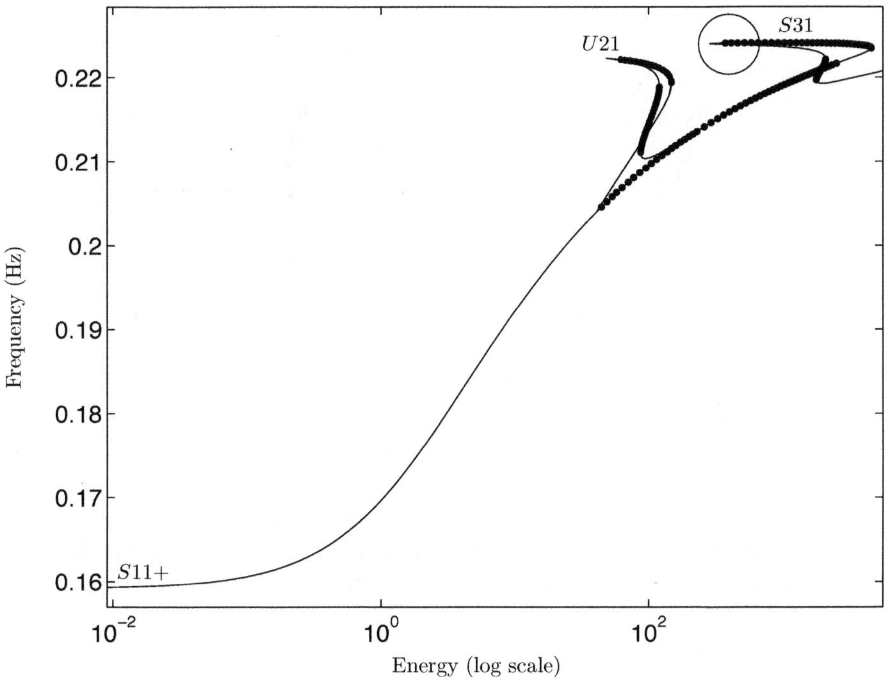

Figure 11. Close-up of $S11+$ for the 2DOF system (30) with stability results (——: stable NNMs; • • • : unstable NNMs).

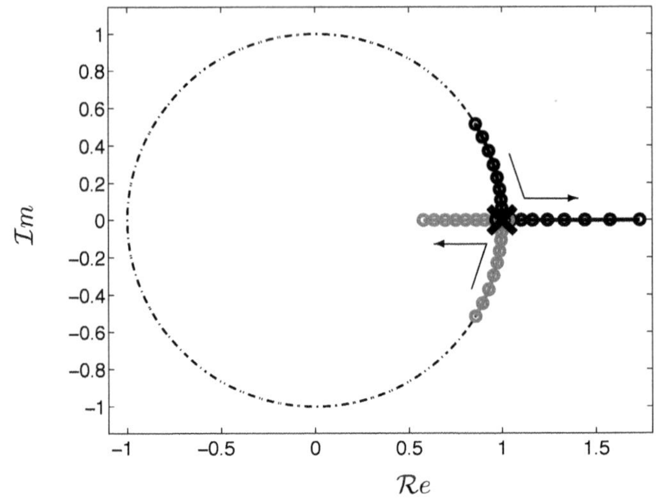

Figure 12. Evolution of the Floquet multipliers in the complex plane corresponding to the circled area in Figure 11. One Floquet multiplier is shown in black, another one in grey. The other two, which stay at +1, are represented by crosses.

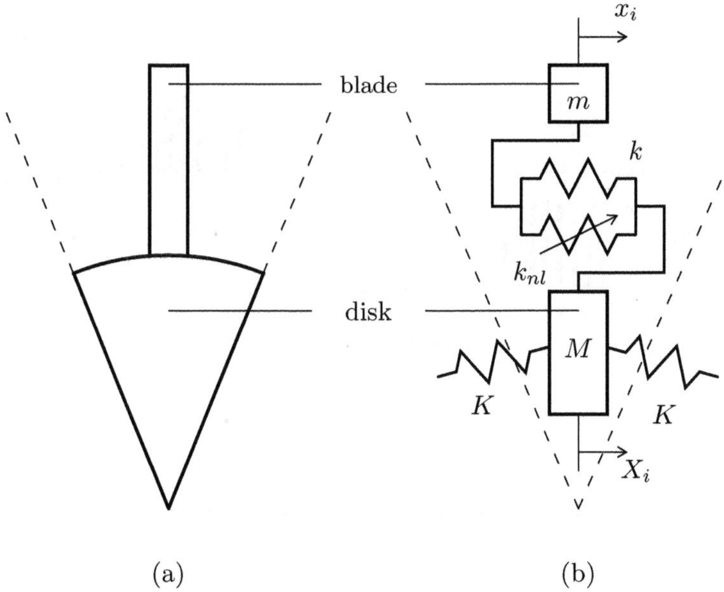

Figure 13. One sector of the nonlinear bladed disk assembly. (a) continuous structure; (b) discrete model.

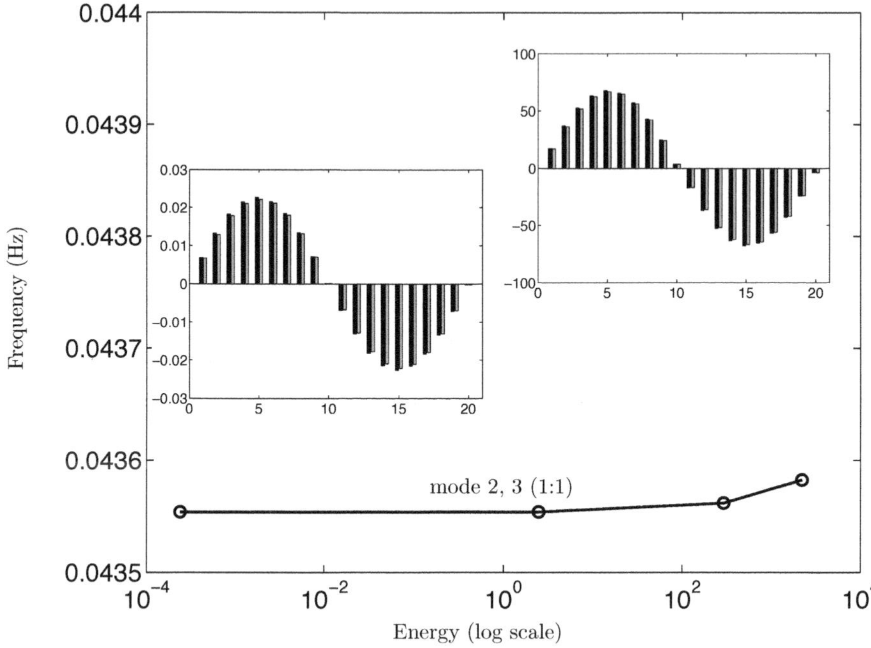

Figure 14. Frequency-energy plot of the 1st and 2nd modes of the bladed disk. The two backbones coincide. The maximum amplitudes of the NNM motions represented by bar graphs are inset; the horizontal axis represents the sector number, and the vertical axis represents the maximum displacements of blade and disk masses (shown in black and grey, respectively).

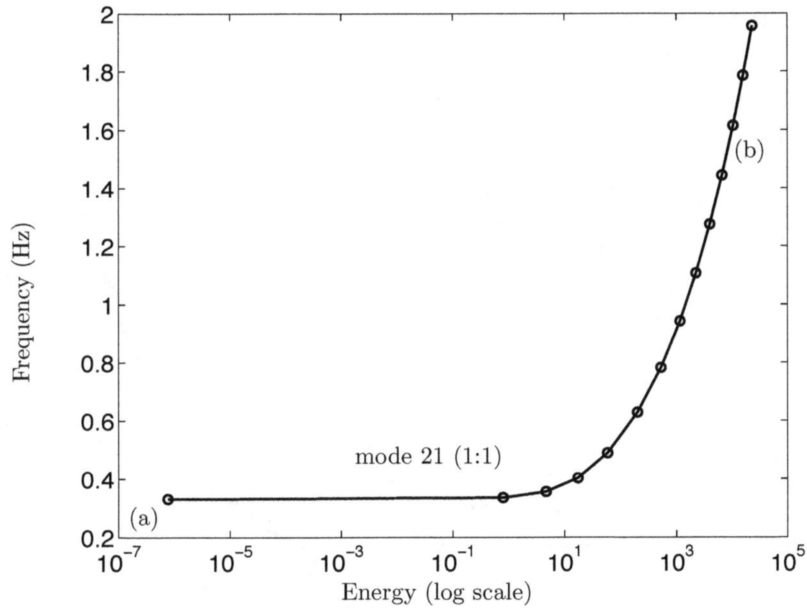

Figure 15. Frequency-energy plot of the 21st mode of the nonlinear bladed disk assembly.

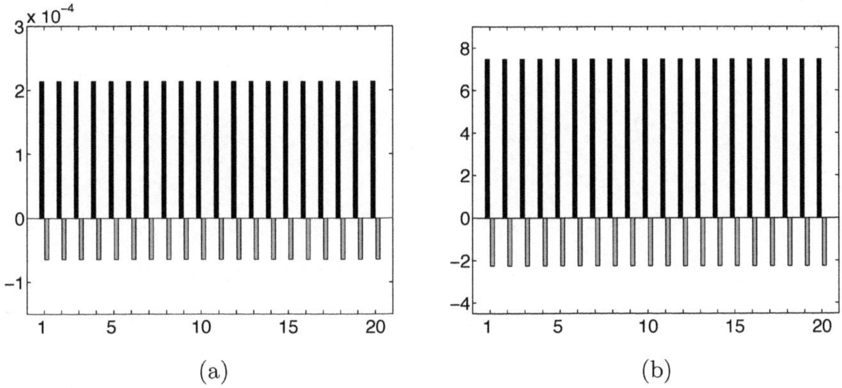

Figure 16. Maximum amplitudes of the 21st NNM of the bladed disk system at (a) low energy and (b) at high energy. The horizontal axis represents the sector number, and the vertical axis represents the maximum displacements of blade and disk masses (shown in black and grey, respectively).

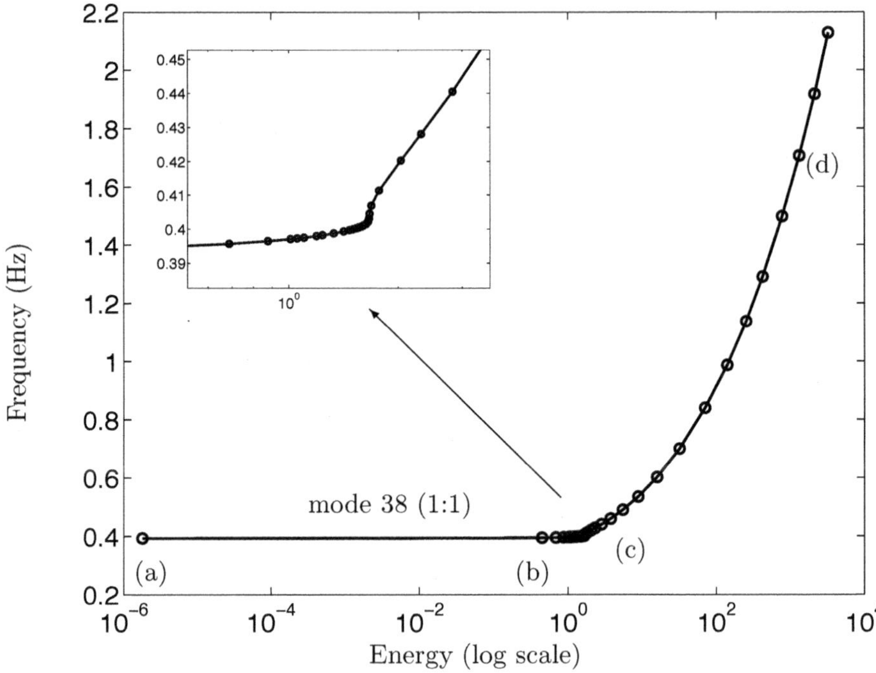

Figure 17. Frequency-energy plot of the 38th mode of the nonlinear bladed disk assembly.

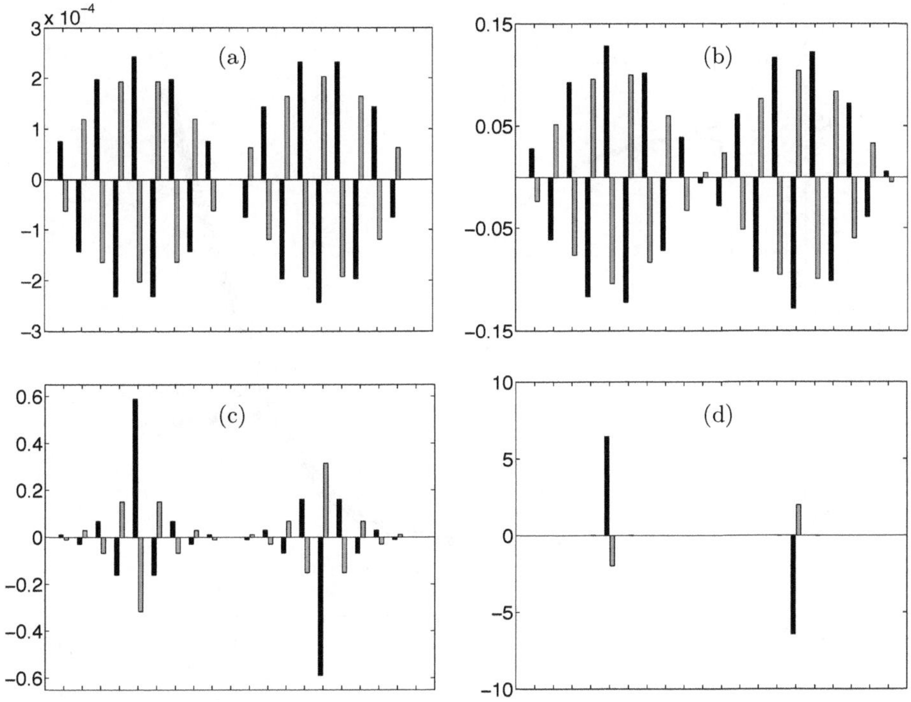

Figure 18. Maximum amplitudes of the 38th NNM of the bladed disk at (a) low energy, (b-c) medium energy and (d) at high energy, corresponding to the points in Figure 17. The horizontal axis represents the sector number, and the vertical axis represents the maximum displacements of blade and disk masses (shown in black and grey, respectively).

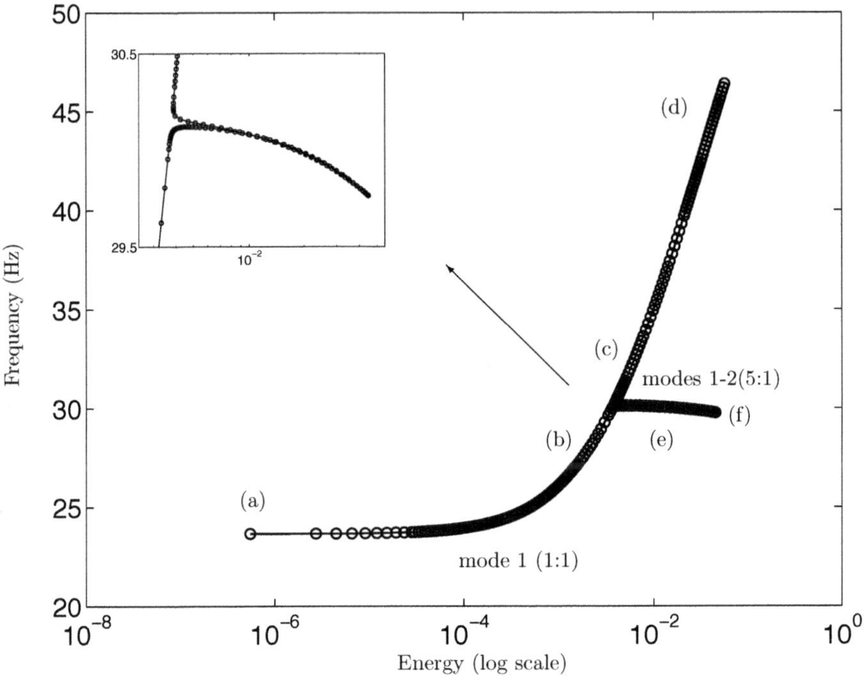

Figure 19. Frequency-energy plot of the first NNM of the nonlinear cantilever beam.

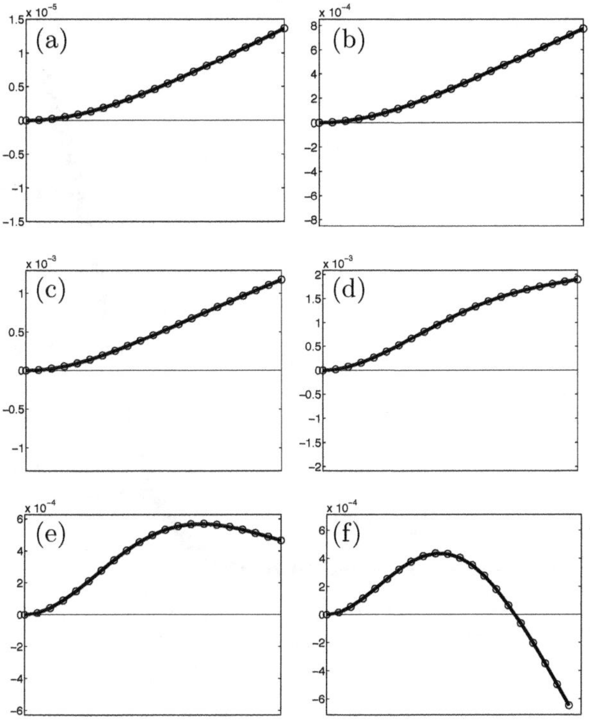

Figure 20. Maximum amplitudes of the first NNM of the nonlinear beam at different energy levels represented in Figure 19. (a-d): fundamental (1:1) NNMs motions; (e,f): internally resonant NNM motions between mode 1 and mode 2.

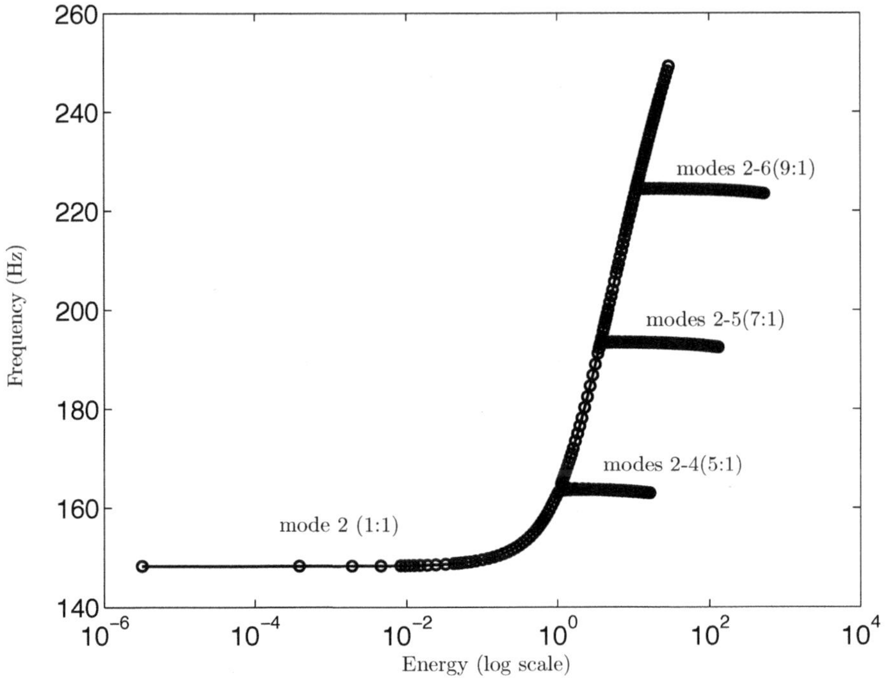

Figure 21. Frequency-energy plot of the second NNM of the nonlinear cantilever beam.

Numerical computation of nonlinear normal modes using HBM and ANM

Bruno Cochelin*

* LMA (Laboratoire de Mécanique et d'Acoustique)
UPR 7051 CNRS, Centrale Marseille
13402 MARSEILLE Cedex 20, France

Abstract These notes correspond to the four lectures given by Bruno Cochelin at the CISM course on "Modal Analysis of Nonlinear Mechanical Systems" held at Udine, from 25 to 29 june 2012. The goal is to describe numerical methods which are complementary to those presented in the previous chapter by Gaetan Kerschen. Once again a conservative nonlinear normal mode will be seen as a one-parameter familly of periodic orbits. The Harmonic Balance Method (HBM) will be used for computing periodic solutions and the Asymptotic Numerical Method (ANM) will be used for the continuation of these periodic solutions. The methods and the results presented above have been obtained with the collaboration of S. Bellizzi, S. Karkar, E.H. Moussi and C. Vergez.

1 Introduction to HBM and ANM on a toy model

Before embarking into details, let us have a quick overview of both numerical techniques on a toy model : the forced Duffing equation

$$\ddot{u} + 2\mu\,\dot{u} + u + u^3 = f\cos(\omega t) \tag{1}$$

where the damping factor μ, the force amplitude f and the angular frequency ω are parameters.

1.1 Harmonic Balance Method (HBM)

The Harmonic Balance approach refers to a class of methods for computing periodic solutions as follows. First, one assumes the solution to be under the form of a truncated Fourier series with selected harmonic index. For the Duffing equation, we choose to take the fundamental harmonic only (index one)

$$u(t) = u_c\cos(\omega t) + u_s\sin(\omega t). \tag{2}$$

G. Kerschen (Ed.), *Modal Analysis of Nonlinear Mechanical Systems*, CISM International Centre for Mechanical Sciences DOI 10.1007/ 978-3-7091-1791-0_6 © CISM Udine 2014

Here, u_c and u_c are unknown amplitudes and the angular frequency ω is assumed to be same as the one of the forcing term $f\cos(\omega t)$. Second, one inserts (2) into the (1), one uses elementary transformations such as $\cos^3(x) = \frac{3}{4}\cos(x) + \frac{1}{4}\cos(3x)$; $\cos^2(x)\sin(x) = \frac{1}{4}\sin(x) + \frac{1}{4}\sin(3x)$, and one collects cosine and sine terms having the same harmonic index, to get

$$\begin{array}{l} \{\ \ (1-\omega^2)u_c + 2\mu\omega u_s + \frac{3}{4}u_c^3 + \frac{3}{4}u_c u_s^2\}\cos(\omega t) \\ +\{(1-\omega^2)u_s - 2\mu\omega u_c + \frac{3}{4}u_s^3 + \frac{3}{4}u_s u_c^2\}\sin(\omega t) \\ +\{\qquad\qquad \dots \qquad\qquad\}\cos(3\omega t) \\ +\{\qquad\qquad \dots \qquad\qquad\}\sin(3\omega t) = f\cos(\omega t) \end{array} \tag{3}$$

Third, in (3) one drops all terms having a Fourier index different from those used in the original assumption (2). Here, terms in $\cos(3\omega t)$ and $\sin(3\omega t)$ are removed. The remaining equations correspond to the balance of the harmonics having the same index as in (2)

$$\begin{array}{l} (1-\omega^2)u_c + 2\mu\omega u_s + \frac{3}{4}u_c^3 + \frac{3}{4}u_c u_s^2 = f \\ (1-\omega^2)u_s - 2\mu\omega u_c + \frac{3}{4}u_s^3 + \frac{3}{4}u_s u_c^2 = 0 \end{array} \tag{4}$$

Finally, one is faced to solve an algebraic system having two equations for two unknowns u_c, u_s and three parameters ω, μ, f.

1.2 Continuation

One now wants to explore the solutions of this 2 degrees of freedom (d.o.f) algebraic system when one of the three parameters varies. For instance, let $\mu = 0.05$ and $f = 0.2$ be fixed parameters and let ω be the free parameter. Let $\mathbf{U} = [u_c, u_s, \omega]$ be the vector containing both the unknowns and the parameter and $\mathbf{R}$ be the vector of equations:

$$\mathbf{R}(\mathbf{U}) = \left\{ \begin{array}{l} (1-\omega^2)u_c + 2\mu\omega u_s + \frac{3}{4}u_c^3 + \frac{3}{4}u_c u_s^2 - f \\ (1-\omega^2)u_s - 2\mu\omega u_c + \frac{3}{4}u_s^3 + \frac{3}{4}u_s u_c^2. \end{array} \right.$$

Since $\mathbf{U} \in \mathbb{R}^3$ and $\mathbf{R} \in \mathbb{R}^2$, the solution set of the algebraic problem

$$\mathbf{R}(\mathbf{U}) = \mathbf{0} \tag{5}$$

is made of one dimensionnal continua of solutions called *solution branches.* Continuation methods (or path following methods) are designed for determining the solution set of (5) by traveling on the solution branches from an initial known solution. In figure 1, we show the result of the contination from the initial solution $\mathbf{U} = [0.195\ 0\ 0]$. This diagram that shows the evolution of u_c and u_s with the parameter ω is called a *bifurcation diagram* [1]. Each solution point on that branch can be the initial solution point for the

[1] Notice that plotting $\sqrt{u_c^2 + u_s^2}$ versus ω would gives the classical backbone curve of the forced Duffing model.

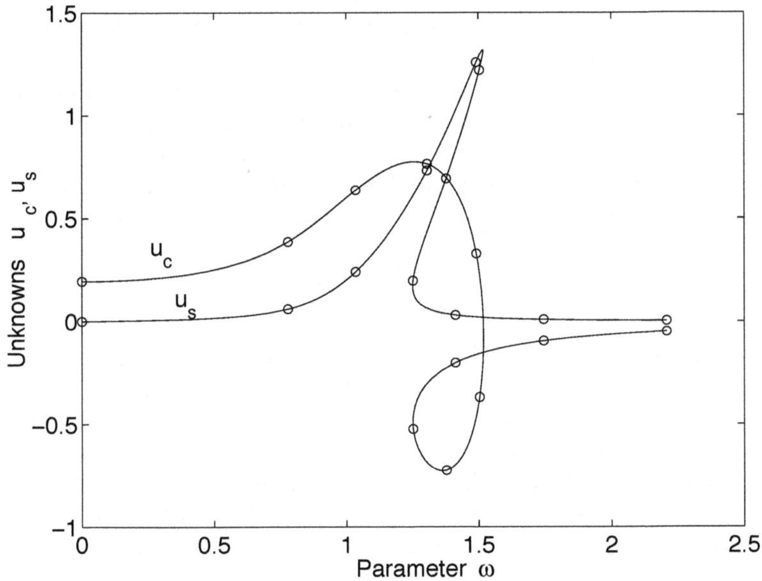

Figure 1. Bifurcation diagram

continuation with respect to another parameter than ω.

We will now go into the detail of both methods, beginning with the continuation. The reason why will be given later on.

2 Continuation with Asymptotic Numerical Method

In this section, we describe a continuation process based on high order Taylor series expansion of the solution branches. It has been originally called the *asympotic numerical method*, Damil and Potier-Ferry (1990); Azrar et al. (1993); Cochelin (1994) .

Let $\mathbf{R}(\mathbf{u}, \lambda) = 0$ be an algebraic system of n equations where $\mathbf{u} \in \mathbb{R}^n$ is a vector of n state variables and $\lambda \in \mathbb{R}$ a single control parameter (or bifurcation parameter). For the sake of compactness of notations, we introduce the extended state vector $\mathbf{U} = \begin{bmatrix} \mathbf{u} \\ \lambda \end{bmatrix} \in \mathbb{R}^{n+1}$ which includes the parameter λ as its last component, and we write the algebraic system as

$$\mathbf{R}(\mathbf{U}) = 0 \tag{6}$$

without distinguishing between state variables and parameter. The jacobian

$\mathbf{R}_{,U} = [\mathbf{R}_{,u}\mathbf{R}_{,\lambda}]$ is a $n \times (n+1)$ matrix composed of the square matrix $\mathbf{R}_{,u}$ and the colum vector $\mathbf{R}_{,\lambda}$.

Let $\mathbf{U}_0$ be a regular solution of (6), that is to say, a solution point where a single branch of solution goes through. In practice, we require $\mathbf{U}_0$ to satisfy (6) only up to a given tolerance ε_R, i.e. , $||\mathbf{R}(\mathbf{U}_0)|| < \varepsilon_R$. By definition of a regular solution, the jacobian $\mathbf{R}^0_{,U}$ at $\mathbf{U}_0$ has maximal rank n. The normalized tangent $\mathbf{U}_1$ at $\mathbf{U}_0$ is one of the two solutions of the following system

$$\begin{aligned} \mathbf{R}^0_{,U}\mathbf{U}_1 &= \mathbf{0} \\ \mathbf{U}_1^T\mathbf{U}_1 &= 1 \end{aligned} \tag{7}$$

having n linear equations and one quadratic equation. Notice that if $\mathbf{U}_1$ is solution, the other solution is $-\mathbf{U}_1$. Let $a = \mathbf{U}_1^T(\mathbf{U} - \mathbf{U}_0)$ be the pseudo-arc length parameter defined as the projection of the branch onto the tangent $\mathbf{U}_1$. According to the Implicit Function Theorem, the unique solution branch passing through $\mathbf{U}_0$ can be represented as a Taylor series in a. The continuation process described hereafter relies on the computation of these Taylor series (power series) at a high order of truncature N, say with $15 < N < 30$,

$$\mathbf{U}(a) = \mathbf{U}_0 + a\mathbf{U}_1 + a^2\mathbf{U}_2 + ... + a^N\mathbf{U}_N. \tag{8}$$

Let a_c be the radius of convergence of the Taylor series, which is in most situations a finite number[2]. When the order of truncature N is high, the truncated power series (8) provides a very accurate representation of the solution branch for a smaller than the radius of convergence a_c (fast convergence zone), and a very bad approximation for a greater than a_c (fast divergence zone). Around a_c, the convergence or the divergence may be slow. Because a_c is finite, the representation (8) gives access only to a limited part of the solution branch around the starting point $\mathbf{U}_0$ as it is shown in figure 2.

We now define the "range of utility" of (8) as the interval $[0, a_{max}]$ for which the truncated series satisfy the equation (6) up to a given required tolerance. Generally, the maximal value a_{max} of the path parameter a is smaller that a_c, so as to have an accurate enough solution. To fix the idea, a_{max} may be somewhere between $0.7a_c$ to $0.9a_c$ and we will see below how this a_{max} can easyly be determined.

In summary, the truncated series (8) with its range of utility $[0, a_{max}]$ provides a continuous representation of the solution branch that starts at $\mathbf{U}_0$ and that ends at the point $\mathbf{U}(a_{max})$. The principle of continuation is

[2]The case of zero radius of convergence or infinite radius of convergence are exceptions rather than the rule.

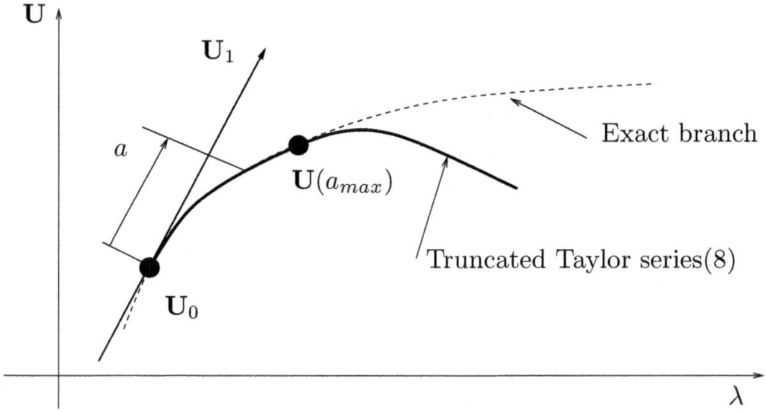

Figure 2. Branch of solution passing through the regular point $\mathbf{U}_0$: exact branch and its Taylor representation

to take the last end point $\mathbf{U}(a_{max})$ as a new starting point $\mathbf{U}_0$, from which a new truncated serie is computed. Finally, as it is shown in figure 2, the solution branch is known as a succession of Taylor series representation (8) having their own start point U_{0_i} and own range of utility a_{max_i}.

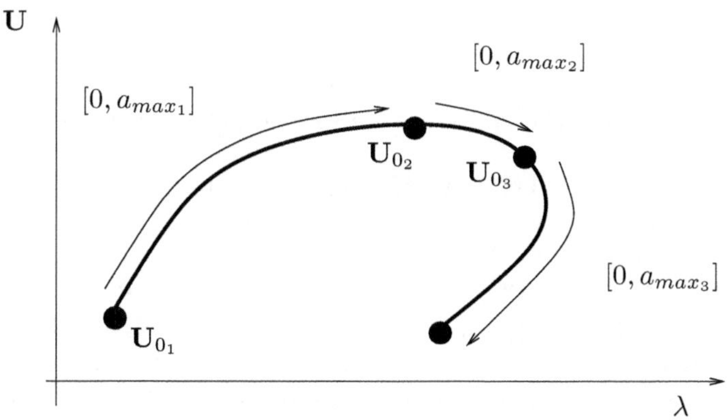

Figure 3. The branch of solution is represented by a succession of Taylor series that can be used inside their "range of utility" $[0, a_{max_i}]$

In the following, we will show how to compute the series (6) and a_{max}. Then we will compare this continuation process with the classical first order predictor-corrector (sometimes referred to as the Newton-Raphson method) that is described in reference papers or textbooks Keller (1987); Seydel (1994); Doedel et al. (1991); Allgower and Georg (1990).

2.1 How to compute the serie efficiently

Two main strategies have been investigated:

- recast the original system of equations into a system with a polynomial quadratic nonlinarity,
- use Automatic Differentiation tools for generating and dealing with the Taylor series.

Hereafter, we focus on the first strategy which is the historical way and we refer to Charpentier et al. (2008); Charpentier (2012) for readers interested in Automatic Differentiation within ANM continuation. We begin by showing that if the system is polynomial and quadratic, then the computation of the series is rather straightforward. Then, we discuss the more subtil question: "can any system be recasted into a polynomial quadratic one"

Computing series for a quadratic system Assuming that $\mathbf{R}(\mathbf{U})$ is quadratic, it can be splitted into three parts representing the constant terms, the linear ones and the quadratic ones as follows:

$$\mathbf{R}(\mathbf{U}) = \underbrace{\mathbf{L}_0}_{\text{constant}} + \underbrace{\mathbf{L}(\mathbf{U})}_{\text{linear}} + \underbrace{\mathbf{Q}(\mathbf{U},\mathbf{U})}_{\text{quadratic}} \tag{9}$$

In (9), $\mathbf{L}(.)$ is a linear operator with respect to its argument and $\mathbf{Q}(.,*)$ is linear with respect to both arguments. Inserting the series (8) into (9) and equating to zero each power of a gives the following succession of linear systems:

- order 0: $\mathbf{L}_0 + \mathbf{L}(\mathbf{U}_0) + \mathbf{Q}(\mathbf{U}_0,\mathbf{U}_0) = \mathbf{0}$, which is obviously satisfied since $\mathbf{U}_0$ is a solution.
- order 1: $\mathbf{R}^0_{,U}(\mathbf{U}_1) = \mathbf{0}$
- order $2 \leq p \leq n$: $\mathbf{R}^0_{,U}(\mathbf{U}_p) + \Sigma_{i=1}^{p-1}\mathbf{Q}(\mathbf{U}_i,\mathbf{U}_{p-i}) = \mathbf{0}$

where $\mathbf{R}^0_{,U}(.) = \mathbf{L}(.) + \mathbf{Q}(\mathbf{U}_0,.) + \mathbf{Q}(.,\mathbf{U}_0)$ is the jacobian at $\mathbf{U}_0$.

The original nonlinear problem has therefore been reduced to a series of N linear systems with n equations. However, at each order, these linear systems are under-determined since they have $n+1$ unknowns. The missing equations are obtained by inserting (8) into the definition of the path parameter a given above. This yields

- order 1 : $\mathbf{U}_1^t\mathbf{U}_1 = 1$
- order $2 \leq p \leq n$: $\mathbf{U}_1^t\mathbf{U}_p = 0$.

It should be noticed that these linear systems share the same jacobian matrix, so this matrix is LU decomposed once for all.

Recasting a given system into a quadratic one The basic idea of the recast process consists in adding new variables and new equations so as to keep only single product between variables. For example, if the expression u^3 appears in the equations, introduce the new variable v defined by the quadratic relation $v = u^2$, and rewrite u^3 as the product uv. For an expression like $\frac{u}{v}$, introduce the new variable $w = \frac{u}{v}$ and the quadratic relation $u = vw$. For $\sqrt{u}$, introduce the new variable v and the quadratic relation $u = v^2$.

Let us now apply the recast on the toy system (4) having cubic expressions. By introducing the new variables $v_c = u_c^2$, $v_s = u_s^2$ and $\beta = \omega^2$, we get the following quadratic system with 5 equations for 6 unknowns,

$$\begin{array}{ccccll}
\overbrace{-f}^{\text{cst}} & + & \overbrace{u_c}^{\text{linear}} & + & \overbrace{(-\beta u_c + 2\mu\omega u_s + \frac{3}{4}u_c v_c + \frac{3}{4}u_c v_s)}^{\text{quadratic}} & = 0 \\
0 & + & u_s & + & (-\beta u_s - 2\mu\omega u_c + \frac{3}{4}u_s v_s + \frac{3}{4}u_s v_c) & = 0 \\
0 & + & v_c & - & u_c u_c & = 0 \\
0 & + & v_s & - & u_s u_s & = 0 \\
0 & + & \beta & - & \omega\omega & = 0.
\end{array} \tag{10}$$

As a second illustration, we consider the simple model with two equations used in Doedel et al. (1991) for continuation illustration

$$\begin{array}{rlll}
r_1(u_1, u_2, \lambda) = & 2u_1 - u_2 + 100\frac{u_1}{1+u_1+u_1^2} & - \lambda & = 0 \\
r_2(u_1, u_2, \lambda) = & 2u_2 - u_1 + 100\frac{u_2}{1+u_2+u_2^2} & - (\lambda + \mu) & = 0.
\end{array} \tag{11}$$

Introducing the four new variables,

$$\begin{array}{rclrcl}
v_1 & = & u_1 + u_1^2 & v_3 & = & \frac{1}{1+v_1} \\
v_2 & = & u_2 + u_2^2 & v_4 & = & \frac{1}{1+v_2}
\end{array} \tag{12}$$

the following quadratic system with 6 equations and 7 unknowns is obtained

$$\begin{array}{ccccccl}
\overbrace{0}^{\text{cst}} & + & \overbrace{2u_1 - u_2 - \lambda}^{\text{linear}} & + & \overbrace{100u_1v_3}^{\text{quadratic}} & = 0 \\
-\mu & + & 2u_2 - u_1 - \lambda & + & 100u_2v_4 & = 0 \\
0 & + & v_1 - u_1 & - & u_1^2 & = 0 \\
0 & + & v_2 - u_2 & - & u_2^2 & = 0 \\
-1 & + & v_3 & + & v_1v_3 & = 0 \\
-1 & + & v_4 & + & v_2v_4 & = 0.
\end{array} \tag{13}$$

It should be noticed that other choises of variables could have been made, for instance $v_1 = 1 + u_1 + u_1^2$ and $v_3 = \frac{1}{v_1}$. So, the recasted form is obviously not unique. The important point is that the recasted system should have the same solutions than the original one. The recast process should lead to a new way of writing the original system and not a modification of the system. For the sake of completness, let us now detail an example where classical mathematical functions appear,

$$r(u,\lambda) = u + e^u + \tan^{-1}(u) - \lambda \tag{14}$$

We introduce two new variables v and w with the relation $v = e^u$, $\tan(w) = u$. Differentiating these relations with respect to the path parameter a permits to get polynomial equations between the variables and their path derivatives

$$\begin{cases} \frac{dv}{da} = v\frac{du}{da} \\ (1+u^2)\frac{dw}{da} = \frac{du}{da} \end{cases} \tag{15}$$

Then, setting $x = u^2$, we can now replace (14) by the following quadratic system

$$\begin{cases} u + v + w - \lambda = 0 \\ \frac{dv}{da} = v\frac{du}{da} \\ (1+x)\frac{dw}{da} = \frac{du}{da} \\ x - u^2 = 0 \end{cases} \tag{16}$$

Notice that the series computation algebra is slightly different in that case because of the derivaties $\frac{du}{da} = u_1 + 2au_2 + 3a^2u_3 + \ldots$.

For concluding, we say that the process of adding suitable variables to recast any system into a polynomial quadratic system is undoubtedly the most difficult and surprizing part of the method. It is however a key point for an efficient computation of the series today. Automatic Differentiation is a promising way of removing this need of a quadratic recast.

2.2 Determining the range of utility of the series

Once the series (8) have been computed up to order N, one has to determine their "range of utility", i.e., the interval $[0, a_{max}]$ inside which the solution $\mathbf{U}(a)$ satisfies the required tolerance $||\mathbf{R}(\mathbf{U}(a))|| < \varepsilon_R$. The residual error of the truncated expression (8) is

$$\begin{aligned}\mathbf{R}(\mathbf{U}_0 + a\mathbf{U}_1 + \cdots + a^N\mathbf{U}_N) = \quad & \mathbf{R}(\mathbf{U}_0) \\ & +a \underbrace{\mathbf{R}_1}_{=0} +a^2 \underbrace{\mathbf{R}_2}_{=0} + \cdots + a^N \underbrace{\mathbf{R}_N}_{=0} \\ & +a^{N+1} \quad \underbrace{\mathbf{R}^{N+1}}_{\text{r.h.s at order} N+1} \quad + \text{h.o.t.}\end{aligned} \tag{17}$$

where $\mathbf{R}(\mathbf{U}_0)$ is the residual error of the start point $\mathbf{U}_0$. The vectors $\mathbf{R}_1, \ldots, \mathbf{R}_N$ correspond to the linear systems solved at each order for computing the $\mathbf{U}_i's$ and they are null. The first significant term is $a^{N+1}\mathbf{R}^{N+1}$ where the vector $\mathbf{R}^{N+1}$ is exactely the right hand side vector of the linear system at order $N+1$. Inside the radius of convergence of the series, the term $a^{N+1}\mathbf{R}^{N+1}$ is greater than the following higher order terms (h.o.t.), and a good approximation of the residual error is given by the simple expression

$$\mathbf{R}(\mathbf{U}_0 + \cdots + a^N\mathbf{U}_N) \simeq \mathbf{R}(\mathbf{U}_0) + a^{N+1}\mathbf{R}^{N+1} \tag{18}$$

Recalling that $||\mathbf{R}(\mathbf{U}_0)|| < \varepsilon_R$, a natural tolerance requirement is that the additional residual error $a^{N+1}\mathbf{R}^{N+1}$ remains with the same order of magnitude as that of the start point $\mathbf{U}_0$. Hence, the requirement $||a^{N+1}\mathbf{R}^{N+1}|| < \varepsilon_R$ gives an explicit formula for the maximal value of the path parameter

$$a_{max} = \left(\frac{\varepsilon_R}{||\mathbf{R}^{N+1}||}\right)^{\frac{1}{N+1}} \tag{19}$$

In practice, the vector $\mathbf{R}^{N+1}$ is easy to compute and the cost for determinig a_{max} is negligeable.

2.3 Comparison with predictor-corrector algorithm

The first-order predictor-corrector algorithm is the most popular method used today to compute solution branches. It works as follows: let $\mathbf{U}_0$ be a regular solution of (6), a prediction point is computed as

$$\mathbf{U}_{pred} = \mathbf{U}_0 + a_p\mathbf{U}_1 \tag{20}$$

where $\mathbf{U}_1$ is again the tangent vector at $\mathbf{U}_0$ and a_p a prescribed scalar which gives the distance from $\mathbf{U}_0$ to $\mathbf{U}_{pred}$. Since $\mathbf{U}_{pred}$ is not an accurate solution, Newton corrections are then applied to $\mathbf{U}_{pred}$ so as to satisfy (6) up to the required tolerance. Applying this process step by step, the method gives the solution branch as a succession of solution points. A crucial issue for the efficiency of the continuation is to have a good step-length control strategy, that is to say, a correct choise of a_p at each step so as to avoid too small steps which loose time unnecessarily and too large steps which involves divergence of the Newton corrections. Second order predictor-corrector methods rely on the improved prediction

$$\mathbf{U}_{pred} = \mathbf{U}_0 + a_p\mathbf{U}_1 + a_p^2\mathbf{U}_2 \tag{21}$$

which limits the error on $\mathbf{U}_{pred}$ and reduces the number of corrections.

We begin the comparison by noticing that both equation (20) and (21) looks like (8) when it is truncated at order one or order two. So, the ANM continuation could be seen as a kind of high-order predictor method which, because the prediction is of high quality, does not need any correction. The most important difference between ANM and prediction-correction is that the path parameter a in (8) is let free and it is determined only after the serie computation by analysing the convergence of the serie. On the contrary, the step length a_p of the prediction has to be fixed to a prescribed value in (20) and (21) before embarking in the Newton corrections whose convergence is not known *apriori*, but only estimated from the story of previous steps.

In summary, the main advantage of the ANM continuation are

- the solution branch is known as a succession of continuous representation given by Eq. (8) with $a \in [0, a_{max}]$ instead of a succession of points.
- since all the linear systems to be solved have the same Jacobian matrix, the computational cost of the series (8) remains low.
- the size a_{max} of each continuous representation is not estimated in advance but calculated *aposteriori* from the convergence properties of the current step. No step-length control strategy is required with ANM continuation.
- a high order Tayor series (8) contains many valuable information on the computed branch. For instance, the presence of a simple bifurcation can be easily detected as we will explained below.

On the other hand, the main drawback of the ANM continuation process (as it has been presented above) is the need to recast the system of equation under a quadratic polynomial form. This is generally the most difficult point for a beginner with ANM. This drawback is now currently being removed by using Automatic Differentiation for the generation of the series.

2.4 MANLAB and DIAMANLAB sofware

MANLAB is an interactive software program that implements ANM continuation for bifurcation analysis of algebraic systems. The first version was programmed in Matlab using an objet-oriented approach by Arquier (2007). The free open source research software can be downloaded at Manlab (2012). MANLAB has a Graphical User Interface (GUI) with buttons, on-line inputs and graphical windows for generating, displaying and analysing the bifurcation diagram and the solution of the system. To enter the system of equations, the user has to provide three vector valued Matlab functions corresponding to the constant, linear and quadratic operators $\mathbf{L}_0$, $\mathbf{L}$ and $\mathbf{Q}$. We gives these inputs for one of the example discussed above, $U = [u_1, u_2, v_1, v_2, v_3, v_4, \lambda]$ and $\mu = 0.05$:

```
function [L0] = L0

L0=zeros(6,1);

L0(1)=   0;
L0(2)=  -0.05;
L0(3)=   0;
L0(4)=   0;
L0(5)=  -1;
L0(6)=  -1;
```

```
function [L] = L(U)

L=zeros(6,1);

L(1)=2*U(1)- U(2) -U(7);
L(2)=2*U(2)- U(1) -U(7);
L(3)=U(3)-U(1);
L(4)=U(4)-U(2);
L(5)=U(5);
L(6)=U(6);
```

```
function [Q] = Q(U,V)

Q=zeros(6,1);

Q(1)=100*U(1)*V(5);
Q(2)=100*U(2)*V(6);
Q(3)=   -U(1)*V(1);
Q(4)=   -U(2)*V(2);
Q(5)=    U(3)*V(5);
Q(6)=    U(4)*V(6);
```

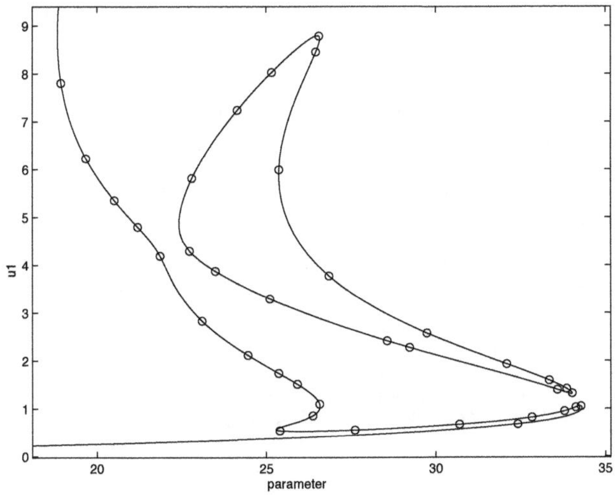

Figure 4. Bifurcation diagram obtained with MANLAB for the two d.o.f. example, showing evolution of u_1 versus λ. The circle correspond to the start points $\mathbf{U}_{0i}$ of each continuation step.

DIAMANLAB is an improved and re-organized version of MANLAB which implements Automatic Differentation for the series computation Char-

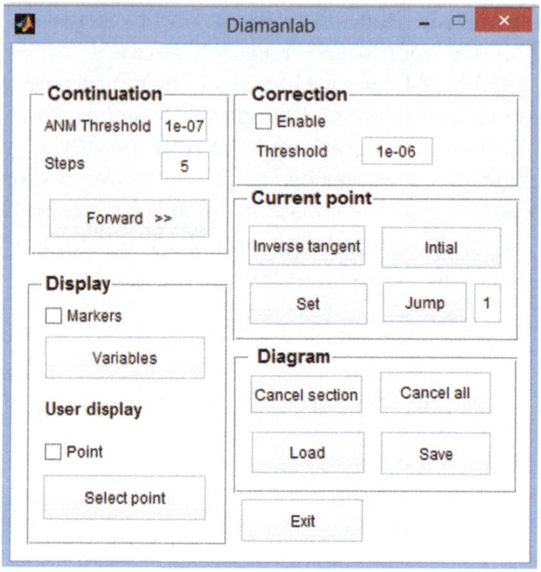

Figure 5. Graphical user interface of DIAMANLAB.

pentier (2012). In DIAMANLAB, the user has to provide its vector of equations under a natural form, as follows

```
function R = R(obj,U)

if isa(U,'Taylor')
 R=Taylor(get(obj,'order'),zeros(2,1));
end

u1=U(1); u2=U(2); lambda=U(3);

R(1) = 2*u1 - u2 + 100*u1/(1+u1+u1*u1) - lambda;
R(2) = 2*u2 - u1 + 100*u2/(1+u2+u2*u2) - (lambda+obj.mu);
```

2.5 Simple bifurcation detection and branch switching using Taylor series

Bifurcation detection and branch switching are classical issues in numerical continuation for which the theoretical backgrounds can be found again in Keller (1987); Seydel (1994); Doedel et al. (1991); Allgower and Georg (1990). This topic is of particular importance for nonlinear normal modes computation in conservative systems because, due to the interactions between modes through internal resonances, the branch of periodic solutions have a great number of bifurcations.

We now address this problem of bifurcation with an approach which is closely linked to the ANM continuation and which differs from classical methods. As stated above, the truncated Taylor series (8) that are computed during the continuation are full of relevant information about the followed branches of solutions. For instance, these series can reveal if there is a bifurcation on the branch, what kind of bifurcation it is and where exactely it is located on the branch. We became aware of this important point by analysing many different problems having simple bifurcations. We clearly and systematically observed that in the vicinity of simple bifurcation point, a geometric series emerges in the representation (8). More precisely, when the start point $\mathbf{U}_0$ is closed to a simple bifurcation point, there exist an order p over which the $\mathbf{U}_p$ vectors of the Taylor series shows the remarkable property

$$\mathbf{U}_{p+1} \simeq \alpha \mathbf{U}_p \qquad \mathbf{U}_{p+2} \simeq \alpha \mathbf{U}_{p+1} \qquad \dots \tag{22}$$

This geometrical series which emerges at a high order is characterized by a common ratio α and an emerging vector $\mathbf{U}_e$. Numerical investigation also reveals that $\frac{1}{\alpha}$ is the distance from $\mathbf{U}_0$ to the bifurcation point $\mathbf{U}_c$ and that the vector $\mathbf{U}_e$ indicates the direction of the bifurcated branch at the bifurcation point. We will explain these finding by using a simplified first order approach and we refer to Cochelin and Médale (2013) for a deeper analysis.

Elements of bifurcation theory Let $\mathbf{U}_c$ be a simple bifurcaton point where two branches of solution of (6) cross each other. Let $\mathbf{U}_{t_1}$ and $\mathbf{U}_{t_2}$ be the two tangents for each branch at $\mathbf{U}_c$. The kernel of the jacobian matrix $\mathbf{R}^c_{,U}$ evaluated at $\mathbf{U}_c$ is two dimensional and we have

$$\mathcal{N}(\mathbf{R}^c_{,U}) = \text{ Span } \{\mathbf{U}_{t_1}, \mathbf{U}_{t_2}\}, \quad \mathcal{N}(\mathbf{R}^{cT}_{,U}) = \text{ Span } \{\psi\}. \tag{23}$$

According to the bifurcation theory, the expression for the branches at $\mathbf{U}_c$ begins like

$$\begin{aligned} &\text{Order } 1 \rightarrow \mathbf{U} = \mathbf{U}_c + a_1\,\mathbf{U}_{t_1} + a_2\,\mathbf{U}_{t_2} + \text{ h.o.t.} \\ &\text{A.B.E.} \rightarrow \underbrace{\psi^T R^c_{,UU}\mathbf{U}_{t_1}\mathbf{U}_{t_2}}_{\text{scalar } \mu_1 \neq 0}\; a_1\,a_2 = 0 \rightarrow \quad a_1\,a_2 = 0 \end{aligned} \tag{24}$$

In the first equation (obtained at order one), a_1 and a_2 are free parameters. The second equation is the so-called Algebraic Bifurcation Equation (second order projected on the left null vector) which finally implies that either a_1 or a_2 should be zero, i.e., the branch begins by following $\mathbf{U}_{t_1}$ or $\mathbf{U}_{t_2}$

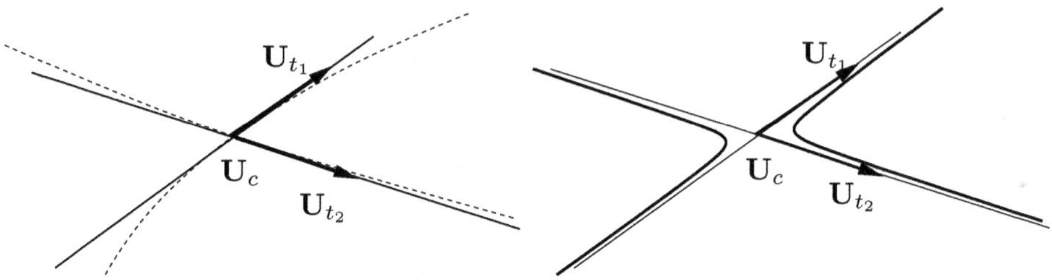

Figure 6. Left : perfect case. Right : perturbed case

Let us now introduce a small perturbation in this analysis and see the change. The reason for paying attention to perturbation is that the start point $\mathbf{U}_0$ is not an exact solution of (6) and that the residual error $\mathbf{R}(\mathbf{U}_0)$ does act as a very small perturbation in the system. Indeed, as we have seen above in (17), the (non truncated) Taylor series $\mathbf{U}(a) = \mathbf{U}_0 + a\mathbf{U}_1 + a^2\mathbf{U}_2 + \ldots$ computed at the start point $\mathbf{U}_0$ is the *exact* solution of

$$\mathbf{R}_p(\mathbf{U}) = \mathbf{R}(\mathbf{U}) - \mathbf{R}(\mathbf{U}_0) = \mathbf{0} \tag{25}$$

which is a *perturbed problem.* Defining $\mu_0 = \psi^T.\mathbf{R}(\mathbf{U}_0)$ which is very small but not zero, the expression of the branches for the perturbed problem are

$$\begin{aligned} &\text{Order } 1 \rightarrow \mathbf{U} = \mathbf{U}_c + a_1\,\mathbf{U}_{t_1} + a_2\,\mathbf{U}_{t_2} + \text{ h.o.t} \\ &\text{A. B. E.} \quad \rightarrow \quad a_1\,a_2 = \tfrac{\mu_0}{\mu_1} \end{aligned} \tag{26}$$

Due to the perturbation the two straight branches shown on the left in figure 6 become the two hyperbola shown on the right in figure 6. We are now ready to give the expression of the branch from a start point which is no more $\mathbf{U}_c$, but a regular point $\mathbf{U}_0$ near $\mathbf{U}_c$. Let $a_1 = -d$ and $a_2 = \frac{-\mu_0}{\mu_1 d}$ so as to define a point $\mathbf{U}_0$ which is on the perturbed branch

$$\mathbf{U}_0 = \mathbf{U}_c - d\,\mathbf{U}_{t_1} - \frac{\mu_0}{\mu_1 d}\,\mathbf{U}_{t_2} \tag{27}$$

Let us make a shift of d for the path parameter $a = a_1 + d$, The perturbed branch can now be written

$$\mathbf{U}(a) = \mathbf{U}_0 + \; a\,\mathbf{U}_{t_1} - \frac{\mu_0}{d\mu_1}\,\frac{\frac{a}{d}}{(1-\frac{a}{d})}\mathbf{U}_{t_2} \tag{28}$$

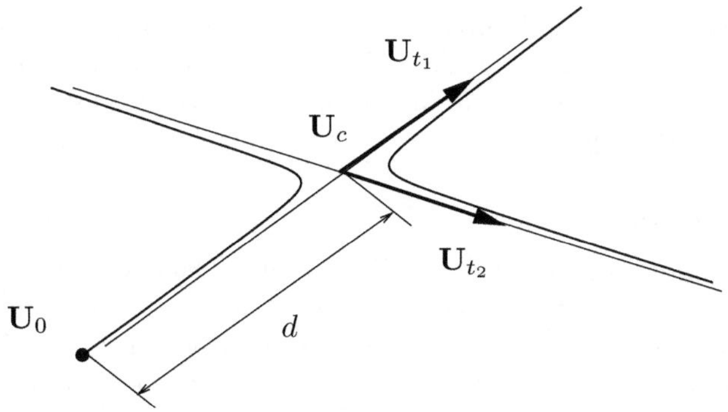

Figure 7. Perturbed branch of solution seen from a start point $\mathbf{U}_0$ on the branch

This simplified first order analysis reveals the key point of the story: the contribution of $\frac{\mu_0}{d\mu_1} \frac{\frac{a}{d}}{(1-\frac{a}{d})}\mathbf{U}_{t_2}$ is very small[3] exept when a is close to d. When expanded into Taylor series, the rational fraction $\frac{\frac{a}{d}}{1-\frac{a}{d}} = \frac{a}{d}+(\frac{a}{d})^2+(\frac{a}{d})^3+\cdots$, is responsible for the emergence of a geometric series with common ratio $\frac{1}{d}$ and emerging vector $\mathbf{U}_{t_2}$.

Practical detection and switching Practically, the presence of a geometrical series is easy to detect by analysing the last term of (6). When a geometrical series is detected, it is extracted additively

$$\mathbf{U}(a) = \mathbf{U}_0 + a\,\hat{\mathbf{U}}_1 + a^2\,\hat{\mathbf{U}}_2 + \cdots + a^{n-1}\,\hat{\mathbf{U}}_{n-1} + \Big(\frac{a}{d} + (\frac{a}{d})^2 + \cdots + (\frac{a}{d})^n\Big)\,\mathbf{U}_e \tag{29}$$

and replaced by a rational fraction as

$$\mathbf{U}(a) = \underbrace{\mathbf{U}_0 + a\,\hat{\mathbf{U}}_1 + a^2\,\hat{\mathbf{U}}_2 + \cdots + a^{n-1}\,\hat{U}_{n-1}}_{\hat{\mathbf{U}}(a)\ \text{cleaned series}} + \frac{a}{d}\left(\frac{1}{1-\frac{a}{d}}\right)\mathbf{U}_e \tag{30}$$

The cleaned series $\hat{\mathbf{U}}(a)$ permits to easily compute the bifucation point as $\mathbf{U}_c = \hat{\mathbf{U}}(d)$ and the two tangents using $\mathbf{U}_{t_1} = \frac{\partial \hat{\mathbf{U}}}{\partial a}(d)$ and $\mathbf{U}_{t_2} = \beta\,\mathbf{U}_{t_1} + \gamma\,\mathbf{U}_e$, so as to satisfy the A.B.E.. The accuracy is very good and the numerical cost is negligible.

[3] We recall that μ_0 is very small.

In MANLAB, simple bifurcation points are automatically detected and marked on the computed branch. When the user places the start point on a bifurcation, the travelling direction is set along the second computed tangent for switching.

2.6 Conclusion

In this section, we have presented the minimal theoretical background for understanding continuation and bifurcation analysis using the asymptotic numerical method. The particularity of this approach is to compute high order Taylor series of the solution branches and to use them at best. Some informations have been given on MANLAB and DIAMANLAB open source softwares which implement the method. We will now go back to the periodic solution computation by the so-called Harmonic Balance Method.

3 High order harmonic balance method

Let us take an autonomous system of differential equations

$$\dot{Y} = f(Y, \lambda) \tag{31}$$

where Y is a vector of unknowns, f a smooth nonlinear vector valued function and λ a real parameter. We assume that this system has branches of periodic solutions when λ varies, and we want to find and follow them by applying the harmonic balance method and the ANM continuation procedure.

3.1 The harmonic balance principle

As recalled in the introduction, the HBM relies in decomposing $Y(t)$ into a truncated Fourier series :

$$Y(t) = Y_0 + \sum_{k=1}^{H} Y_{c,k} \cos(k\omega t) + \sum_{k=1}^{H} Y_{s,k} \sin(k\omega t) \tag{32}$$

This ansatz is put into Eq. (31) and $f(Y, \lambda)$ is expanded into Fourier series. By balancing the first $2H + 1$ harmonic terms, one obtains an algebraic system with $2H + 1$ vector equations for the $2H + 1$ vector unknowns Y_i, the unknown pulsation ω, and the parameter λ. Adding a phase condition (Seydel (1994)) yields a system with N equations and $N + 1$ variables. The branches of solutions of this algebraic system are then followed using a continuation technique. This procedure provides only approximate periodic solutions, since in the expansion of Eq. (31), all the harmonic terms greater than H remain unbalanced. However, if the number of harmonics H is large enough, accurate solutions can be obtained. We note that the name "harmonic balance" seems to have first appeared in 1936 Krylov and Bogoliubov (1947) and the first convergence results are attributed to Urabe (1965). Today, HBM is still under progress, see Krack et al. (2013); Grolet and Thouverez (2012); Guskov et al. (2008) among others, and the method is also being adapted for Computational Fluid Mechanics Ekici et al. (2008).

One crucial point in the HBM is the expansion of the vector $f(Y, \lambda)$ into Fourier series. This can be quite a straighforward procedure if $f(Y, \lambda)$ is a polynomial of degree two or three, and if the number of harmonics H is sufficiently small. In this case, the expansion can be carried out by hand. But in the most general situation, where $f(Y, \lambda)$ shows nonlinearities of any kind, this computation can be very cumbersome, even with the help of a symbolic software program. This drawback has been the starting-point in the developpment of many revisited HBM in the literature such as the incremental harmonic balance by Lau and Cheung (1981) and the alternating frequency/time domain harmonic balance by Ling (1987).

Hereafter, we shall follow a simple but powerful procedure that overcomes the drawback mentioned above: with the same spirit as for the ANM continuation, we shall first recast the original system (31) into a new system where the nonlinearities are polynomial and quadratic.

3.2 A key point: the quadratic recast

The quadratic system replacing (31) will be written as follows

$$m(\dot{Z}) = c + l(Z) + q(Z, Z) \tag{33}$$

and it will contain both differential and algebraic equations. The unknown vector Z (of size N_e) contains the original components of the vector Y and some new variables which are added to get the quadratic form. The right hand side of (33) is written as follows: c is a constant vector with respect to the unknown Z, $l(.)$ is a linear vector valued operator with respect to the vector entry, and $q(.,.)$ is a quadratic vector valued operator, which is linear with respect to both entries. We assume that only the vectors c, and $l(.)$ depend on the real parameter λ. On the left hand side, $m(.)$ is a linear vector valued operator with respect to the vector entry. The algebraic equations correspond to zero values of $m(.)$.

Since the nonlinearities are quadratic, the HBM can obviously be easily and systematically applied on Eq. (33), even with a large number of harmonics. The question of how a system can be recasted into the form (33) will be again illustrated with some elementary examples where the transformation from (31) to (33) will be performed explicitly. We will deal first with the classical Van der Pol oscillator, the Rössler system and a model for clarinet-like musical instruments. Other examples, with nonlinearities of others kinds such as rational fraction, will be given later on.

Example 1: The Van der Pol oscillator. We take the following second order autonomous system, known as the Van der Pol oscillator, where the parameter λ governs the amplitude of the nonlinear damping term.

$$\ddot{u} - \lambda(1 - u^2)\,\dot{u} + u = 0 \tag{34}$$

This equation can be classically recast into a first order system by introducing the velocity $v(t) = \dot{u}(t)$ as a new unknown. We obtain

$$\begin{array}{rcl} \dot{u} & = & v \\ \dot{v} & = & \lambda(1 - u^2)\,v - u \end{array} \tag{35}$$

If we now introduce the auxiliary variables $w(t) = 1 - u^2(t)$ and $r(t) = v(t)w(t)$, the system can be written in the form of Eq. (33) with $Z = [u, v, w, r]^t$. The various terms are arranged below so that it will be clear to the readers how the different functions m, c, l and q are formed.

$$\begin{array}{ccccc} \dot{u} & = & 0 & +v & \\ \dot{v} & = & 0 & -u + \lambda r & \\ 0 & = & 1 & -w & -u^2 \\ \underbrace{0}_{m(\dot{Z})} & = & \underbrace{0}_{c} & \underbrace{+r}_{l(Z,\lambda)} & \underbrace{-vw}_{q(Z,Z)} \end{array} \tag{36}$$

We finally obtain two first order differential equations and two algebraic equations with quadratic polynomial nonlinearities. Only the operator l depends here on λ.

Example 2: The Rössler system. We take the following first order autonomous system of three equations known as the Rössler system.

$$\begin{array}{rcl} \dot{x} & = & -y - z \\ \dot{y} & = & x + ay \\ \dot{z} & = & b + z(x - \lambda) \end{array} \tag{37}$$

where x, y, z are functions of time and a, b, c are parameters. This system can be written in the form of Eq. (33) without any auxiliary variables (i.e. $Z = [x, y, z]^t$):

$$\begin{array}{ccccc} \dot{x} & = & 0 & -y - z & \\ \dot{y} & = & 0 & +x + ay & \\ \underbrace{\dot{z}}_{m(\dot{Z})} & = & \underbrace{b}_{c} & \underbrace{-\lambda z}_{l(Z,\lambda)} & \underbrace{+zx}_{q(Z,Z)} \end{array} \tag{38}$$

Here again, only the operator l depends on λ. This example will be used here to show that the approach presented in this paper can be used to find period-doubling and bifurcations with period 4.

Example 3: A model for clarinet-like musical instruments. In the case of small amplitude oscillations, a simple model for reed instruments (clarinet, saxophone, etc) can be written using a modal formulation as in Silva et al. (2008); Karkar et al. (2012):

$$\begin{array}{ll} \ddot{x} + q_r \omega_r \dot{x} + \omega_r^2 x = \omega_r^2 p & \\ \ddot{p}_n + 2\alpha_n c\, \dot{p}_n + \omega_n^2 p_n = \frac{2c}{l}\, \dot{u} & \forall n \in [1 \dots N] \\ u = \zeta(1 - \lambda + x)\sqrt{\lambda - p} & \\ p = \sum_{n=1}^{N} p_n & \end{array} \tag{39}$$

where the unknowns x, $p_{n=1..N}$, p, u are functions of time and q_r, ω_r, α_n, ω_n, c, l, ζ are given parameters describing either the physics or the player's action. N is the number of acoustic modes. λ stands here for the blowing pressure.

It is worth noting that even if this system contains a square root, it can be recast in the form of Eq. (33): if we introduce the auxiliary variables $y = \dot{x}$, $z_n = \dot{p}_n$ and $v = \sqrt{\lambda - p}$, system (39) can be rewritten, with $Z = [x, y, p_1, \dots, p_N, z_1 \dots z_N, u, v]^t$ as follows:

$$\begin{array}{ccccccc} \dot{x} & = & 0 & +y & & \\ \dot{y} & = & 0 & +\omega_r^2 p - q_r \omega_r y - \omega_r^2 x & & \\ \dot{p}_n & = & 0 & +z_n & & \forall n \in [1 \dots N] \\ \dot{z}_n - \frac{2c}{l}\dot{u} & = & 0 & -2\alpha_n c\, z_n - \omega_n^2 p_n & & \forall n \in [1 \dots N] \\ 0 & = & 0 & -u + \zeta(1-\lambda)v & +\zeta x v & \\ 0 & = & 0 & -p + p_1 + \dots + p_N & & \\ \underbrace{0}_{m(\dot{Z})} & = & \underbrace{\lambda}_{c(\lambda)} & \underbrace{-p}_{l(Z,\lambda)} & \underbrace{-v^2}_{q(Z,Z)} & \end{array} \tag{40}$$

Here, the linear term l, but also the constant term c, are functions of λ.

3.3 The harmonic balance method applied to a quadratic system

In this section, the harmonic balance method is applied to the system (33). The unknown (column) vector Z is decomposed into Fourier series with H harmonics:

$$Z(t) = Z_0 + \sum_{k=1}^{H} Z_{c,k} \cos(k\omega t) + \sum_{k=1}^{H} Z_{s,k} \sin(k\omega t) \tag{41}$$

The components of the Fourier series are collected into a large (column) vector U, with size $(2H+1)\times N_e$, where N_e is the number of equations in (33).

$$U=[Z_0^t,Z_{c,1}^t,Z_{s,1}^t,Z_{c,2}^t,Z_{s,2}^t,\dots,Z_{c,H}^t,Z_{s,H}^t]^t \tag{42}$$

By introducing the expansion (41) into the set of Eqs. (33), collecting the terms of the same harmonic index, and neglecting the higher order harmonics, one obtains a large system of $(2H+1)\times N_e$ equations for the unknown vector U,

$$\omega M(U)=C+L(U)+Q(U,U) \tag{43}$$

The new operators $M(.)$, C, $L(.)$, and $Q(.,.)$ that apply to U depend only on the operators $m(.)$, c, $l(.)$ and $q(.,.)$ of Eq. (33) and on the number of harmonics H. We refer to Cochelin and Vergez (2009) for the explicit formulas giving $M(.)$, $L(.)$ and $Q(.,.)$ as functions of $m(.)$, c, $l(.)$ and $q(.,.)$. The final system (43) contains $(2H+1)\times N_e$ equations for the $(2H+1)\times N_e$ unknowns U plus the angular frequency ω and the continuation parameter λ. Since the original system (31) is autonomous, a phase condition has to be added to Eq. (43) to define a unique orbit. Indeed, the time t does not appear in Eq. (33) and if $U(t)$ is a solution of Eq. (43) then $U(t+\tau)$ is also a solution, for any τ. We refer here to textbooks dealing with periodic solution continuation (Seydel (1994); Doedel (2007); Nayfeh and Balachandran (1995)) for the choice of the phase condition. $Z_{s,1}=0$ is a frequently encountered additional phase condition, for example.

In the three examples presented here, we have organised the terms of equations so that the parameter λ only appears in the operators c and l. In addition, we have managed to make this dependence linear. Once again, this formulation is not limited to the three examples presented here, but it can be obtained for a very large class of dynamical systems provided suitable additional variables are introduced. We recall that the parameter λ should not appear in the operator q, for which the HBM algebra is the more complex. For example, if there is a term λu^2 in Eq. (33), it should be rewritten as λv with the auxiliary variable $v=u^2$, and put into l instead of q.

In what follows, it is therefore assumed that c and l can be written:

$$\begin{aligned} c &= c_0+\lambda c_1 \\ l(.) &= l_0(.)+\lambda l_1(.) \end{aligned} \tag{44}$$

where c_0, c_1, l_0 and l_1 are independent of λ. Under this assumption, the operators C and L simply become

$$\begin{aligned} C &= C_0+\lambda C_1 \\ L(.) &= L_0(.)+\lambda L_1(.) \end{aligned} \tag{45}$$

and the final algebraic system becomes

$$\mathbf{R}(\mathbf{U}) = \mathbf{L0} + \mathbf{L}(\mathbf{U}) + \mathbf{Q}(\mathbf{U},\mathbf{U}) \tag{46}$$

with $\mathbf{U} = [U^t, \lambda, \omega]^t$ and

$$\begin{array}{ll} \mathbf{L0} & = C_0 \\ \mathbf{L}(\mathbf{U}) & = L_0(U) + \lambda C_1 \\ \mathbf{Q}(\mathbf{U},\mathbf{U}) & = Q(U,U) + \lambda L_1(U) - \omega M(U) \end{array} \tag{47}$$

In Eq. (46), $\mathbf{L0}$ is a constant vector, $\mathbf{L}(.)$is a linear vector valued operator and $\mathbf{Q}(.,.)$ is a bilinear vector valued operator.

Implementation in MANLAB The high order harmonic balance method has been implemented in MANLAB. The user has to provide the operator $m(.)$, c_0, c_1, $l(.)$ $l_1(.)$ and $q(.,.)$ and the number of harmonics. The functions $\mathbf{L0}$, $\mathbf{L}$ and $\mathbf{Q}$, which are the actual input for the continuation are generated automatically. The examples presented in this paper are available online (Manlab (2012)).

3.4 The case of a periodically forced system

We now focus on periodically forced (non-autonomous) systems:

$$\dot{Y} = f(t, Y, \lambda) \tag{48}$$

where f is periodic in t, with the period T (forcing period). We look for periodic solutions (responses) with a period pT or $\frac{T}{p}$, where p is an integer. Here, the periodic forcing term should be also expanded into harmonics and resulting term should be taken into account in the balance of each harmonic. The approach is illustrated below with two further examples.

Example 4: Forced Duffing oscillator.

$$\ddot{u} + 2\mu\dot{u} + u + u^3 = f\cos(\lambda t) \tag{49}$$

We take the damping coefficient μ and the force amplitude f constant, and use the forcing angular frequency as the varying parameter λ. By using $v(t) = \dot{u}(t)$ and $w(t) = u^2(t)$, this equation can be recast as follows

$$\underbrace{\begin{array}{c} \dot{u} \\ \dot{v} \\ 0 \end{array}}_{m(\dot{Z})} \begin{array}{c} = \\ = \\ = \end{array} \underbrace{\begin{array}{c} \\ f\cos(\lambda t) \\ 0 \end{array}}_{c(t,\lambda)} \underbrace{\begin{array}{c} v \\ -2\mu v - u \\ w \end{array}}_{l(Z)} \underbrace{\begin{array}{c} \\ -uw \\ -uu \end{array}}_{q(Z,Z)} \tag{50}$$

where $Z = [u, v, w]^t$, and the forcing term is deliberately put into c. The forcing frequency is now related to the response frequency by putting $\lambda = \omega$ or possibly, $\lambda = p\omega$ (p is an integer). The term $c(t)$ is then expanded into harmonics with respect to ω.

This results in slight changes in the procedure:

- because of the synchronization of the response and the forcing, the phase condition has to be removed. Note that the parameter λ is no longer an unknown, since it was chosen as a multiple of ω. In comparison with the case of an autonomous system, both the number of equations and the number of unknowns have decreased by one.
- in the final system (46)(47), the operators C_1 and L_1 disappear and the forcing amplitude f has to be accounted for in L_0.

Lastly, we take the Raylegh-Plesset equation used to model large amplitude vibrations of gas bubbles in fluids. The forcing is handled slightly differently and this example also shows how to cope with a minus three power.

Example 5: The forced Rayleigh-Plesset equation (Plesset (1949)).

Let R be the radius of the vibrating bubble, and R_0 the radius at rest. The equation of motion of the bubble is

$$R\ddot{R} + \frac{3}{2}\dot{R}^2 = A\{(\frac{R_0}{R})^3 - 1\} + B\cos(\lambda t) \tag{51}$$

where A and B are fixed. We introduce the normalized radius $u = \frac{R}{R_0}$ and the (normalized) velocity $v = \frac{\dot{R}}{R_0}$. Dividing Eq (51) by R_0^2 and defining $a = \frac{A}{R_0^2}$, $b = \frac{B}{R_0^2}$, we get the first order system

$$\begin{aligned} \dot{u} &= v \\ u\dot{v} &= a(u^{-3} - 1) - \tfrac{3}{2}v^2 + b\cos(\lambda t) \end{aligned} \tag{52}$$

We now introduce the following auxiliary variables $x = \frac{1}{u}$, $y = x^2$, and $z = v^2$, and get

$$\begin{aligned} \dot{u} &= v \\ \dot{v} &= a(y^2 - x) - \tfrac{3}{2}xz + bx\cos(\lambda t) \end{aligned} \tag{53}$$

After undergoing this transformation, the system is quadratic but the forcing term has now been multiplied by the unknown function x, and it is no longer a constant term with respect to the unknown. We introduce another auxilliary variable $r(t) = \cos(\lambda t)$, and replace the term $bx\cos(\lambda t)$ by bxr. Finally, we obtain the following system, where the first two equations stand for (53) and the last four equations define the auxilliary variables.

$$\begin{array}{ccccccccc}
\dot{u} & = & 0 & + & v & + & \\
\dot{v} & = & 0 & + & (-ax) & + & ay^2 - \frac{3}{2}xz + bxr \\
0 & = & 1 & + & 0 & + & (-ux) \\
0 & = & 0 & + & y & + & (-xx) \\
0 & = & 0 & + & z & + & (-vv) \\
\underbrace{0}_{m(\dot{Z})} & = & \underbrace{-\cos(\lambda t)}_{c(t)} & + & \underbrace{r}_{l(Z)} & + & \underbrace{\qquad\qquad}_{q(Z,Z)}
\end{array} \tag{54}$$

Here the unknown vector is $Z = [u, v, x, y, z, r]^t$, and the forcing term is clearly present in the operator c.

3.5 The case of conservative nonlinear normal modes

The toy models that we have used above belong to the class of general dissipative systems for which periodic orbits are generically isolated. The limit cycle attractor of the Van der Pol oscillator is an example of such isolated periodic orbit. In such systems, the control parameter *lambda* appears explicitely in the equation.

The situation is quite different for conservative system, i.e., systems having a first integral. We restrict here the discussion to mechanical systems having an integral correponding to the conservation of the total energy. For example, the undamped free Duffing oscillator

$$\ddot{u} + u + u^3 = 0 \tag{55}$$

for which there is conservation of $E = \frac{1}{2}\dot{u}^2 + \frac{1}{2}u^2 + \frac{1}{4}u^4$. In these systems, periodic orbits belong to one-dimensional [4] families that can be parametrized by the values of the first integral, here the value of the energy E. The difference is that the continuation parameter is not explicitely available in the equations and this causes the numerical framework presented above to fail. The solution to this problem has been given by Sepulchre and MacKay (1997) and extended by Munoz-Almaraz et al. (2003) for general conservative system. The key idea is to embed the conservative system into a more general system by adding a one-parameter dissipative term in such a way that periodic orbits can exist only for zero values of the parameter. Therefore, the dissipative and the conservative systems have the same periodic orbits.

[4] Equation (55) has a two-dimensional family of a periodic orbits, but (55) with a phase condition has a one-dimensional family of periodic orbits

For mechanical system having total energy conservation, one can simply add a linear disspation multiplied by λ. For instance,

$$\ddot{u} + \lambda \dot{u} + u + u^3 = 0 \tag{56}$$

for the Duffing example. It is obvious that because the energy is monotonic, here increasing if $\lambda < 0$ and decreasing if $\lambda > 0$, a periodic solution can exist only if $\lambda = 0$. A periodic solution of (56) with $\lambda = 0$ is also a periodic orbit for (55). The advantage of (56) over (55) is that the continuation parameter is explicit and the system can be treated numerically with the same methodology as for the five examples presented above. Notice that during the continuation, the control parameter λ remains equal to zero.

NNMs of a two d.o.f. spring-mass example Let us consider the two d.o.f. spring mass system already used by Cyril Touze in this book.

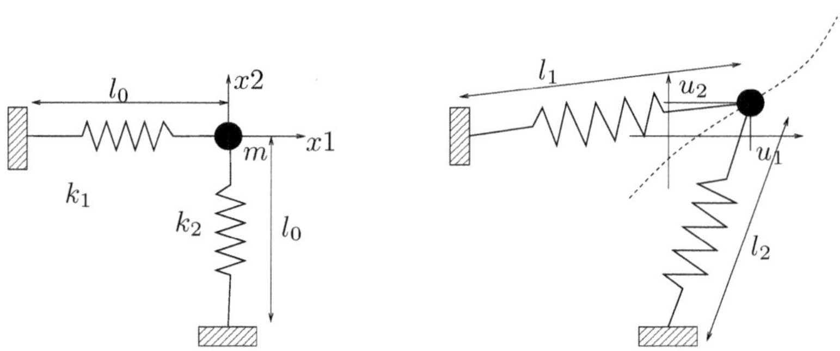

Figure 8. A two d.o.f. spring mass system

Let $u_1(t)$ and $u_2(t)$ be the (dimensionless) displacements, $e_1 = u_1 + 1/2(u_1^2 + u_2^2)$ and $e_2 = u_2 + 1/2(u_1^2 + u_2^2)$ the Green-Lagrange strain, k_1 and k_2 the spring stiffnesses, $W = 1/2k_1e_1^2 + 1/2k_2e_2^2$ the strain energy and m the mass. The equation of motion reads

$$\begin{aligned} m\ddot{u}_1 + k_1e_1(1+u_1) + k_2e_2(u_1) &= 0 \\ m\ddot{u}_2 + k_1e_1(u_2) + k_2e_2(1+u_2) &= 0, \end{aligned} \tag{57}$$

and the final system which is ready for applying the method (first order, quadratic, explicit λ) is

$$
\begin{array}{lll}
\dot{u}_1 & = v_1 & \\
\dot{u}_2 & = v_2 & \\
m\dot{v}_1 & = -\lambda v_1 - k_1 e_1 & -u_1(k_1 e_1 + k_2 e_2) \\
m\dot{v}_2 & = -\lambda v_2 - k_2 e_2 & -u_2(k_1 e_1 + k_2 e_2) \\
0 & = e_1 - u_1 & -1/2(u_1^2 + u_2^2) \\
0 & = e_2 - u_2 & -1/2(u_1^2 + u_2^2).
\end{array}
\tag{58}
$$

NNMs of elastic structure Large amplitute vibrations of elastic structures can be modelized using nonlinear elasticity theory with a finite element discretization. This mainly concerns thin structures made of beams, plates, shells and trusses which undergo large displacements but small strains and small stresses. One speaks of geometrically nonlinear elastic problem. Let $\mathbf{U}$ be the vector of nodal displacement,$\mathbf{V}$ the velocity, $\mathbf{S}$ the vector of the second Piola-Kirchhoff stress component at each Gauss point, the problem can be easily formulated so as to be treated by the numerical methods presented above

$$
\begin{array}{rcl}
\dot{\mathbf{U}} & = & \mathbf{V} \\
\mathbf{M}\dot{\mathbf{V}} & = & -\int_\Omega (B_L + B_{NL}(\mathbf{U}))^t \mathbf{S} dv - \lambda \mathbf{V} \\
\mathbf{S} & = & \mathbf{D}(B_L + \frac{1}{2} B_{NL}(\mathbf{U}))\mathbf{U}
\end{array}
\tag{59}
$$

where $\mathbf{M}$ is the mass matrix, $\mathbf{D}$ is the Hooke operator, B_L and B_{NL} are the linear and nonlinear strain-displacement operators, see Crisfield (1991) for instance. Under this form, the governing equations consist in a group of first order quadratic differential equations (velocity definition, equations of motion) and a group of algebraic quadratic equations (constitutive law at finite element's Gauss point).

3.6 Stability

Once a periodic solution has been computed, its stability can be analyzed by using any classical method described in text-book, for instance in Nayfeh and Balachandran (1995); Seydel (1994), and we do not enter into details for the sake of brevity of the presentation. However, a purely harmonic-based stability analysis has been developped by A. Lazarus and O. Thomas as an ideal companion to the purely harmonic based method presented in this chapter. The method is described in Lazarus and Thomas (2010) and it has been implemented in the Manlab software by these authors.

3.7 Numerical results on selected examples

In the following, the numerical results obtained using HBM and ANM are either compared with time domain simulations or used to illustrate particular features: the influence of the number of harmonics or the ability to deal with bifurcation for instance.

Example 1: The Van der Pol oscillator.

Numerical results obtained on the Van der Pol oscillator (34) are presented in figure 9 for $H = 10$ harmonics and in figure 10 for $H = 50$ harmonics.

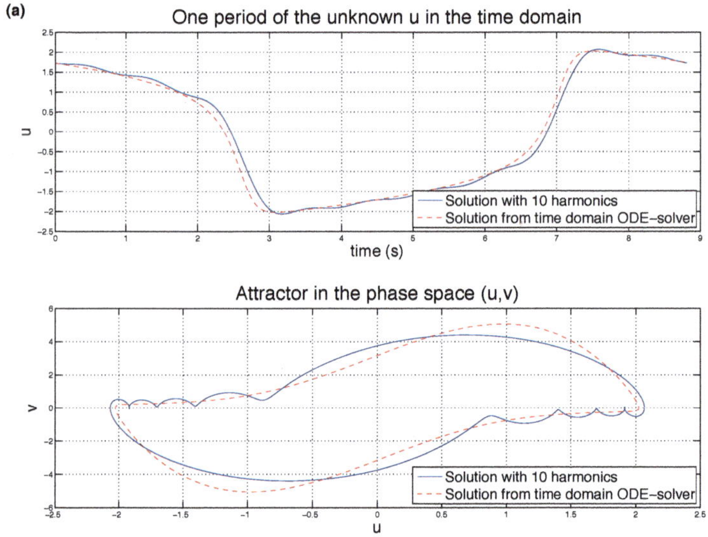

Figure 9. The Van der Pol oscillator: harmonic balance results with 10 harmonic compared to time domain simulation.

Good agreement is obtained with the results of a time domain simulation, performed using Matlab ODE solvers. As expected, the number of harmonics H in the solution sought by the harmonic balance method had to be adapted to span the bandwith of the solution obtained by direct integration. We can see that for $\lambda = 3$, the value for this test, a relatively high number of harmonics H is required to match the reference solution.

Example 2: The Rössler system.

The ability to follow period-doubling bifurcations is illustrated in figure

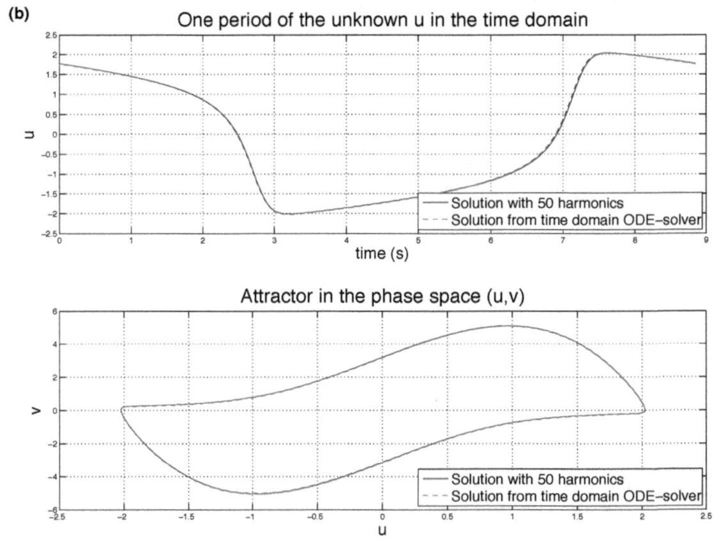

Figure 10. The Van der Pol oscillator: harmonic balance results with 50 harmonics compared to time domain simulation.

11. The strategy used here therefore consists in introducing subharmonic amplitudes as additional unknowns, i.e., the fundamental oscillation is not on the first harmonic. To detect K period-doubling bifurcations, the solution is then sought as:

$$Z(t) = Z_0 + \sum_{k=1}^{H} Z_{c,k} \cos(\frac{k}{2K}\omega t) + \sum_{k=1}^{H} Z_{s,k} \sin(\frac{k}{2K}\omega t) \qquad (60)$$

When the solution belongs to the T-periodic solution branch, $Z_{c,k}|_{k=1..K-1} = 0$ and $Z_{s,k}|_{k=1..K-1} = 0$. When the solution belongs to the $2T$-periodic solution branch, $Z_{c,k}|_{k=1..K-1,k\neq 2^{K-1}} = 0$ and $Z_{s,k}|_{k=1..K-1,k\neq 2^{K-1}} = 0$. One practical consequence in terms of the computational cost is that in order to span the same bandwidth, the number of harmonics has to be mulitplied by 2^K. This case is illustrated in figure 11 where $H = 10$ from the T-periodic solution to $H = 2^2 \times 10 = 40$ after two period-doubling bifurcations.These figure show the ability of the method to follow the sub-harmonic cascade from (a) $\lambda = 2.5$ (period T), to (b) $\lambda = 3.5$ (period $2T$) and (c) $\lambda = 4$ (period $4T$).

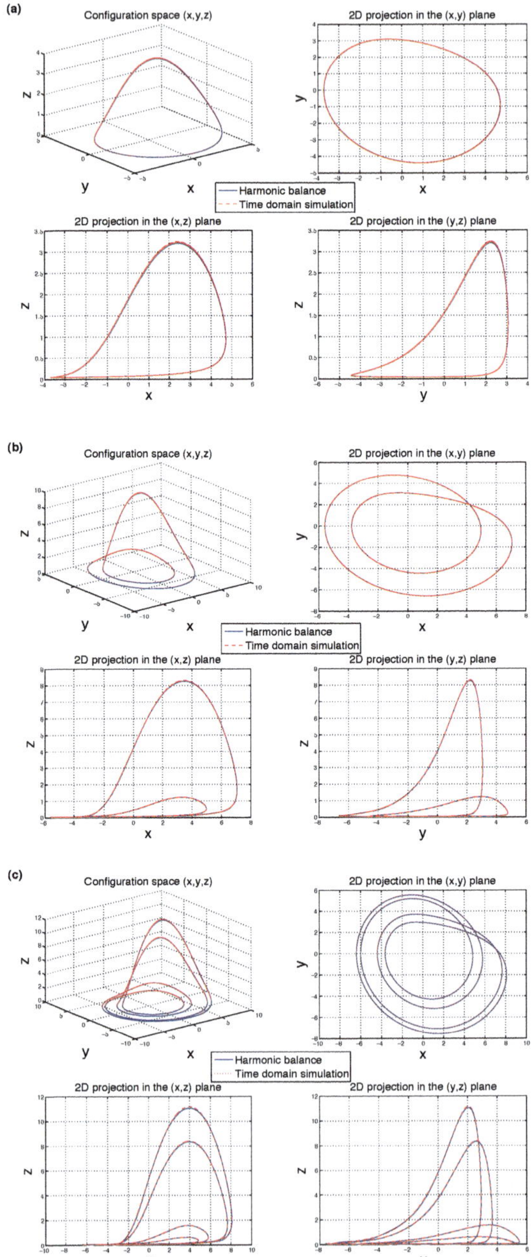

Figure 11. The Rössler system (Eq. (37) with $a = b = 0.2$): Comparison between harmonic balance method and time domain simulation.

Example 3: The clarinet model.

The following results are due to Sami Karkar Karkar (2012). Various periodic regimes of a clarinet musical instrument model have been computed by using and developping the MANLAB software. The bifurcation parameter is here the blowing pressure inside the mouth of the player normalized by the pressure for which the flow is zero. Figure 12 shows the stationary solution of the model with the position of the Hopf bifurcations.

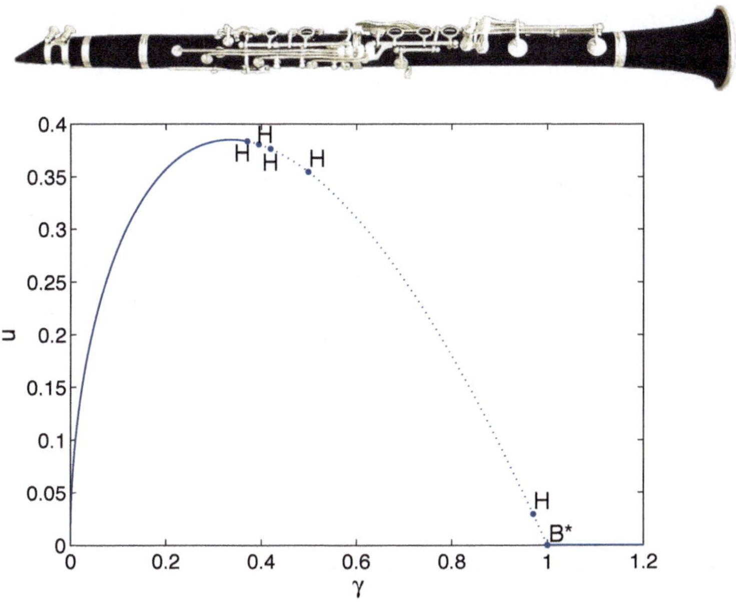

Figure 12. Stationary solution: flow u through the clarinet versus the (dimensionless) blowing pressure γ. Thick line: stable stationary solution. Dot line: unstable stationary solution. Mark H: Hopf bifurcations .

The branch of periodic solutions emerging from the first hopf bifurcation is presented in figure 13 . Various characteristics of the clarinet instrument are retrieved :

- for $\gamma = 0.505$, the reed begins to impact the mouthpiece during the period. This results in a change of the slope in the diagram.
- the saturation threshold is obtained for $\gamma = 1.742$ (maximal energy pressure inside the mouthpiece).
- the sound cut-off occurs here for $\gamma = 1.914$

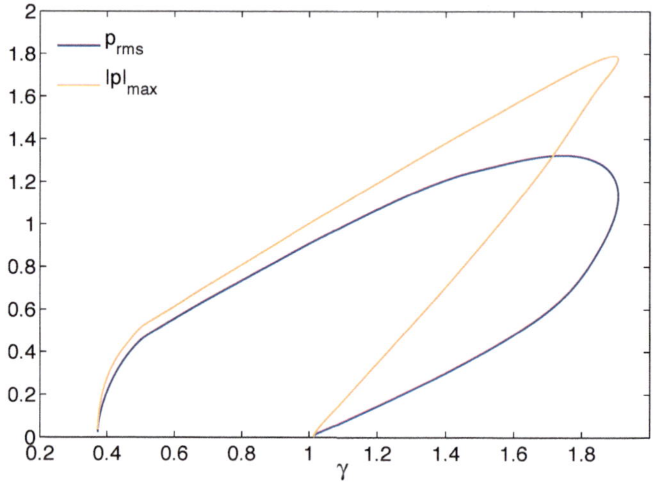

Figure 13. Branch of periodic solution emerging from the first Hopf bifurcation (25 Harmonics). Total pressure value at the end of the resonator versus the blowing pressure γ (blue : rms value, red : max value)

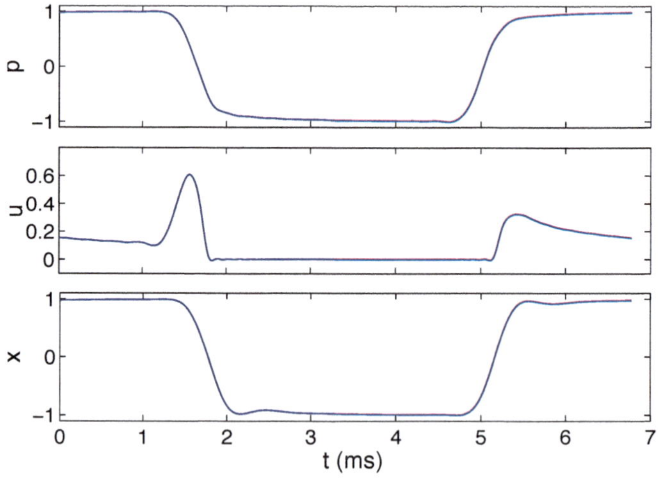

Figure 14. Time serie for the pressure at the end of the resonator $p(t)$, the flow $u(t)$, and the motion of the reed $x(t)$. Periodic solution for $\gamma = 1$ with 25 harmonics.

Example 5: Forced Rayleigh-Plesset equation

The softening effect of the Rayleigh-Plesset bubble model is illustrated in figure 15 showing the backbone curve of forced response. Convergence with respect to harmonic number is highlighted by using from 2 to 6 harmonics. We refer to Pauzin et al. (2011) for a detailed analysis of the model and results.

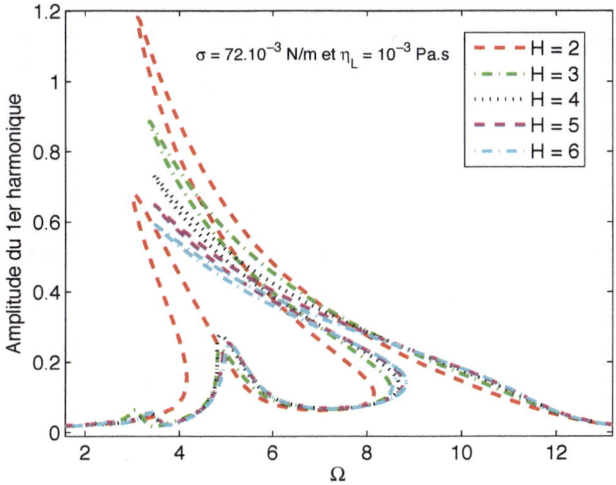

Figure 15. Rayleigh Plesset bubble model: amplitude of the first harmonics versus forcing frequency Ω .

Example 6: NNMs of the 2 d.o.f. spring mass system

Figure 16 shows a bifurcation diagram and a screenshot of the MANLAB software during the computation of NNMs for the 2 d.o.f. spring mass system. One can recognize the Rosenberg modal line in the phase space which looks like the ones presented by Cyril Touzé in this book. Due to many possible subharmonic resonances and internal resonances between the two modes, the bifurcation diagram is very elaborate. Only some branches are shown in the figure.

More example

Other periodic solution continuation examples using Manlab can be found in Kacem et al. (2011); Lazars et al. (2012) dealing with MEMS.

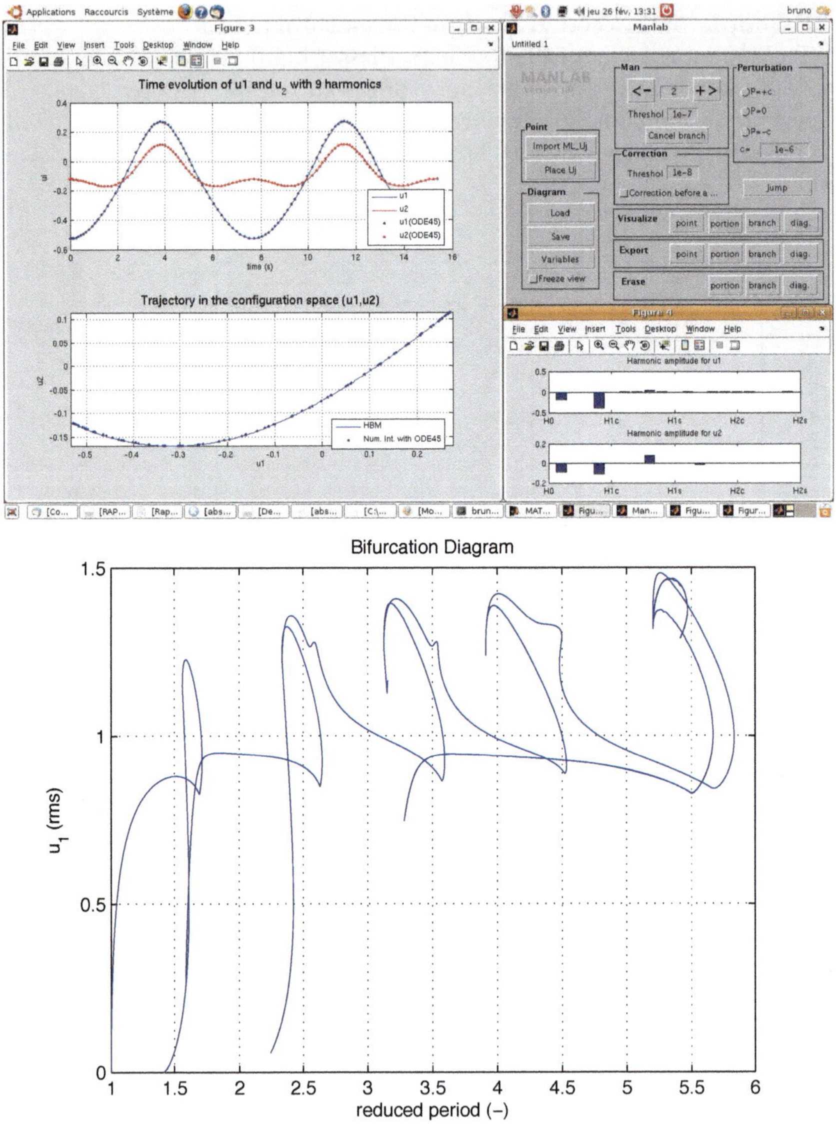

Figure 16. (top) Screenshot of the MANLAB-1.0 software during the computation of the 2 d.o.f. spring mass system NNMs; (bottom) Bifurcation diagram: displacement u_1 versus the reduced period. Branches of periodic solution computed from the first linear mode. Eventhough the system is simple, the diagram of periodic solutions is very elaborate due to many bifurcations. Only some branches are represented in this diagram.

4 Application : NNMs of elastic structures with contactors

This section addresses the computation of NNMs for linear elastic structures having localized nonsmooth nonlinearities. Typical applications are concerned with thin structures such as beams and plates whose vibration amplitudes are limited by localized elastic contactors.

As compared to the other models used in this presentation, the challenge is here that the nonlinearity is nonsmooth. Another difficulty is that the finite element model for the elactic structure may be of a large size. An improved version of the HBM+ANM for large size systems with localized nonsmooth nonlinearities has been developped in Moussi (2013).

- the nonsmooth contact law are regularized and written under quadratic polynomial form.
- the computing time is greatly optimized. For instance, an FFT/FFT^{-1} process is used for reporting heavy frequency computation in the time domain where these computations are much faster.
- a different harmonic truncature is used for the displacement (rather smooth) and for the nonlinear forces (not smooth). This permits to significantly reduce the size of the algebraic system provided by the HBM.

Notice that for systems having a small number of nonlinear equations and a large number of linear equations, dynamic model reduction could be efficiently used. A general presentation of the HBM method that includes such model redution can be found for instance in Sarrouy and Sinou (2011). No reduction has not been considered in the following.

4.1 NNMS of a bar with an unilateral contactor

We consider an elastic bar which is clamped at one end and restrained at the other end by an unilateral contactor made of a linear spring and a gap.

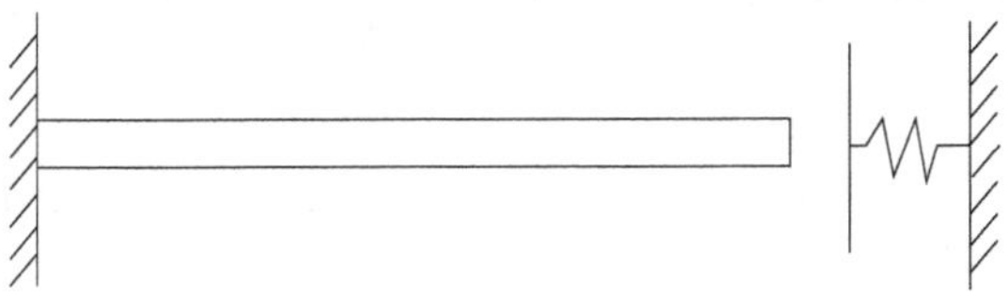

Figure 17. Clamped-free bar with an unilateral contactor at one end.

The governing equations for the axial displacement $u(x,t)$ are

$$\begin{aligned}
&\rho A \frac{\partial^2 u}{\partial t^2}(x,t) - EA\frac{\partial^2 u}{\partial x^2}(x,t) = 0 \qquad x \in]0,L[\\
&u(0,t) = 0 \\
&EA\frac{\partial u}{\partial x}(L,t) = -f(t)
\end{aligned} \tag{61}$$

where ρ is the density , E is the Young modulus and A the cross section area. The end $x = L$ of the bar is limited by an elastic contactor with a gap g and a stiffness α. This contactor applies a piecewise linear force on the beam $-f(t)$ given by

$$\begin{cases} f(t) = 0 & \text{if} \quad u(L,t) < g \\ f(t) = \alpha(u(L,t) - g) & \text{if} \quad u(L,t) > g. \end{cases} \tag{62}$$

In order to apply HBM and ANM methods, the piecewise linear contact law has to be regularized. We assume that $f(t)$ and $u(L,t)$ are related by the polynomial and quadratic expression

$$f(t)(f(t) - \alpha(u(L,t) - g)) = \alpha\eta \tag{63}$$

where η is a small parameter. For $\eta = 0$, the solution to (63) is either $f = 0$ or $f = \alpha(u - g)$. For η very small, (63) has two solutions, one of which is very closed to "exact" law (62) and the other one being non physical.

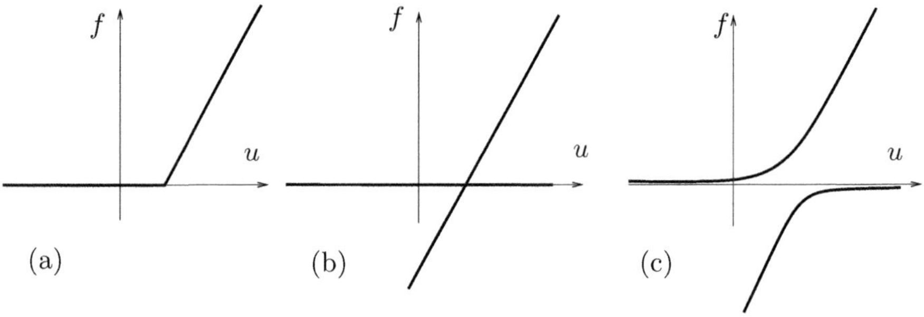

Figure 18. Contact law of the elastic contactor: (a) exact law. (b) solution of (63) with $\eta = 0$. (c) solution of (63) with $\eta > 0$ and small.

Classical 2 d.o.f. finite elements with linear displacements are used for the spatial discretization

$$\boldsymbol{M}_e = \frac{\rho A l_e}{6}\begin{pmatrix} 2 & 1 \\ 1 & 2 \end{pmatrix} \text{ et } \boldsymbol{K}_e = \frac{EA}{l_e}\begin{pmatrix} 1 & -1 \\ -1 & 1 \end{pmatrix}, l_e = \frac{L}{n} \tag{64}$$

The final system to be treated by HBM and ANM method is

$$\begin{cases} \mathbf{0} & = \boldsymbol{M}\ddot{\boldsymbol{U}}(t) + \lambda\dot{\boldsymbol{U}}(t) + \boldsymbol{K}\boldsymbol{U}(t) + \begin{pmatrix} \mathbf{0} \\ \boldsymbol{f}(t) \end{pmatrix} \\ 0 & = f(t)(f(t) - \alpha(U_L(t) - g)) - \alpha\eta \end{cases} \tag{65}$$

In the following, we use $\rho = 7800$ kg/m^3, $E = 2.1 \times 10^{11}$ Pa, $L = 1$ m, $A = 0.03^2$ m^2 , $g = 10^{-4}$ m, $\alpha = 810^7$ N/m, $\eta = 2.510^{-7}$. The first three linear frequencies of the beam are then $f_1 = 1297.5$ Hz , $f_2 = 3900.6$ Hz and $f_3 = 6527.7$ Hz.

4.2 First nonlinear mode

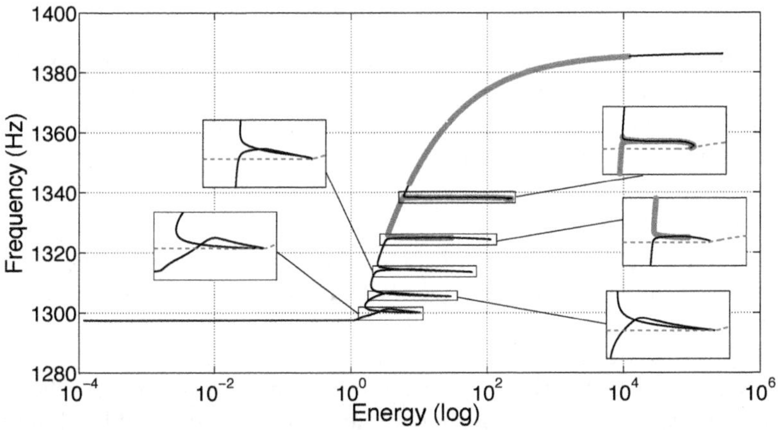

Figure 19. Frequency-energy plot of the first NNM for the bar with an elastic end stop. Number of harmonics $H_u = 11$ for the dsiplacement.

The frequency-energy plot of the first NNM is presented in figure 19 using 11 harmonics for the displacement and 151 harmonics for the force. For low energy level, the bar does not enter in contact with the contactor and the NNM looks like the first linear mode with a constant frequenecy and a fixed modal shape. This corresponds to the horizontal line at $f = 1297$ Hz. For energy level between 1 and 100, the frequency increases from 1297 Hz to around 1370 Hz due to the stiffening effect provided by the contactor. One can observe, five "tongues" on the main curve which corespond to five internal resonances $3:1$, $5:1$, $7:1$, $9:1$ and $11:1$ respectively with the NNMs 2, 3, 4, 5 and 6.

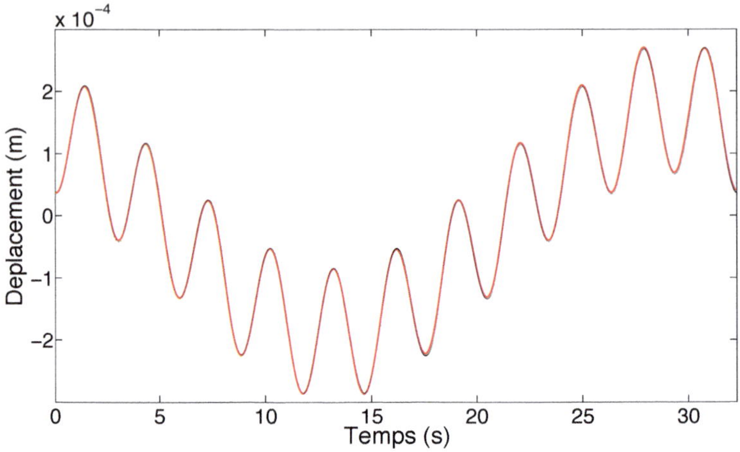

Figure 20. Time serie on a period for the end displacement $u(L,t)$ of the bar. Periodic solution $E = 144.06$, $f = 1338.3$ correpond to a stable solution on the 11:1 tongue. HBM+ANM solution (in black) compared to time integration solution (in red).

The HBM solution with 11 harmonics for the displacement is compared to a reference solution (time integration) on figure 20 showing a very good agreement.

Finally, the frequency-energy plot obtained by using only 1 or 3 or 7 harmonics instead of 11 for the displacement is shown on figure 21. The main branch remains the same with almost the same asymptot. The difference is the presence or not of the tongues corresponding to internal resonances. For example, with 3 harmonis only the first 3:1 tongue is described, and with 7 only the first three one $3:1$, $5:1$, $7:1$.

One can see that the use of a small number of harmonics on the displacement permits to limit the number of internal resonances and greatly simplifies the frequency-energy plot.

4.3 Beam with a bilateral contactor

We consider a clamped-simply supported beam with a bilateral contactor as shown in figure 22. This system is modeled by the following boundary value problem:

$$\rho A \frac{\partial^2 u}{\partial t^2}(x,t) + EI\frac{\partial^4 u}{\partial x^4}(x,t) + \delta_{x_s}(x) f_{nl}(u(x_s,t)) = 0, \ \forall x \in [0,L], \quad (66)$$

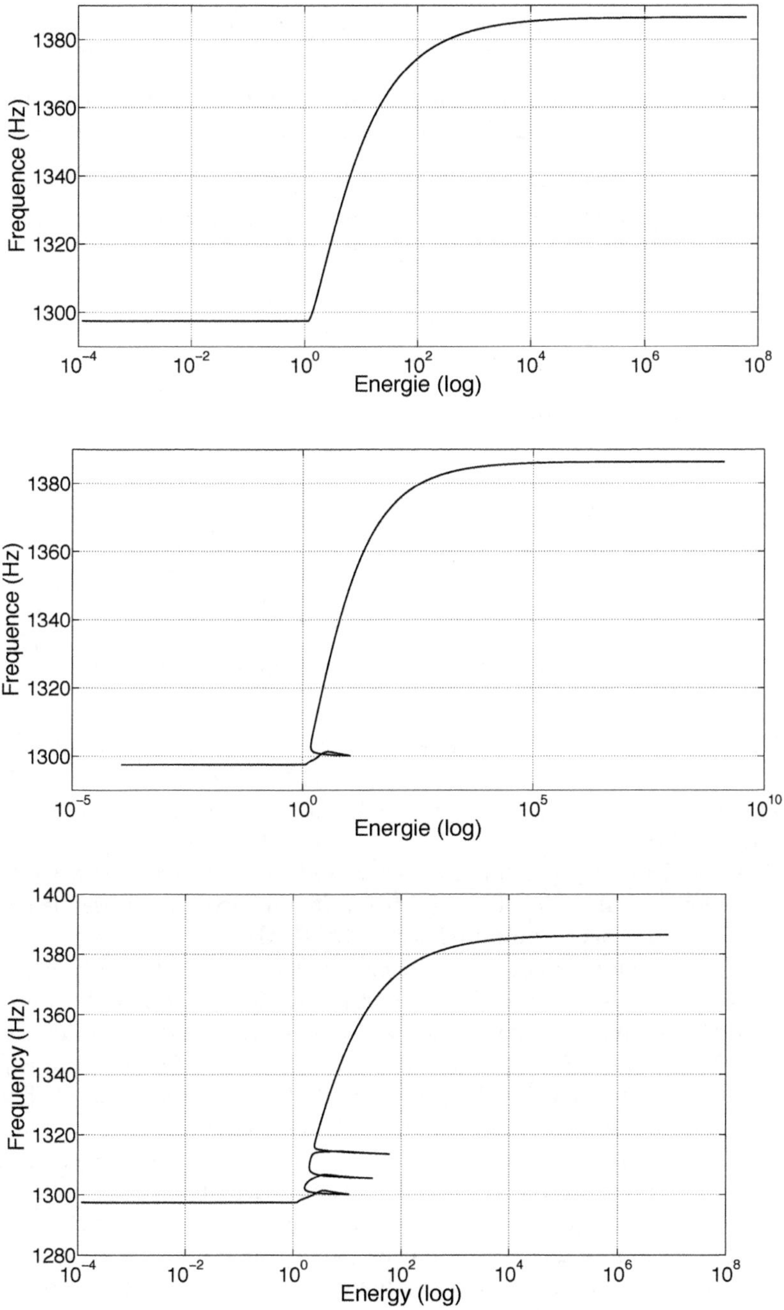

Figure 21. Frequency-energy plot of the first NNM. $H_u = 1$ (top), $H_u = 3$ (middle) and $H_u = 7$ (bottom).

$$u(0,t) = 0, \ \frac{\partial u}{\partial x}(0,t) = 0, \ u(L,t) = 0, \ \frac{\partial^2 u}{\partial x^2}(L,t) = 0, \tag{67}$$

where I denotes the moment of inertia and x_s the position of the contactor.

The localized nonlinear force f is defined by

$$\begin{cases} f(t) = \alpha(w(x_s) + g) & \text{if} \quad w(x_s) < -g. \\ f(t) = 0 & \text{if} \quad -g < w(x_s) < g \\ f(t) = \alpha(w(x_s) - g) & \text{if} \quad w(x_s) > g. \end{cases} \tag{68}$$

and regularized as

$$f(t)(f(t) - \alpha(w(x_s) + g))(f(t) - \alpha(w(x_s) - g)) = \alpha^2 \eta w(x_s) \tag{69}$$

where η is a small parameter.

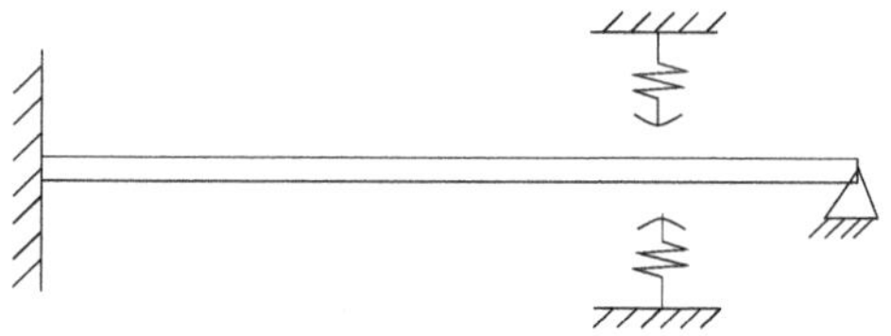

Figure 22. A clamped-simply supported beam with a bilateral contactor.

The following numerical values have been used: $\rho = 8357$ kg/m^3, $E = 2.06 \times 10^{11}$ Pa, $A = \pi r^2$ m^2, $I = \frac{\pi}{64} \times ((0.02222^4 - 0.00127^4)$ m^4, $r = 11.11 \times 10^{-3}$ m, $L = 1.59$ m, $x_s = 1.3515$ m, $\alpha = 8 \times 10^5$ N/m, $g = 0.00029$ m and $\eta = 0.005$.

A finite element procedure has been used to discretize (66) with 20 Hermite elements giving $n = 40$ d.o.f. with only one involving possible contact. The NNMs have been computed with MANLAB.

The four first resonance frequencies of the clamped-simply supported beam are $f_1 = 26.817$ Hz, $f_2 = 86.907$ Hz, $f_3 = 181.330$ Hz, and $f_4 = 310.113$ Hz. Using the associated normal modes as starting point, the four first NNMs were computed using the following orders of truncation $H_u = 11$ for the d.o.f not involved in the contact and $H_f = 201$ for the d.o.f. involved in the bilateral contact. The behavior of the first NNM is shown in figure 23 in terms of frequency energy plot. At low energy level, the frequency is constant corresponding to modal motions without contact. When contact occurs, the frequency increases with the energy showing four tongues associated to four internal resonances (square markers). The first tongue corresponds to a 11 : 1 internal resonance between the fourth and the first NNMs. The fourth NNM is also reported (dashed curve) using the scale

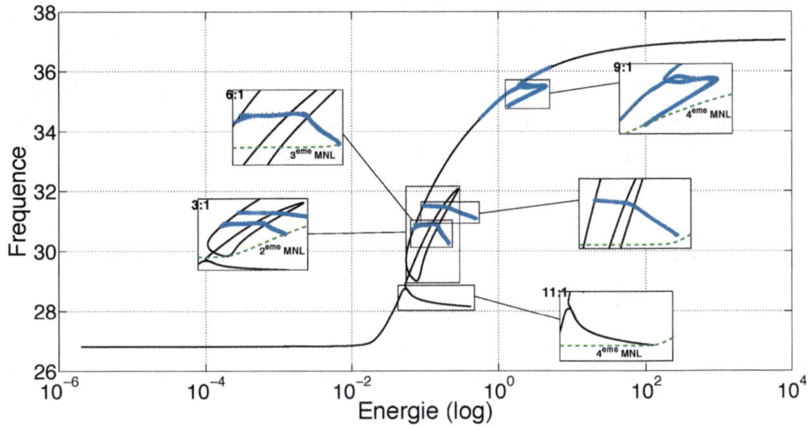

Figure 23. FEP of the first NNM (continuous line). The square markets correspond to internal resonances.

factor $\frac{1}{11}$ in frequency axis. For this NNM, the frequency is also constant $(=\frac{f_4}{11})$ at low energy level. The second tongue corresponds to a $3:1$ internal resonance between the second and the first NNMs and so on. At high energy level, the frequency of the first NNM increases up to a resonance frequency of the linear system associated to a nul gap $(g=0)$.

4.4 Conclusion

The two academic examples described in this section show that the combination of HBM and ANM is operant for the computation of NNMs of elastic structures having localized nonsmooth nonlinearities, provided that the contact law are regularized. This result is not evident since the use of a Fourier serie is probably not the best way to represent a nonsmooth function, eventhough it is regularized. However, as it was already claimed in Karkar et al. (2013), the continuation is robust provided that a large number of harmonic is used, up to 1000 for instance in Karkar et al. (2013).

The vibro-impact of the U-tube of nuclear power plant steam generator is adressed in Moussi (2013) with the same methodoly as for the bar and the beam example. The stiffness of contactor is set to high values to modelize the impact between the tube and the supporting plate. Accordingly, a high number of harmonics is used to capture the nonsmooth response due to impact. The computation of the NNMs with the internal resonance tongues explain particular vibrating regimes which are in line with experimental results conducted by EdF R&D compagny. These regimes are of course not predicted by the linear theory.

This last industrial experience confirms, if needed, that the definition and the calculation of NNM is not only an academic topic for researchers, but it becomes progressively an effective tool to describe and explain the behaviour of practical applications.

Bibliography

E. L. Allgower and K. Georg. *Numerical Continuation Methods: an Introduction.* Springer-Verlag, New-York, 1990.

R. Arquier. *Calcul des orbites priodiques en mcanique : applications aux modes de vibrations non linaires de structures.* Phd thesis, , Aix-Marseille University, 2007.

L. Azrar, B. Cochelin, N. Damil, and M. Potier-Ferry. An Asymptotic Numerical Method to compute the post-buckling behaviour of elastic plates and shells. *International Journal for Numerical Methods in Engineering*, 36:1251–1277, 1993.

I. Charpentier. On higher-order differentiation in nonlinear mechanics. *Optimization Methods and Software*, 27:221–232, 2012.

I. Charpentier, A. Lejeune, and M. Potier-Ferry. *The Diamant approach for an efficient Automatic Differentiation of the Asymptotic Numerical Method*, volume 64 of *Lecture Note in Computational Science and Engineering*, pages 139–149. Springer, 2008.

B. Cochelin. A path following technique via an Asymptotic Numerical Method. *Computers and Structures*, 53:1181–1192, 1994.

B. Cochelin and M. Médale. Power series analysis as a major breakthrough to improve the efficiency of asymptotic numerical method in the vicinity of bifurcations. *Journal of Computational Physics*, 236:594–607, 2013.

B. Cochelin and C. Vergez. A high order purely frequency-based harmonic balance formulation for continuation of periodic solutions. *Journal of Sound and Vibration*, 324:243–262, 2009.

M. Crisfield. *Non-linear finite element analysis of solids and structures.* John Wiley and Sons, New York, 1991.

N. Damil and M. Potier-Ferry. A new method to compute perturbed bifurcations: application to the buckling of imperfect elastic structures. *International Journal of Engineering Sciences*, 28:943–957, 1990.

E. Doedel. Lecture notes on numerical analysis of nonlinear equations. In B. Krauskopf, H.M. Osinga, and J. Galan-Vioque, editors, *Numerical Continuation Methods for dynamical systems*, pages 1–49. Springer, 2007.

E. Doedel, H. B. Keller, and J. P. Kernevez. Numerical analysis and control of bifurcation problems (i) bifurcation in finite dimensions. *Int. J. Bifurcation and Chaos*, 3:493–520, 1991.

K. Ekici, K.C. Hall, and E.H. Dowell. Computationally fast harmonic balance methods for unsteady aerodynamic predictions of helicopter rotors. *Journal of Computational Physics*, 227:6206–6225, 2008.

A. Grolet and F. Thouverez. On a new harmonic selection technique for harmonic balance method. *Mechanics Systems and Signal Processing*, 30:43–60, 2012.

M. Guskov, J-J. Sinou, and F. Thouverez. Multi-dimensional harmonic balance applied to rotor dynamics. *Mechanics Research Communications*, 35:537–545, 2008.

N. Kacem, S. Baguet, S. Hentz, and R. Dufour. Computational and quasi-analytical models for non-linear vibrations of resonant mems and nems sensors. *International Journal of Non-Linear Mechanics*, 46(3):532 – 542, 2011.

S. Karkar. *Mthodes numriques pour les systmes dynamiques non linaires : application aux instruments de musique auto-oscillants.* Phd thesis, Aix-Marseille University, 2012.

S. Karkar, C. Vergez, and B. Cochelin. Oscillation threshold of a clarinet model : a numerical continuation approach. *Journal of Acoustical Society of America*, 131:698–707, 2012.

S. Karkar, B. Cochelin, and C. Vergez. A high-order purely frequency based harmonic balance formulation for continuation of periodic solutions: The case of non-polynomial nonlinearities. *Journal of Sound and Vibration*, 332:968–977, 2013.

H.B. Keller. *Lectures on numerical methods in bifurcation problems.* Springer-Verlag, Berlin, 1987.

M. Krack, L. Panning von Scheidt, and J. Wallaschek. A high-order harmonic balance method for systems with distinct states. *Journal of Sound and Vibration*, 332:5476–5488, 2013.

Krylov and Bogoliubov. *Introduction to Nonlinear Mechanics, English translation of Russian edition from 1936.* Princeton University Press, New Jersey, 1947.

S.L. Lau and Y.K. Cheung. Amplitude incremental variational principle for nonlinear vibration of elastic systems. *ASME Journal of Applied Mechanics*, 28:959–964, 1981.

A. Lazars, O. Thomas, and J.F. Deü. Finite elementreduced order models for nonlinear vibrations of piezoelectric layered beams with applications to nems. *Finite Elements in Analysis and Design*, 49:35–51, 2012.

A. Lazarus and O. Thomas. A harmonic-based method for computing the stability of periodic solutions of dynamical systems. *Comptes Rendus Mécaniques*, 338:510–517, 2010.

Wu X.X. Ling, F.H. Fast galerkin method and its application to determine periodic solutions of non-linear oscillators. *International Journal of Non-linear Mechanics*, 22:89–98, 1987.

Manlab. An interactive path-following and bifurcation analysis software, http://manlab.lma.cnrs-mrs.fr/, 2012. URL `http://manlab.lma.cnrs-mrs.fr/`.

El-Hadi Moussi. *Analyse de structures vibrantes dotes de non-linarits localises jeu l'aide des modes non-linaires.* Phd thesis, Aix-Marseille University, 2013.

F. J. Munoz-Almaraz, E. Freire, J. Galn, E. Doedel, and A. Vanderbauwhede. Continuation of periodic orbits in conservative and hamiltonian systems. *Physica D: Nonlinear Phenomena*, 181(1-2):1 – 38, 2003.

A. H. Nayfeh and B. Balachandran. *Applied Nonlinear Dynamics: Analytical Computational and Experimental Method.* Wiley, 1995.

Marie-Christine Pauzin, Serge Mensah, Bruno Cochelin, and Jean-Pierre Lefebvre. High order harmonic balance formulation of free and encapsulated microbubbles. *Journal of Sound and Vibration*, 330(5):987 – 1004, 2011.

Plesset. The dynamics of cavitation bubbles. *Journal of Applied Mechanics*, 16:277–28, 1949.

E. Sarrouy and J.J. Sinou. *Non-linear periodic and quasi-periodic vibrations in mechanical systems - On the use of the harmonic balance methods*, pages 419–434. Advances in Vibration Analysis Research. Intech, 2011.

J.A. Sepulchre and R. MacKay. Localized oscillations in conservative or dissipative nteworwks of weakly coupled autonomous oscillators. *Nonlinearity*, 10:679–713, 1997.

R. Seydel. *Practical bifurcation and stability analysis.* Springer-Verlag, 1994.

F. Silva, J. Kergomard, C. Vegez, and J. Gilbert. Interaction of reed and acoustic resonator in clarinet-like systems. *Journal of Acoustical Society of America*, 124:3284–3295, 2008.

M. Urabe. Galerkin's procedure for nonlinear periodic systems. *Archives for Rational Mechanics and Analysis*, 20:120–152, 1965.

Elements of Nonlinear System Identification of Broad Applicability

A.F. Vakakis
Department of Mechanical Science and Engineering
University of Illinois at Urbana-Champaign
Urbana, USA

1 Introduction

Linear modal analysis in the time and frequency domains is well established. Yet, as mechanical systems become increasingly more complex, incorporating electromechanical components or biological and biomimetic elements, the likelihood exists that their dynamics will be strongly nonlinear and nonstationary. Examples of sources of strong nonlinearity, some of which are realized even for small amplitudes of vibration) include local buckling, plastic deformations, clearance and backlash, hysteresis, friction-induced oscillations, and vibro-impact motions. Such effects cannot be accurately addressed by linear modal analysis, since standard techniques, such as the classical FT, and well-established concepts such as normal mode, natural frequency, and modal space cannot be applied to the identification of non-linear and nonstationary dynamic regimes.

In this section we discuss the basic elements of a nonlinear system identification (NSI) methodology with promise of broad applicability that has the potential to overcome such limitations of linear modal analysis. Ideally, such an NSI methodology would be as utilitarian as experimental linear modal analysis, and would be applicable to a broad class of nonlinear systems. Moreover, desirable attributes of this methodology would be to be,

- applicable to direct analysis of measured time series
- based on a solid physics-based, theoretical foundation
- able to identify complex nonlinear resonance interactions, and provide interpretation and modeling of such interactions
- able to address the dependence of the nonlinear dynamics on energy, type of excitation, and, perhaps, more important, initial conditions
- considered as natural extension of classical linear modal analysis, extending, if possible well-established linear concepts to the nonlinear regime.

G. Kerschen (Ed.), *Modal Analysis of Nonlinear Mechanical Systems*, CISM International Centre for Mechanical Sciences DOI 10.1007/ 978-3-7091-1791-0_7 © CISM Udine 2014

This is a very demanding list, indeed, given the highly individualistic nature of nonlinear dynamical systems which restricts the unifying dynamical features that are amenable to system identification. This is further highlighted by the fact that existing NSI methods (1) are either applicable to relatively simple systems (e.g., one or two-DOF), or are computationally intensive and, hence, difficult to implement in practical settings. Moreover, no general methods exist for systems with non-smooth nonlinearities (e.g., clearances and dry friction) or with strong nonlinearities (e.g., ultra-flexible wings where geometric nonlinearities are prominent).

The proposed NSI methodology can potentially address these challenges. After reviewing the basic analytical, computational and post-processing components of the proposed NSI methodology, and its local and global aspects we provide some examples of applications to nonlinear mechanical systems.

2 Slow flows and empirical mode decomposition

The basic elements of the NSI methodology have been presented elsewhere (2; 3; 4; 5; 6; 7; 8; 9), so here we provide only a brief overview of its most important components and refer the reader to these references for more technical details. The methodology adopts an integrated global / local approach to NSI, whereby global features of the dynamics are identified in the frequency-energy domain by constructing frequency-energy plots (FEPs) directly from measured data, while local features are identified by constructing appropriate local slow-flow models. An advantage of the methodology is based on direct analysis of measured time series which contain complete information of the nonlinear dynamics to be identified, and holds promise for broad applicability to a wide range of dynamical systems, including systems with strong or even non-smooth nonlinearities, as well as systems with time-varying properties. The basic elements of the methodology consists of, (i) analytical constructions of slow flow models, (ii) numerical empirical mode decompositions combined with Hilbert transforms, (iii) correspondence between the analytical and numerical parts, (iv) FEP reconstructions (global aspect of NSI), and (v) time series reconstructions (local aspect of NSI).

A slow-flow model of the dynamics of an oscillating system captures its essential (important) dynamics, after the non-essential (fast) dynamics is averaged out. This is performed by slow/fast partitions of measured time series. We consider a general n-degree-of-freedom (DOF) nonlinear dynamical system,

$$\dot{\underline{X}} = \underline{f}(\underline{X}, t),\ \underline{X} = \{\underline{x}^T\ \ \dot{\underline{x}}^T\}^T \in R^{2n},\ t \in R \qquad (1)$$

where $\underline{x}$ is an n-response and $\underline{f}$ is an $2n$-vector function (underlines denote vectors), and apply the complexification-averaging (CX-A) technique first introduced in (10) under the central assumption that the measured time series are in the form of modulated oscillatory signals. Assuming that the dynamics possesses N distinct components at well-separated frequencies $\omega_1, ..., \omega_n$ (which will be labeled fast frequencies) the response of each DOF of the system can be expressed as a summation of N independent components,

$$x_k(t) = x_k^{(1)}(t) + ... + x_k^{(N)}(t), \; k = 1, ..., n \tag{2}$$

where $x_k^{(m)}(t)$ indicates the response of the $k-th$ coordinate of (1), associated with the fast frequency ω_m with $\omega_1 > ... > \omega_n$. To each component of (2) we assign a complex variable defined as,

$$\Psi_k^{(m)}(t) = \dot{x}_k^{(m)}(t) + j\omega_m x_k^{(m)}(t) = \underbrace{\phi_k^{(m)}(t)}_{\text{Slow component}} \; \underbrace{e^{j\omega_m t}}_{\text{Fast component}} \tag{3}$$

where $j = (-1)^{1/2}$. A slow/fast partition of the dynamics in terms of a slow (complex) amplitude $\phi_k^{(m)}(t)$ and a fast oscillation $e^{j\omega_m t}$ was assumed. In essence the slow complex amplitudes are considered as slowly varying modulations of the fast oscillation at frequency ω_m. That such a partition (which is by no means unique or universal) holds is a central assumption in our methodology. Substituting (2) and (3) into (1) and performing multi-phase averaging (11) for each of the fast frequencies we construct the slow flow of (1) in the form,

$$\dot{\underline{\phi}}_k = \underline{F}_k(\underline{\phi}_1, ..., \underline{\phi}_n), \; \underline{\phi}_k \in C^N \tag{4}$$

where $\underline{\phi}_k = \{\phi_k^{(1)}, ..., \phi_k^{(N)}\}^T$, $k = 1, ..., n$ is the N-vector of complex amplitudes or modulations associated with the $k - yh$ DOF of (1). Clearly, *the number of fast frequencies N determines the dimensionality of the slow flow (4), and (in contrast to linear systems) it is not necessarily equal to the number of degrees of freedom, n, of (1).* This is due to the possibility of higher harmonics entering in the response due to bifurcations, nonlinear modal interactions or other nonlinear phenomena occurring in the dynamics.

The slow flow model (4) captures the slow evolution of the N harmonic components of (1), or the essential (important) dynamics after the fast frequencies (or un-important dynamics) are averaged out. We make the remark here that the previous slow flow formulation might seem artificial, since it is based on direct analysis of the equations of motion (1) which, in general, are not known in experimental NSI (in fact, the derivation of a reduced-order model such as (1) is one of the outcomes of NSI!). This theoretical construction, however, is important since it provides the theoretical foundation for the NSI methodology. Indeed, *even if the governing equations of motion of a measured dynamical process are not known, the underlying important dynamics is captured by a slow flow;* it follows that the construction of such a (reduced order) model directly from measured time series data through NSI will be one of the primary tasks of the current NSI methodology. The slow formulation provides the basic analytical element in the methodology and it will be complemented by a numerical part which we will now briefly discuss.

The second basic element of the NSI methodology is *Empirical Mode Decomposition (EMD)* combined with the numerical *Hilbert Transform (HT).* EMD was conceived as a numerical post- processing technique for analyzing nonstationary and nonlinear time series (12), and has been applied to numerous applications. This decomposition method is based on identifying the characteristic time scales in measured oscillatory time series in an adaptive and highly efficient way, and is especially suitable for nonlinear and nonstationary processes. EMD yields a complete and nearly (but not thoroughly) orthogonal basis of intrinsic mode functions IMFs, which are oscillatory modes embedded in the time series, each with its own characteristic time scale; the linear superposition of IMFs reconstructs the measured time series. Hence, *EMD is a multi-scale decomposition of a measured time series in terms of embedded oscillatory modes at different time scales of the dynamics.* The main loop of the EMD algorithm (the sifting algorithm) for extracting the IMFs from a signal $x(t)$ is described in (12) and (2; 3; 4; 5; 6; 7; 8; 9) (together with some extensions that improve its effectiveness), so it will not be repeated here.

By this construction process the superposition of the K leading IMFs reconstructs approximately the measured time series. However, due to the ad hoc nature of the sifting algorithm, only a subset of these IMFs is physically meaningful with the rest being of spurious nature (13). As discussed in (14; 15), however, the dominant (and physically meaningful) IMFs can be identified by comparing their instantaneous frequencies to the *wavelet transform (WT) spectra* of the original time series. The instantaneous frequencies of the dominant IMFs coincide with the dominant harmonics of the wavelet spectra. This process also identifies the dominant time scales (or frequencies) of the dynamics in the time series. It follows that EMD provides the following ad hoc numerical decomposition of the dynamical response of the $k-th$ coordinate of the general system (1) in the form,

$$x_k(t) = c_k^{(1)}(t) + ... + c_k^{(N)}(t),\ k = 1, ..., n \tag{5}$$

where $c_k^{(m)}(t)$ is the $m-th$ dominant IMF of the response $x_k(t)$, associated with the fast frequency ω_m, with $\omega_1 > ... > \omega_N$. We note at this point the similarity between the numerical construction (5) and the theoretical slow flow construction (2); this similarity is a key point in NSI methodology.

Although EMD is a powerful signal decomposition method for oscillatory measured data, it has certain deficiencies. First, it may lead to spurious (i.e., non-physically meaningful) IMFs due to the lack of orthogonality of the IMFs; as mentioned above this is addressed by identifying a set of dominant IMFs through the use of wavelet transforms of the experimental measured timeseries. Second, there are concerns regarding the frequency content of the IMFs, since in order to obtain meaningful results when applying HT to the IMFs, it is necessary that these are mono-component or, at least, narrowband (otherwise we obtain mixed-mode IMFs in the form of beat phenomena with closely spaced frequencies). Related to this, there are issues concerning the uniqueness of the EMD results; EMD does not result in a unique decomposition of a measured time series since it is applied in an ad hoc manner and depends on a free stopping parameter; that is, EMD is not robust in practice. The set of extracted IMFs can be considered as a basis for reconstructing the original measured time series if it satisfies (or nearly satisfies) the basic conditions of completeness and orthogonality. By virtue of the EMD algorithm, completeness of the IMFs is guaranteed by construction. It is the lack of orthogonality between IMFs, however, that generates spurious features in the results and prevents uniqueness of the decomposition. These issues have been addressed in (2; 3; 4; 5) through the use of masking and mirror-image signals that lead to well-decomposed, nearly orthogonal sets of IMFs. A final issue concerns the ad hoc nature of EMD and the lack of a theoretical foundation for the derived near-orthogonal basis of the IMFs of the decomposed time series. This very issue is addressed by reconciling the analytical and numerical decompositions (2) and (5), respectively. This is discussed below.

The matching of the slow flow decomposition and EMD is performed by shifting the analysis to the complex plane. For the slow flow construction this was already performed through the definition (3). A similar complexification for the IMFs defined in (5) can be performed by numerical Hilbert transform. The Hilbert transform (HT) $h(t)$ of a (mono-component) signal

$y(t)$ is defined as

$$h(t) \equiv H[y(t)] = \frac{PV}{\pi} \int_{-\infty}^{\infty} \frac{y(s)}{t-s}\, ds \tag{6}$$

where PV stands for Cauchy principal value. Moreover, a basis result in complex analysis states that a complex function whose imaginary part is the Hilbert transform of its real part is analytic. Motivated by this result we proceed complexify the $m-th$ IMF $c_k^{(m)}(t)$ of the time series $x_k(t)$ in (5) by defining the analytic complex function,

$$\hat{\Psi}_k^{(m)}(t) \equiv c_k^{(m)}(t) + jH[c_k^{(m)}(t)] \tag{7}$$

Then we can compute the *instantaneous amplitude and phase of the $m-th$ IMF* as,

$$\hat{A}_k^{(m)}(t) = \{c_k^{(m)^2}(t) + H[c_k^{(m)}(t)]^2\}^{1/2}, \tan\hat{\theta}_k^{(m)}(t) = H[c_k^{(m)}(t)]/c_k^{(m)}(t) \tag{8}$$

and its *instantaneous frequency* as $\hat{\omega}_k^{(m)}(t) = \dot{\hat{\theta}}_k^{(m)}(t)$. This leads to the slow-fast representation of the complexified IMF (7) in a form which is similar to the analytical slow flow decomposition (3):

$$\hat{\Psi}_k^{(m)}(t) \equiv \hat{A}_k^{(m)}(t)e^{j\hat{\theta}_k^{(m)}(t)} = \underbrace{\hat{A}_k^{(m)}(t)\, e^{j[\hat{\theta}_k^{(m)}(t)-\omega_k^{(m)}t]}}_{Slow\ component} \quad \underbrace{e^{j\omega_k^{(m)}t}}_{Fast\ component} \tag{9}$$

Hence, complexifying the IMFs provides a way to relate the EMD results to the underlying slow flow dynamics, and *to physically interpret the (previously ad hoc) dominant IMFs in terms of the underlying slow flow dynamics.* Indeed, from the previous derivations, the response of the $k-th$ DOF of (2) can be expressed in two different ways,

Slow flow (analytical)

$$x_k(t) = x_k^{(1)}(t) + \dots + x_k^{(N)}(t),\ \Psi_k^{(m)}(t) = \dot{x}_k^{(m)}(t) + j\omega_m x_k^{(m)}(t) \equiv \underbrace{\phi_k^{(m)}(t)}_{Slow\ component} \underbrace{e^{j\omega_m t}}_{Fast\ component} \tag{10}$$

EMD (numerical)

$$x_k(t) = c_k^{(1)}(t) + \dots + c_k^{(N)}(t),\ \hat{\Psi}_k^{(m)}(t) = c_k^{(m)}(t) + jH[c_k^{(m)}(t)] = \underbrace{\hat{A}_k^{(m)}(t)\, e^{j[\hat{\theta}_k^{(m)}(t)-\omega_k^{(m)}t]}}_{Slow\ component} \quad \underbrace{e^{j\omega_k^{(m)}t}}_{Fast\ component} \tag{11}$$

where we have expressed the complex envelope (9) in polar form as $\hat{\Psi}_k^{(m)}(t) = \hat{A}_k^{(m)}(t)\ e^{j\hat{\theta}_k^{(m)}(t)}$. Given, that the time series is decomposed in terms of dominant IMFs, it follows that $\omega_k^{(m)} \approx \omega_m$ for $m = 1, ..., N$ where N is the number of dominant harmonic components in the slow flow decomposition of the dynamics (i.e., the number of fast frequencies which defines the dimensionality, or the number of significant frequency-time scales in the dynamics). It follows that the *above partitions can be directly related, since they represent identical theoretical and numerical multi-scale slow-fast decompositions of the measured time series,*

$$\underbrace{x_k^{(m)}(t)}_{Theoretical model} \quad \rightarrow \quad \underbrace{c_k^{(m)}(t)}_{Numerical model} \quad , \ k = 1, ..., n, \ \ m = 1, ..., N \tag{12}$$

We note that the complexification (7) based on the Hilbert Transform is different than the corresponding complexification (3) employed in the analytical slow flow analysis, so we can consider the following alternative complexification,

$$\begin{aligned} \hat{\hat{\Psi}}_k^{(m)}(t) = \dot{c}_k^{(m)}(t) + j\omega c_k^{(m)}(t) &\equiv \underbrace{\hat{\hat{\phi}}_k^{(m)}(t)}_{Slow\ component} \quad \underbrace{e^{j\omega_m t}}_{Fast\ component} \\ &= \underbrace{\hat{\hat{A}}_k^{(m)}(t) e^{j[\hat{\hat{\theta}}_k^{(m)}(t) - \omega_m t]}}_{Slow\ component} \quad \underbrace{e^{j\omega_m t}}_{Fast\ component} \end{aligned} \tag{13}$$

where we introduced the polar representation $\hat{\hat{\Psi}}_k^{(m)}(t) = \hat{\hat{A}}_k^{(m)}(t) e^{j\hat{\hat{\theta}}_k^{(m)}(t)}$. Then, simple manipulations can relate the two IMF complexifications (7) and (13) through the relationship:

$$\hat{\Psi}_k^{(m)}(t) = \frac{1}{j\omega_m} \hat{\hat{\Psi}}_k^{(m)}(t) \tag{14}$$

Given the correspondence (12) the correspondence between the analytical and numerical complexifications yields:

$$\begin{aligned} &\Psi_k^{(m)}(t) \rightarrow \hat{\hat{\Psi}}_k^{(m)}(t) = j\omega_m \hat{\Psi}_k^{(m)}(t) \Rightarrow \\ &\underbrace{\phi_k^{(m)}(t) \rightarrow j\omega_m \hat{A}_k^{(m)}(t) e^{j[\hat{\theta}_k^{(m)}(t) - \omega_m t]}}_{Equivalence\ of\ slow\ complex\ amplitudes} \end{aligned} \tag{15}$$

The results (15) provides a *physics-based theoretical foundation for EMD*, whereby the dominant IMFs represent the underlying slow flow of the dynamics and, hence, capture all the important multi-scale dynamics. This leads us to the important conclusion that *dominant IMFs extracted by means of EMD of experimentally measured time series represent the underlying slow flow dynamics of the measured dynamical system.* This important result is central to the NSI methodology of broad applicability. Note that no assumptions have been made regarding the type or dimensionality of the system and the type and strength of the nonlinearity. Hence, the slow-fast partitions discussed above and the physical interpretation of the results of EMD should hold for a broad class of dynamical systems. We reemphasize at this point, however, that this approach is valid for oscillatory measured time series possessing a finite number of well separated fast frequencies (for a discussion of how the methodology can be extended to systems with internal resonances see (16)).

As an example consider the following two-DOF strongly nonlinear system (i.e., n=2 in terms of the previous notation), composed of a linear oscillator (LO) coupled to a lightweight nonlinear energy sink (NES) (3):

$$\begin{aligned} \ddot{x}_1 + \omega_0^2 x_1 + \epsilon\lambda_1\dot{x}_1 + \epsilon\lambda_2(\dot{x}_1 - \dot{x}_2) + C(x_1 - x_2)^3 &= 0 \\ \epsilon\ddot{x}_2 + \epsilon\lambda_2(\dot{x}_2 - \dot{x}_1) + C(x_2 - x_1)^3 &= 0 \end{aligned} \tag{16}$$

For parameters $\omega_0 = 1$, $C = 1$, $\epsilon = 0.05$, $\lambda_{1,2} = 0.03$ and initial conditions $\dot{x}_1(0) = -0.059$, $\dot{x}_2(0) = 0.015$, $x_1(0) = x_2(0) = 0$, the responses are depicted in Figure 1, and the realization of 1:3 transient resonance capture (1:3 TRC) (17) is clear; that is, the NES oscillates with a dominant frequency equal to (1/3) of the frequency of the LO. The corresponding dominant IMFs of each of the two responses are depicted in Figure 1b. In terms of the previous notation, we have the following analytical and empirical decompositions:

$$x_1(t) \approx x_1^{(1)}(t) \approx c_1^{(1)}(t) \ \text{ and } \ x_2(t) \approx x_2^{(1)}(t) + x_2^{(2)}(t) \approx c_2^{(1)}(t) + c_2^{(2)}(t) \tag{17}$$

The IMFs $c_1^{(1)}(t)$ and $c_2^{(1)}(t)$ are nearly monochromatic at frequency $\omega_1 = \omega_0$, whereas the IMF $c_2^{(2)}(t)$ is nearly monochromatic at frequency $\omega_2 = \omega_1/3$.

In Figure 2 we present the correspondence between the slow flow and the empirical mode decomposition for the responses of system (16). The response of the LO possesses a dominant IMF at frequency $\omega_1 = \omega_0$, whereas the response of the NES possesses two IMFs are frequencies $\omega_1 = \omega_0$ and $\omega_2 = \omega_1/3$. Good correspondence between the analytical slow flow predictions and the numerical EMD results is noted for the IMFs (only the second IMF of the response of the NES is considered in Fig. 2 as it is the dominant one).

3 Frequency-energy plots, wavelet transforms and damped transitions

Given the dependence of the responses of nonlinear dynamical systems on energy (e.g., initial conditions or forcing level), it is necessary to study the dynamics in the frequency energy plane. The corresponding representation of the sets of periodic (or quasi-periodic (18)) orbits of an undamped and unforced model of a dynamical system in that plane will be referred to as frequency energy plot (FEP), and will provide a synoptic picture of the underlying intrinsic nonlinear dynamics.

We will demonstrate this plot by considering the two-DOF oscillator (19; 20) composed a linear oscillator coupled to a strongly nonlinear oscillating attachment:

$$\begin{aligned} \ddot{y} + \omega_0^2 y + \lambda_1 \dot{y} + \lambda_2(\dot{y} - \dot{v}) + C(y - v)^3 &= F(t) \\ \epsilon \ddot{v} + \lambda_2(\dot{v} - \dot{y}) + C(v - y)^3 &= 0 \end{aligned} \tag{18}$$

In Figure 3 we depict the FEP for this system for $\lambda_1 = \lambda_2 = F(t) = 0$, where for a given frequency and energy a periodic orbit is represented by a point in the plot, and a branch, represented by a solid line, is the set of periodic orbits possessing the same qualitative features. For instance, branch $S11+$ depicts a branch of periodic orbits for which the linear and nonlinear components oscillate in a symmetric fashion and with the same frequencies (hence the 'S' and '11' designations) and in an in-phase fashion (hence the '+' sign). Similarly, branch $U21$ depicts a branch of periodic orbits where the oscillations of the two components of the system are unsymmetric (hence the 'U' designation) and the frequency of the nonlinear oscillator is twice that of the linear one (hence the '21' designation). Symmetric S-solutions correspond to orbits that satisfy the initial conditions, $\dot{v}(0) = \pm\dot{v}(T/2)$, $\dot{y}(0) = \pm\dot{y}(T/2)$, $v(0) = y(0) = 0$, where T is the period of the oscillation, whereas, unsymmetric U-solutions are periodic orbits that fail to satisfy these conditions. Regarding the frequency designation of each periodic orbit, we assign a frequency index equal to the ratio of its indices, e.g., $S21$ is represented by the frequency index $\omega = 2/1 = 2$. This convention rule holds for every branch except some of the branches $S11\pm$, which, however, are particular branches, forming the basic backbones of the entire FEP. Moreover, on the energy axis we depict the (conserved) total energy of the system when it oscillates in the corresponding periodic motion. Transitions between certain branches seem to involve 'jumps', but this is only due to the frequency convention adopted, and no actual discontinuities in the dynamics occur (e.g., by the previous designations, a branch $S(kn)(km)$, k integer is identified with Snm).

There is a sequence of higher- and lower-frequency periodic solutions bifurcating or emanating from the backbone branches $S11\pm$, which are designated as subharmonic tongues. Each tongue occurs in the neighbourhood of an internal resonance between the linear oscillator and the nonlinear attachment, and corresponds to either symmetric (e.g., $S13\pm$) or unsymmetric (e.g., $U21\pm$) subharmonic motion of the system. Certain periodic orbits (termed impulsive orbits (18)] and depicted by dots in Figure 3) satisfy the special initial conditions $v(0) = \dot{v}(0) = y(0) = 0$ and $\dot{y}(0) \neq 0$, i.e., correspond to application of an impulse of magnitude $\dot{y}(0)$ to the linear oscillator; these orbits play an important role in targeted energy transfers in the damped and impulsively forced system (18) (21).

The FEP is especially useful when considering damped responses of the weakly damped responses system (18) either under transient excitations or due to nonzero initial conditions. In particular, one can systematically interpret the complex multi-frequency nonlinear transitions by relating them to the different branches of solutions in the FEP. To show this we consider the system parameters $\omega_0 = 1$, $C = 1$, $\epsilon = 0.05$, $\lambda_1 = 0$, $\lambda_2 = 0.0015$, $F(t) = 0$, and numerically integrate the equations (18) subject to initial conditions $v(-T/4) = \dot{v}(-T/4) = x(-T/4) = 0$ and $\dot{x}(-T/4) = -0.1039$ (these correspond to excitation of a stable impulsive orbit on the subharmonic branch $U76$). In Figure 4 we depict the transient responses of the linear and nonlinear oscillator, and in Figure 5 the corresponding wavelet transform (WT) spectra of these motions.

The WT can be viewed not only as a basis for functional representation, but at the same time as a useful technique for time-frequency analysis. In contrast to the Fast Fourier Transform (FFT) which assumes signal stationarity, the WT involves a windowing technique with variable-sized regions, by constructing a series of wavelet functions derived from a basic 'mother wavelet' function. Small time intervals are considered for high frequency components whereas the size of the interval is increased for lower frequency components, thereby giving better time and frequency resolutions than the FFT. Different types of mother wavelets can be considered, e.g., Morlet wavelets, which are Gaussian-windowed complex sinusoids of frequency ω_0, $\Psi_M(t) = e^{-t^2/2}e^{j\omega_0 t}$, or Cauchy wavelets. These two mother wavelets provide similar results when applied to the signals considered here. The WT spectra presented in Figure 4 depict contour plots of the amplitude of the WT as a function of frequency (vertical axis) and time (horizontal axis). Heavy shaded areas correspond to regions where the amplitude of the WT is high, whereas lightly shaded regions correspond to low amplitudes. Such plots enable one to deduce the temporal evolutions of the dominant frequency components of the signals analysed.

In Figure 6 we depict the WT spectra of the damped responses Figure 5 superimposed on the undamped FEP of Figure 3, by replacing the time variable with the corresponding instantaneous energy at a given instant of time. Whereas this is only a phenomenological comparison, it helps us gaining an understanding of the different resonance dynamics that occur in the damped transient dynamics. We make the general observation that the highly complex nonlinear dynamics can be fully interpreted in the FEP. Indeed, initially a stable impulsive orbit on branch $U76$ is excited, but as energy decreases due to damping there occurs a transition (jump) to the stable branch $S13$-, where the oscillation locks into a 1:3 transient resonance capture with the linear oscillator possessing a harmonic component with frequency three times higher than that of the nonlinear attachment; as energy decreases there occurs escape from this resonance capture, and the motion evolves along the lower frequency branches $S15$, $S17$,...

These results demonstrate that depictions of damped responses on the FEP can be a powerful tool for interpreting highly complex, multi-frequency nonlinear transitions. The usefulness of FEP depictions is further highlighted that the fact that it provides both the global synoptic picture of the intrinsic nonlinear dynamics of (both discrete and continuous) dynamical systems, as well as local interpretations of specific damped transitions (i.e., of oscillations initiated by specific forcing or initial conditions). As such, the FEP will prove to be central in the integrated NSI methodology discussed in the next section.

4 Global and local aspects of the NSI methodology

The NSI methodology combines the previous concepts and techniques into an integrated scheme with local and global aspects of identification. The central assumption of the methodology is that the measured dynamics can be decomposed in terms of slowly modulated fast oscillations, which is a reasonable assumption for non-chaotic measured vibration data. The basic elements are as follows:

Under transient shock excitation (similar to hammer tests of linear modal analysis), perform measure transient time series from a number of sensing positions throughout the test item, and perform EMD of the measured time series. This will lead to a set of experimentally extracted intrinsic mode functions (IMFs) at each sensing location. By Hilbert-transforming the IMFs we compute their instantaneous frequencies and compare them to wavelet transform spectra of the measured time series; this determines the dominant IMFs and the corresponding dominant fast frequencies in the measured dynamics at each sensing location. This procedure identifies the basic (dominant) time scales and the dimensionality of the measured dynamics, and provides the foundation for the nonlinear system identification.

Based on the theoretical correspondence of the experimentally measured dominant IMFs and the underlying intrinsic slow-flow dynamics of the system, we relate the slow components of the dominant experimental IMFs to the intrinsic slow flow dynamics. This is the fundamental underlying principle of the NSI. Indeed, using the dominant IMFs we can reconstruct the measured time series and depict them in an FEP assuming that the mass distribution of the system is known (an assumption that is not restrictive). Under the assumption of weak dissipation this will reconstruct a portion of the FEP of the intrinsic dynamics of the system under investigation and provides the global aspect of the NSI methodology. Note that no a priori model is assumed for this FEP reconstruction, so this procedure is non-parametric. By considering different time series we can construct the global picture of the intrinsic dynamics of the system and gain an estimate of the strength of the nonlinear effects in the dynamics (e.g., curved branches in the FEP indicate strongly nonlinear dynamics, whereas horizontal branches linear or weakly nonlinear dynamics).

In addition, by considering a specific measured damped transition we can define a parametric reduced-order slow-flow model of the system with the dimensionality of the dynamics, and identify its parameters, thus constructing a local slow flow model of the dynamics. This provides the local parametric aspect of the NSI methodology. The so constructed slow flow models fully reconstruct the specific measured time series. By varying the excitation and/or the initial conditions of the system we can consider different nonlinear transitions of the system over different frequency and energy ranges, and construct the corresponding portions of the FEP of the system together with the associated local slow flow models.

The final outcomes of the proposed NSI methodology are the construction an FEP of the global dynamics depicting the possible coexisting families of periodic solutions (nonlinear modes) over the frequency and energy ranges of interest, and the corresponding local slow flow models of the dynamics describing nonlinear transitions on the FEP. Hence, the proposed approach addresses in a systematic way a fundamental limitation of other nonlinear system identification methods which fail to account for the fact that the responses of nonlinear systems may depend crucially on the initial conditions and/or the applied excitations. The added flexibility of 'probing' the dynamics over different frequency and energy ranges in order to extract different local models is important when identifying systems capable of strongly nonlinear dynamical behavior.

5 Applications

As a first application we apply the local aspect of the NSI methodology to identify the strongly nonlinear dynamical interaction of a viscously damped dispersive finite rod on distributed elastic support, with an essentially nonlinear attachment at its right end (cf. Figure 7 (22)). The essential nonlinearity of the attachment is due to the lack of a linear component in the stiffness connecting it to the rod, whereas the attachment is lightweight (its mass is scaled by the small parameter $0 < \epsilon << 1$). Denoting by $v(t)$ and $u(x,t)$ the responses of the nonlinear attachment and the rod, respectively, the equations of motion are given by:

$$\frac{\partial^2 u(x,t)}{\partial t^2} + \omega_0^2 u(x,t) + \epsilon\lambda_1 \frac{\partial u(x,t)}{\partial t} - \frac{\partial^2 u(x,t)}{\partial x^2} = 0, \quad 0 \le x \le L$$

$$u(0,t) = 0, \quad \frac{\partial u(L,t)}{\partial x} = -\epsilon \ddot{v}(t)$$

$$\epsilon \ddot{v}(t) + \epsilon\lambda_2 \left[\dot{v}(t) - \frac{\partial u(L,t)}{\partial t}\right] + C\left[v(t) - u(L,t)\right]^3 = 0$$

$$u(x,0) = r(x), \; \frac{\partial u(x,0)}{\partial t} = s(x), \; v(0) = v, \; \dot{v}(0) = \dot{v}_0 \qquad (19)$$

where $r(x)$ and $s(x)$ are the initial displacement and velocity distributions of the rod, and the viscous dissipative terms of the system are assumed to be of $O(\epsilon)$.

Assuming proportional viscous damping distribution for the rod (its discretized viscous damping matrix is chosen as $D = \alpha_1 M$, where M is the mass matrix of the discretized finite element model of the rod with $\alpha_1 = 0.005$ (22)), and parameters $\epsilon = 0.05$, $C = 1$, $L = 1$, $\omega_0 = 1$, $\lambda_2 = 0.2$, a specific damped transition of this system is depicted in Figure 8. The initial conditions for the damped response are given by:

$$v(0) \approx -0.1650$$

$$u(x,0) \approx 0.1052 \sin(x\sqrt{\omega^2 - \omega_0^2}) \cos \omega t + 0.000988 \times$$

$$\sin(x\sqrt{9\omega^2 - \omega_0^2}) \cos 3\omega t + 0.17536 \sin(x\sqrt{25\omega^2 - \omega_0^2}) \cos 5\omega t \,|_{t=0} \qquad (20)$$

This specific set of initial conditions corresponds to initiation of the transient dynamics at the point of the FEP with dominant frequency $\omega = 2.214\,\text{rad/s} \approx \omega_4/5$ (where ω_4 is the fourth eigenfrequency of the rod with no attachment) on a subharmonic tongue (cf. Fig. 8c). Hence, the damped transition initially (Regime I cf. Fig. 8c) possesses two dominant harmonic components at frequencies ω_4 (the dominant harmonic of the rod) and $\omega_4/5$ (the dominant harmonic of the nonlinear attachment), but as energy decreases due to damping dissipation, multi-frequency transitions occur. Indeed, there is a high-energy transition from the initially excited subharmonic tongue to a prolonged 1:3 subharmonic transient resonance capture (Regime II), as signified by the presence of strong harmonics at frequencies $\omega_1/3$ and ω_1. This high- to low-frequency transition results in relatively large amplitudes of the nonlinear attachment as targeted energy transfer (21) from the rod to the attachment takes place. Finally, there is a final low-energy transition to a linearized state of the dynamics (Regime III), where the response of the nonlinear attachment is negligible, and the dynamics is dominated by the response of the linear rod.

In Figure 9 we present the EMD analysis of the rod end response. There are two dominant, near-orthogonal mono-component IMFs. The superposition of these IMFs accurately reconstructs the rod end response (cf. Fig. 9c). Moreover, the instantaneous frequencies of the dominant IMFs coincide with the dominant harmonics of the wavelet transform spectra of the rod end response. The two dominant IMFs possess nearly constant frequencies identical to the first and fourth eigenfrequencies of the rod with no attachment, which correlates with the representation of the wavelet spectrum of this damped transition on the FEP (cf. Fig. 8c).

In Figure 10 we depict the IMFs together with their instantaneous frequencies of the response of the nonlinear attachment. Similar to the response of the end of the rod, this response possesses two dominant IMFs that accurately reconstruct the computed time series. In this case, however, the first dominant IMF possesses a nearly constant frequency equal to one-fifth of the fourth eigenfrequency of the rod with no attachment, whereas the second IMF possesses a slowly varying eigenfrequency which ends up 'locking' to one third of the first eigenfrequency of the rod in the time interval $250 < t < 500$. Again, these results correlate with the damped transitions depicted in the FEP of Figure 8c.

Based on the EMD results we wish to perform system identification and reduced order modeling of the described strongly nonlinear modal interactions between the rod and the nonlinear attachment, by relying solely on direct analysis of the computed time series. Considering first the rod end response we note that this response can be modeled by a system of uncoupled, forced linear oscillators, termed intrinsic modal oscillators IMOs:

$$\begin{aligned} \ddot{u}_1 + \lambda_4^u \dot{u}_1 + \omega_4^2 u_1 &\approx \Lambda_4^u(t) e^{j\omega_4 t} + cc \\ \ddot{u}_2 + \lambda_1^u \dot{u}_2 + \omega_1^2 u_2 &\approx \Lambda_1^u(t) e^{j\omega_1 t} + cc \end{aligned} \tag{21}$$

where cc denotes denotes complex conjugate. The inhomogenous (forcing) terms in (21) are in the form of 'fast' oscillating terms $e^{j\omega_k t}$, $k = 1, 4$

modulated by the 'slowly' varying complex amplitudes $\Lambda_k^u(t)$, $k = 1, 4$; moreover, we have assumed that the modulations $\Lambda_k^u(t)$ vary much slower than the corresponding carrying signals $e^{j\omega_k t}$. We considered this form of reduced-order model for the rod end response, so that *the response of each of the two IMOs (21) reproduces approximately one of the dominant IMFs of the dynamics of the rod end (cf. Figure 9).* Since the superposition of the IMFs reconstructs the original time series, the same result should hold for the combined response of the IMOs in (21). The specific forms of the inhomogeneous (forcing) terms in (21), i.e., as modulated periodic signals, are dictated by the realization that any alternative type of excitation would be off-resonance and, hence, their effects on the rod dynamics would be small, of a secondary nature.

Following similar reasoning and taking into account that the damped response of the nonlinear attachment possesses two dominant fast frequencies approximately equal to $\omega_4/5$ and $\omega_1/3$, we construct the following approximate reduced-order model for the damped dynamics of the nonlinear attachment:

$$\begin{aligned} \ddot{v}_1 + \lambda_4^v \dot{v}_1 + (\omega_4/5)^2 v_1 &\approx \Lambda_4^v(t) e^{j(\omega_4/5)t} + cc \\ \ddot{v}_2 + \lambda_1^v \dot{v}_2 + (\omega_1/3)^2 v_2 &\approx \Lambda_1^v(t) e^{j(\omega_1/3)t} + cc \end{aligned} \tag{22}$$

Again, each of the two IMOs in (22) reproduces approximately a dominant IMF of the response of the nonlinear attachment (cf. Fig. 10). In particular, the first IMO models the initial stage of the response of the nonlinear attachment, i.e., the initial 1:5 transient resonance capture of the damped dynamics (Regime I), whereas the second IMO models the second (delayed) 1:3 transient resonance capture (Regime II).

Considering now to the reduced-order models (21) and (22), we compute their damping coefficients and inhomogeneous terms by imposing the requirement that each of the IMOs in (21) and (22) reproduces a dominant IMF of the rod end or the nonlinear attachment responses, respectively. To demonstrate this analytical computation we consider an IMO with the general form:

$$\ddot{y} + \lambda \dot{y} + \omega^2 y \approx \Lambda(t) e^{j\omega t} + cc \tag{23}$$

where $\Lambda(t)$ is a slow (complex) modulation of the fast periodic signal $e^{j\omega_k t}$. We analyze the dynamics of (23) by performing a slow/fast partition of the dynamics (21), introducing the new complex variable $g(t) = \dot{y}(t) + j\omega y(t)$, and expressing the real dependent variable and its derivatives as,

$$y(t) = \frac{g - \overline{g}}{2j\omega}, \quad \dot{y}(t) = \frac{g + \overline{g}}{2}, \quad \ddot{y}(t) = \dot{g} - \frac{j\omega}{2}(g + \overline{g}) \tag{24}$$

where overbar denotes complex conjugate. Then, (23) is expressed as,

$$\dot{g}(t) - j\omega g(t) + (\lambda/2)\,[g(t) + \overline{g}(t)] = \Lambda(t) \tag{25}$$

a relation that is exact up to this point. We now introduce a slow-fast partition of the response of (25) in the form $g(t) = \phi(t)e^{j\omega t}$, where $\phi(t)$ is a slow complex modulation (amplitude). Substituting this expression into (25) and performing averaging with respect to the fast frequency ω we derive the following approximate complex modulation equation governing the slow evolution of the amplitude $\phi(t)$ and relating it to the forcing modulation $\Lambda(t)$:

$$\dot{\phi}(t) + (\lambda/2)\phi(t) = \Lambda(t) \tag{26}$$

This key result provides a means of for estimating (identifying) the slowly varying forcing term $\Lambda(t)$ and the damping coefficient λ in the IMO (23), if the slow modulation of its response is experimentally or computationally given. Indeed, assuming that the response of the IMO (23) is approximately equal to one of the corresponding dominant IMFs $c(t)$ derived by EMD of the measured time series, we may regard $\phi(t)$ as the slow component of the complexification $\hat{\Psi}(t) \equiv c(t) + jH[c(t)]$ of that IMF (where $H[\bullet]$ denotes Hilbert transform). The slow component of the complexification $\hat{\Psi}(t)$ is then computed by means of the previously discussed expressions (7) and (8) providing its amplitude and phase. Assuming that the IMF possesses the dominant ('fast') frequency ω, we may then express the complexification of the IMF in the polar form,

$$\hat{\Psi}(t) = \underbrace{\hat{A}(t)e^{j[\hat{\theta}(t)-\omega t]}}_{Slow\ component}\ \underbrace{e^{j\omega t}}_{Fast\ component} \tag{27}$$

where $\hat{A}(t) = \{c^2(t) + H[c(t)]^2\}^{1/2}$, $\tan\hat{\theta}(t) = H[c(t)]/c(t)$ and $\hat{\omega}(t) = \dot{\hat{\theta}}(t)$. It follows that the slow modulation $\phi(t)$ can be approximated as,

$$\phi(t) \approx j\omega\hat{A}(t)e^{j[\hat{\theta}(t)-\omega t]} \tag{28}$$

where the factor $j\omega$ accounts for the slightly different definitions of the complex variables $g(t)$ and $\hat{\Psi}(t)$. Relations (26-28) can be used to determine the complex modulation $\Lambda(t)$ of the forcing term of the IMO (23), once its damping coefficient λ is estimated through an optimization procedure (i.e., by requiring that the response of the IMO best approximates its corresponding IMF).

By the outlined procedure we can derive local reduced-order models identifying a specific such as the one presented in Fig. 8). Indeed, applying it to the IMOs (21) and (22) we estimated the slow forcing modulations $\Lambda_{1,4}^{u,v}(t)$. The results are depicted in Figures 11 and 12 for corresponding optimized damping coefficients $\lambda_4^u = 0.18$, $\lambda_1^u = 0.4$, $\lambda_4^v = 1$, and $\lambda_1^v = 1.2$. In essence, the forcing terms $\Lambda_p^{u,v}(t)e^{j\omega_p t}$ in (21) and (22) represent nonlinear modal interactions quantifying transient energy exchanges between the dominant harmonics of the rod end and attachment responses, as well as the time windows where these exchanges occur. The responses of the IMOs reproduce all the dominant IMFs, so by superimposing we can reproduce the measured nonlinear time series. As an example, the comparison of the IMFs to the responses of the repespective IMOs for the nonlinear attachment are shown in Figure 13. The early- and later-time discrepancies can be attributed to the temporal variations of the dominant frequencies of the nonlinear attachment in these regimes of the dynamics in Regimes I and III (cf. the wavelet spectrum of Fig. 8c), which was not taken into account in the IMOs (22), since, for simplicity, fixed eigenfrequencies were assumed.

In summary, employing the local aspect of the NSI methodology we are able to identify the nonlinear temporal energy exchanges between the rod and the attachment and the precise time scales (frequencies) at which these interactions occur. This represents *multi-scale system identification* of the nonlinear dynamics.

We now provide an application for the global aspect of the NSI methodology, by showing partial reconstruction the main features of the FEP of a dynamical system under the assumption of weak dissipation. This is performed by considering different damped nonlinear transitions initiated at different energy levels which resemble 'hammer tests' used in traditional experimental modal analysis.

Considering a specific weakly damped transition, we first perform EMD and identify the dominant IMFs corresponding to that transition. The instantaneous frequency of an identified dominant IMF, say $c_k^{(m)}$, can be computed directly through relations (7-9) as, $\hat{\omega}_k^{(m)}(t) = \dot{\hat{\theta}}_k^{(m)}(t)$. The corresponding instantaneous 'energy' of the IMF (i.e., at each time instant of the damped transition) can be expressed as a sum of kinetic and potential energies, $\hat{E}_k^{(m)}(t) = 1/2\left[\dot{c}_k^{(m)^2}(t) + \omega_m^2 c_k^{(m)^2}(t)\right]$. If the mass distribution for the system is known, then the instantaneous mechanical energy of the system can be estimated as a summation of the 'energies' of the IMFs multiplied by appropriate mass factors,

$$E_{\text{total}}(t) = \eta \sum_{k=1}^{n} \sum_{m=1}^{N} m_k \hat{E}_k^{(m)}(t) \tag{29}$$

where the weighting coefficients m_k correspond to the mass distribution of the system among components (it can de deduced from the physical configuration of the system), and η is a factor used to match the exact initial conditions of the damped transition with the approximate initial conditions satisfied by the IMFs (this can be directly deduced from the measured time series). If the system is linear then $\eta = 1$, so $\eta > 1$ accounts for the energy of the nonlinear terms. Using these expressions a partial construction of the FEP can be made (corresponding to the considered transition) and a global picture of the dynamics can be gained. By considering different nonlinear transitions we can construct different regions of the FEP and perform global identification of the dynamics over broad frequency and energy ranges.

As a demonstration in Figure 14 we provide a partial reconstruction of the FEP for the strongly nonlinear system of coupled oscillators (18). In this case, the mass distribution of the system is $m_1 = 1$ (linear oscillator) and $m_2 = \epsilon = 0.05$ (nonlinear attachment), and the correction factor is computed as $\eta = 1.5$. The FEP reconstructions are performed using two specific damped transitions (6) and the instantaneous frequency and energy estimates derived above.

6 Conclusion

The discussed NSI methodology is based on post processing of measured time series and holds promise of broad applicability to dynamical systems with weak or strong nonlinearities and time-invariant or time-variant parameters. In addition, it appears as conceptual extension of classical experimental modal analysis EMA. Indeed, whereas in EMA one constructs reduced-order models in the frequency or temporal domains in terms of bases of orthogonal vibration modes, in the discussed NSI methodology one deals with measured time series (which eliminates the need of frequency analysis a basic limitation when one performs nonlinear dynamic analysis) and decomposes them in terms of nearly orthogonal, approximately monochromatic IMFs resulting from EMD. Instead of classical Fast Fourier Transforms used in EMA, the NSI considers wavelet transforms that provide not only the frequency components of the measured signals but also the temporal variation of these components due to non-stationary effects. Moreover, instead of considering the dynamics in the frequency domain (as in EMA), in NSI one considers frequency energy plots which enable study of the dynamics for changes in forcing levels and initial conditions. The use of such tools addresses the basic feature of nonlinear systems that their dynamics depend on the energy level (which does not occur in linear theory).

Among the current challenges that need to be addressed concern the development of data-driven reduced order models in multi-physics problems using the NSI methodology, of identifying and modeling non-smooth structural dynamics including the effects of clearance and friction, to extend the NSI methodology to systems with closely spaced modes exhibiting beat phenomena (a first step towards addressing the issue is considered in (9) or modes with slowly varying fast frequencies, and to wave-related applications. Finally, there is the need to extend the discussed discrete reduced-order models to distributed-parameter systems by adding a additional spatial dimension to the analysis.

Bibliography

[1] G. Kerschen, K. Worden, A.F. Vakakis, J.C. Golinval, Past, present and future of nonlinear system identification in structural dynamics, Mechanical Systems and Signal Processing 20 (2006), 505-592.

[2] Y.S. Lee, S. Tsakirtzis, A.F. Vakakis, D.M. McFarland, L.A. Bergman, Physics-based foundation for empirical mode decomposition: correspondence between intrinsic mode functions and slow flows, AIAA Journal 47 (2009), 2938-2963.

[3] Y.S. Lee, A.F. Vakakis, D.M. McFarland, L.A. Bergman, A global-local approach to nonlinear system identification: A review, Journal of Structural Control and Health Monitoring 17 (2010), 742-760.

[4] Y.S. Lee, S. Tsakirtzis, A.F. Vakakis, D.M. McFarland, L.A. Bergman, A time-domain nonlinear system identification method based on multi-scale dynamic partitions, Meccanica (2010).

[5] Y.S. Lee, A.F. Vakakis, D.M. McFarland, L.A. Bergman, Non-linear system identification of the dynamics of aeroelastic instability suppression based on targeted energy transfers, The Aeronautical Journal 114 (2010), paper No. 3437.

[6] A.F. Vakakis, L.A. Bergman, D.M. McFarland, Y.S. Lee, M. Kurt, Current efforts towards a nonlinear system identification methodology of broad applicability, Proceedings of the Institution of Mechanical Engineers, Part C: Journal of Mechanical Engineering Science, 225 (2011), 2497-2515.

[7] M. Kurt, H. Chen, Y.S. Lee, D.M. McFarland, L.A. Bergman, A.F. Vakakis, Nonlinear system identification of the dynamics of a vibro-impact beam: Numerical results, Archive of Applied Mechanics 82 (2012), 1461-1479.

[8] M. Eriten, M. Kurt, G. Luo, D.M. McFarland, L.A. Bergman, A.F. Vakakis, Nonlinear system identification of frictional effects in a beam with a bolted joint connection, Mechanical Systems and Signal Processing 39 (2013), 245-264.

[9] M. Kurt, M. Eriten, D.M. McFarland, L.A. Bergman, A.F. Vakakis, Strongly nonlinear beats in the dynamics of a cantilever beam with a nonlinear spring: Analysis and identification, Journal of Sound and Vibration, in press.

[10] L.I. Manevitch, The description of localized normal modes in a chain of nonlinear coupled oscillators using complex variables, Nonlinear Dynamics, 25 (2001), 95-109.

[11] F. Verhulst, Singular perturbation methods for slow-fast dynamics, Nonlinear Dynamics, 50 (2007), 747-753.

[12] N. Huang, Shen Z., Long S., Wu M., Shih H., Zheng Q., Yen N.-C., Tung C., Liu, H., The Empirical Mode Decomposition and the Hilbert Spectrum for nonlinear and nonstationary time series analysis, Proceedings of the Royal Society of London A 454 (1998), 903-995.

[13] G. Kerschen, A.F. Vakakis, Y.S. Lee, D.M. McFarland, L.A. Bergman, Toward a fundamental understanding of the Hilbert-Huang transform in nonlinear structural dynamics, Journal of Vibration and Control, 14 (2008), 77-105.

[14] S. Tsakirtzis, Passive Targeted Energy Transfers from Elastic Continua to Essentially Nonlinear Attachments for Suppressing Dynamical Disturbances, PhD Thesis, Dept. of Applied Mathematical and Physical Sciences, National Technical University of Athens, Athens, Greece, 2006.

[15] F. Georgiades, Nonlinear Localization and Targeted Energy Transfer Phenomena in Vibrating Systems with Smooth and Non-smooth Stiffness Nonlinearities, PhD Thesis, Dept. of Applied Mathematical and Physical Sciences, National Technical University of Athens, Athens, Greece, 2006.

[16] M. Kurt, M. Eriten, D.M. McFarland, L.A. Bergman, A.F. Vakakis, Strongly Nonlinear Beats in the Dynamics of a Cantilever Beam with a Nonlinear Spring: Analysis and Identification, Journal of Sound and Vibration (in review).

[17] V.I. Arnold (Ed.), Dynamical Systems III, Encyclopaedia of Mathematical Sciences, Vol. 3, Springer Verlag, 1988.

[18] G. Kerschen, O. Gendelman, A.F. Vakakis, L. Bergman, D.M. McFarland, Impulsive Periodic and Quasiperiodic Orbits in a System of Coupled Oscillators with Essential Stiffness Nonlinearity, Communications in Nonlinear Science and Numerical Simulation 13 (2008), 959-978.

[19] Y.S. Lee, G. Kerschen, A.F. Vakakis, P.N. Panagopoulos, L.A. Bergman, D.M. McFarland, Complicated Dynamics of a Linear Oscillator with a Light, Essentially Nonlinear Attachment, Physica D, 204 (2005), 41-69.

[20] G. Kerschen, Y.S. Lee, A.F. Vakakis, D.M. McFarland, L.A. Bergman, Irreversible Passive Energy Transfer in Coupled Oscillators with Essential Nonlinearity, SIAM Journal on Applied Mathematics 66 (2006), 648-679.

[21] A.F. Vakakis, O. Gendelman, L.A. Bergman, D.M. McFarland, G. Kerschen, Y.S. Lee, Passive Nonlinear Targeted Energy Transfer in Mechanical and Structural Systems: I and II, Springer Verlag, New York and Berlin, 2008.

[22] S. Tsakirtzis, Y.S. Lee, A.F. Vakakis, L.A. Bergman, D.M. McFarland, Modeling of nonlinear modal interactions in the transient dynamics of an elastic rod with an essentially nonlinear attachment, Communications in Nonlinear Science and Numerical Simulation (2010), 2617-2633.

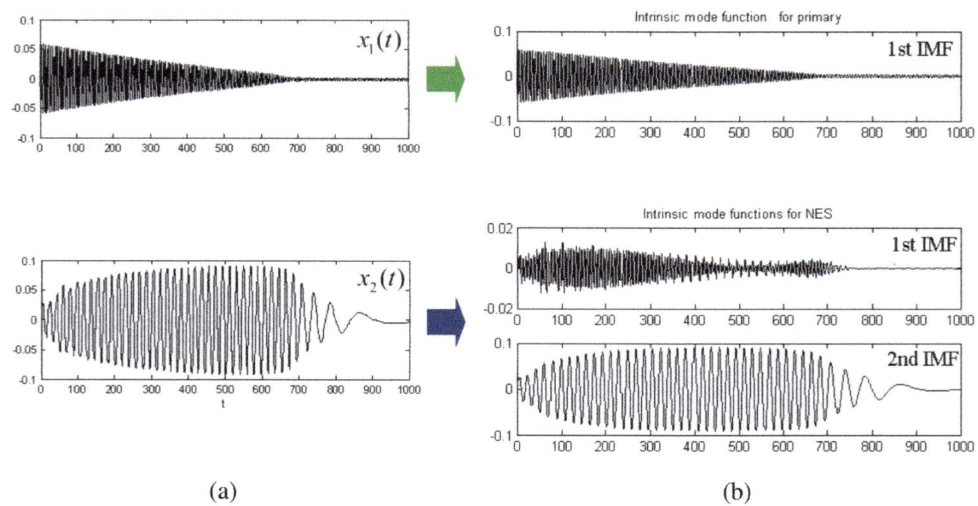

Figure 1. 1:3 TRC in the strongly nonlinear system 16: (a) Transient responses, (b) dominant IMFs of the responses.

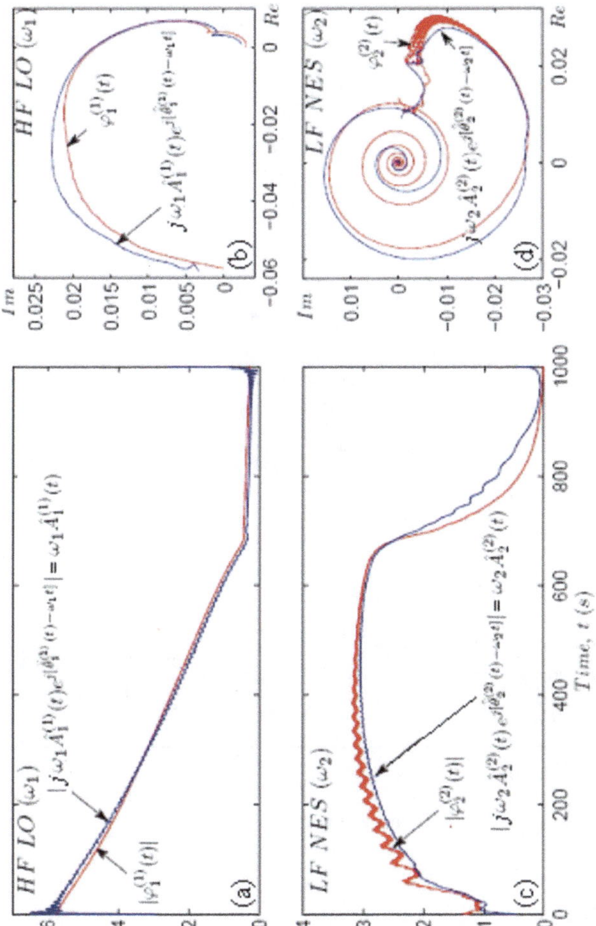

Figure 2. Correspondence between slow flow dynamics and EMD results for the strongly nonlinear system (16): (a,b) Response of the LO at its dominant frequency ω_1, (c,d) response of the NES at $\omega_2 = \omega_1/3$ (3).

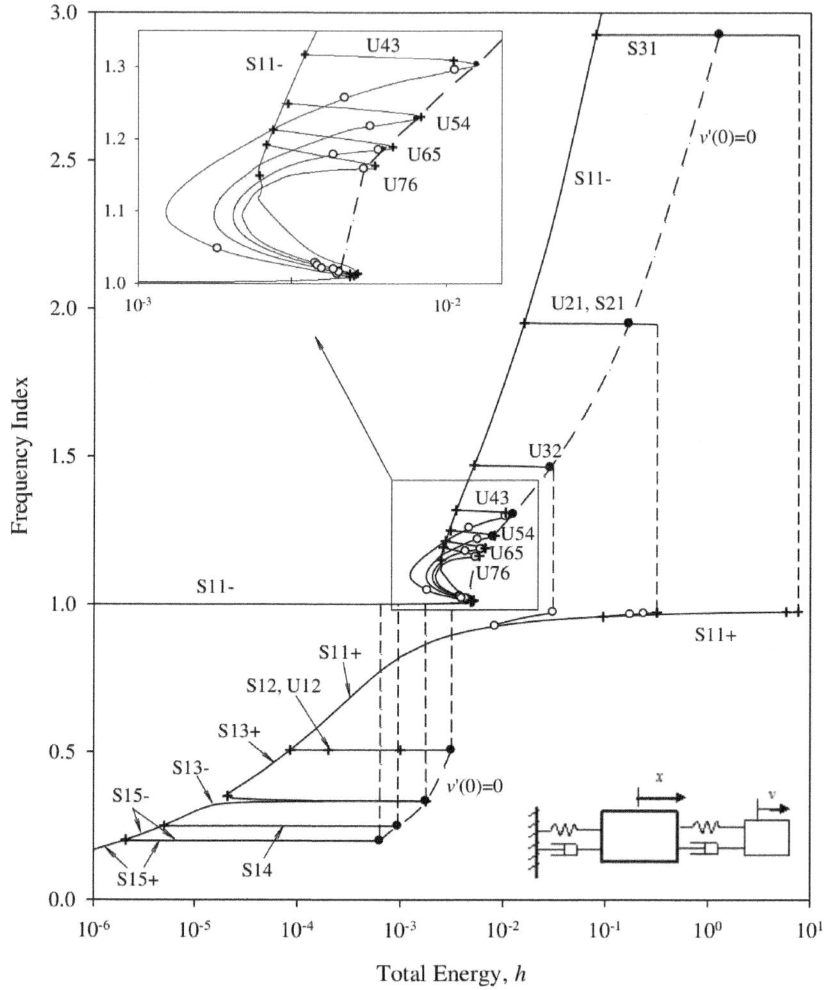

Figure 3. FEP of system (3) (19; 20; 21).

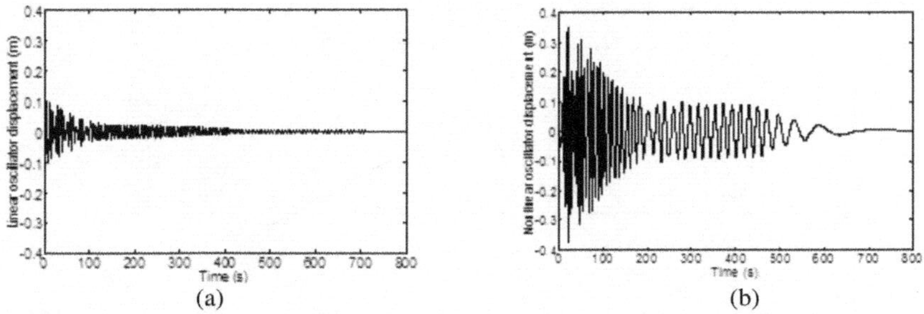

Figure 4. Transient responses of the linear oscillator (a), and the nonlinear oscillator (b) (19).

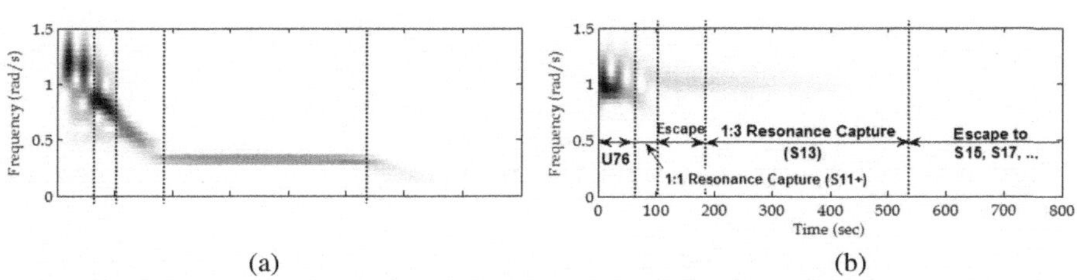

Figure 5. Contours of wavelet transforms (WT spectra) of the transient responses of the linear oscillator (a), and the nonlinear oscillator (b) (19).

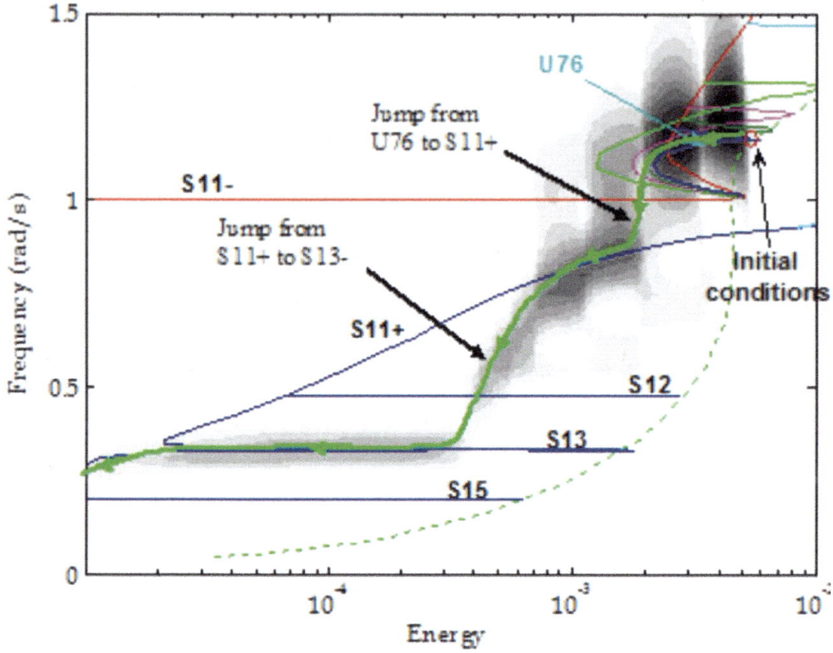

Figure 6. WT spectra of the relative response $y(t) - v(t)$ superimposed on the FEP (19).

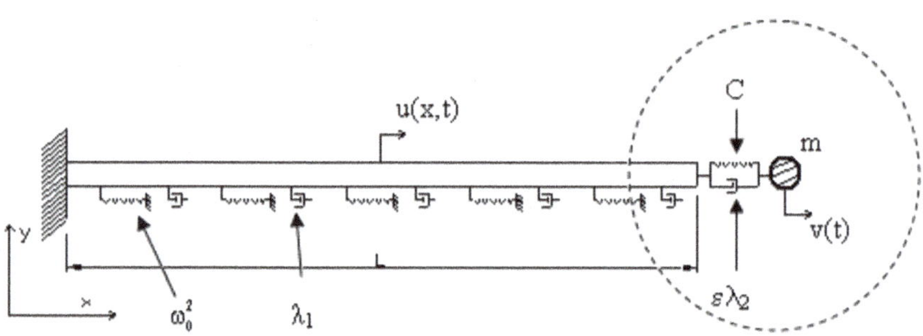

Figure 7. Elastic dissipative rod with essentially nonlinear end attachment (22).

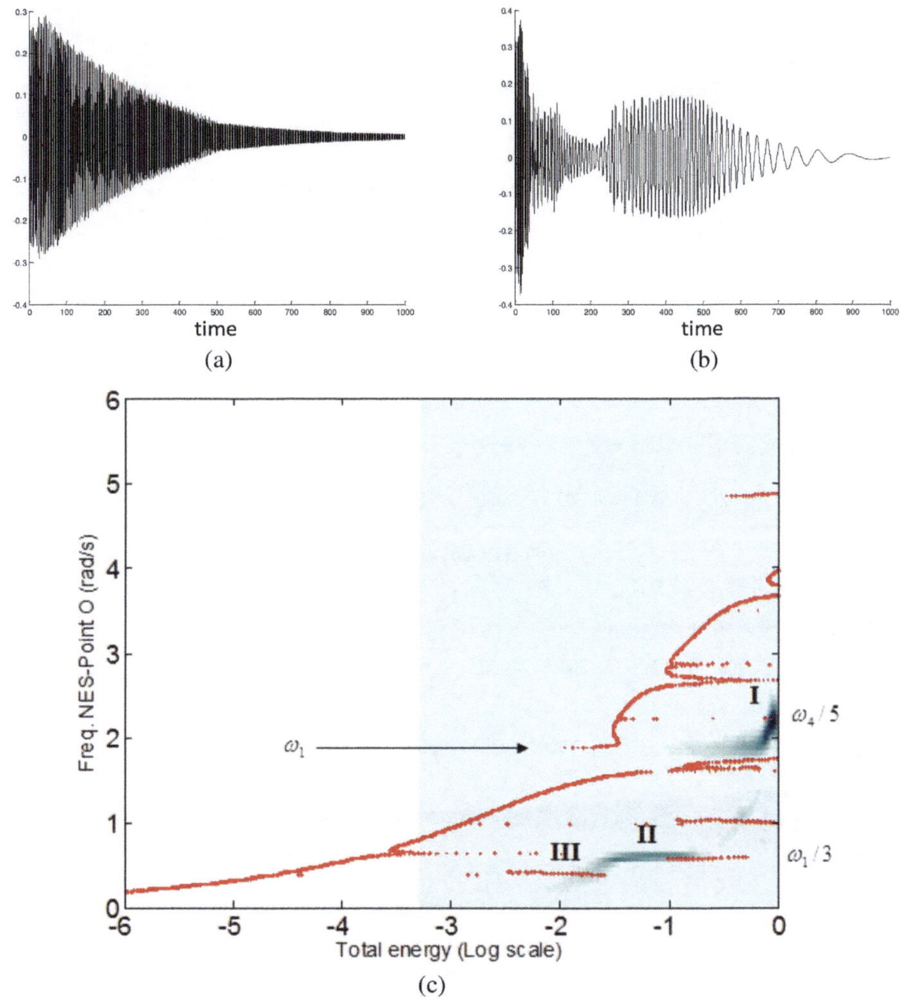

Figure 8. Highly complex damped transition: (a) Response of the rod, (b) response of the attachment, (c) depiction on the FEP of the relative rod end-attachment response (22).

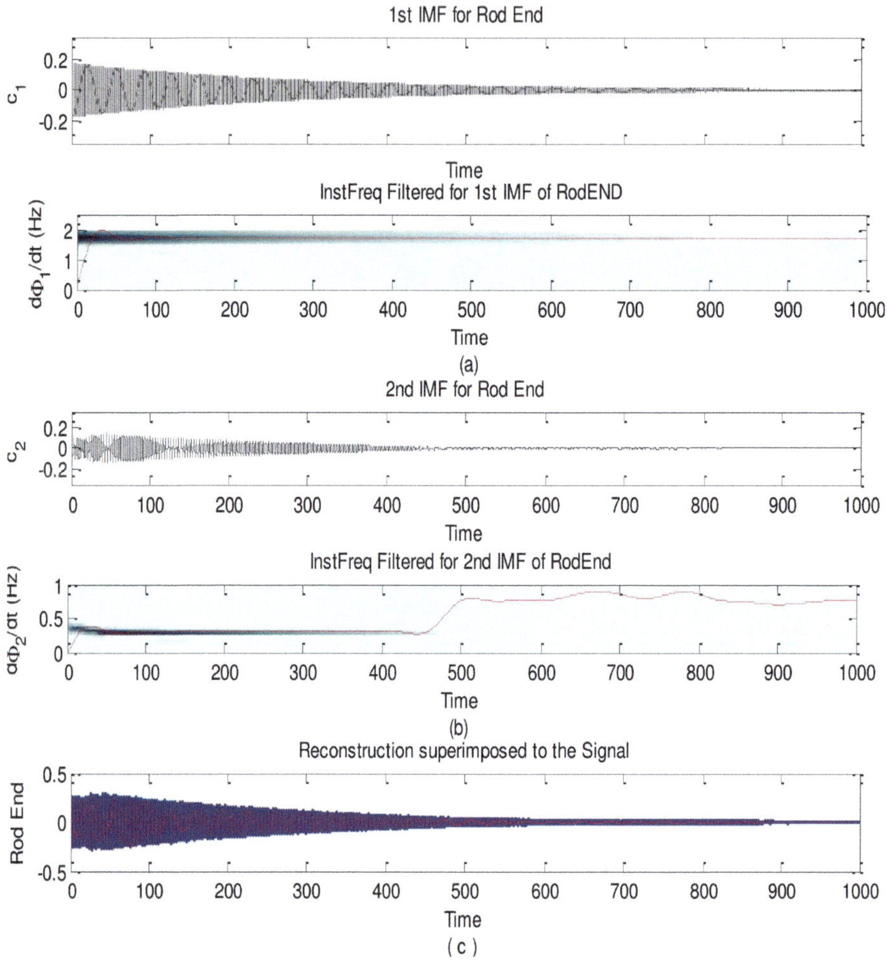

Figure 9. EMD analysis of rod response: (a) 1st IMF and its instantaneous frequency superimposed on the WT spectrum, (b) 2nd IMF and its instantaneous frequency superimposed on the WT spectrum, (c) reconstruction of response by superimposing the two leading (dominant) IMFs (22).

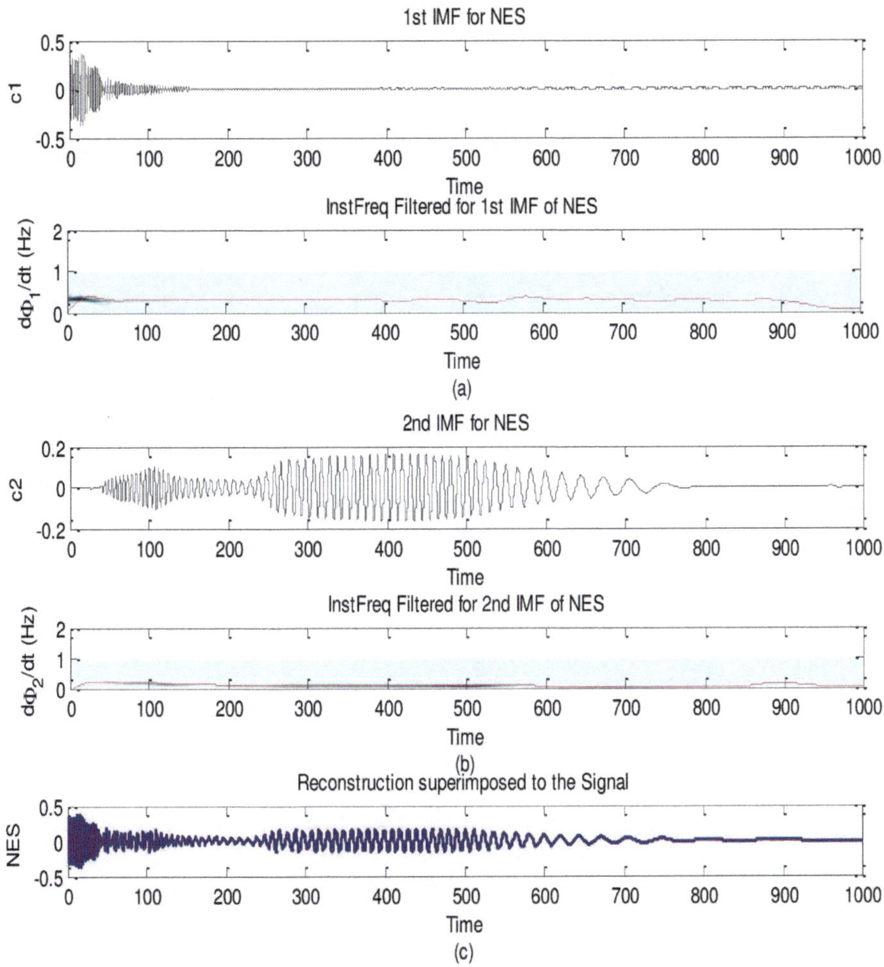

Figure 10. EMD analysis of the response of the nonlinear attachment: (a) 1st IMF and its instantaneous frequency superimposed on the WT spectrum, (b) 2nd IMF and its instantaneous frequency superimposed on the WT spectrum, (c) reconstruction of response by superimposing the two leading (dominant) IMFs.

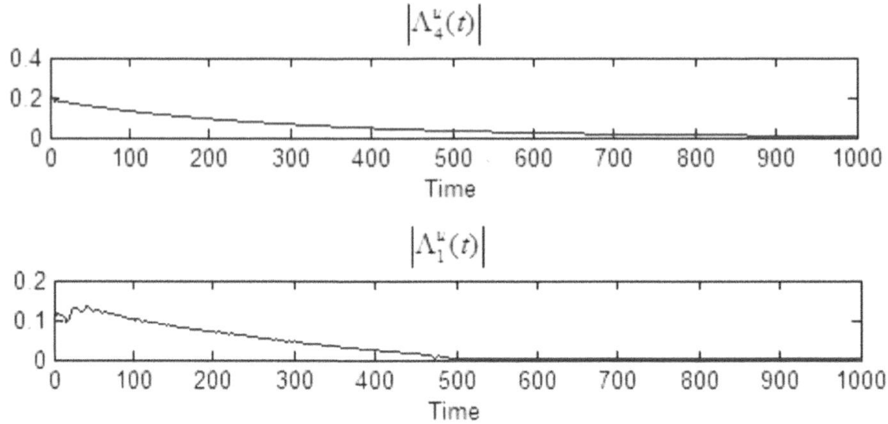

Figure 11. Magnitudes of slow forcing modulations of the IMOs of the rod end response at frequencies ω_4 and ω_1 (22).

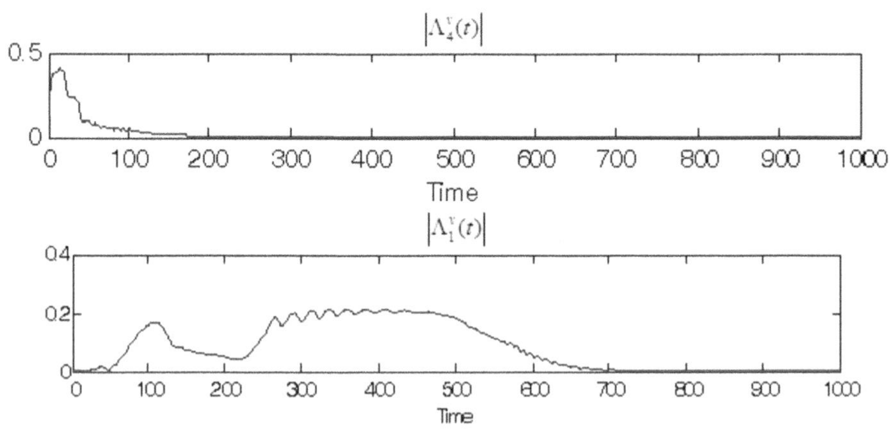

Figure 12. Magnitudes of slow forcing modulations of the IMOs of the nonlinear attachment response at frequencies $\omega_4/5$ and $\omega_1/3$ (22).

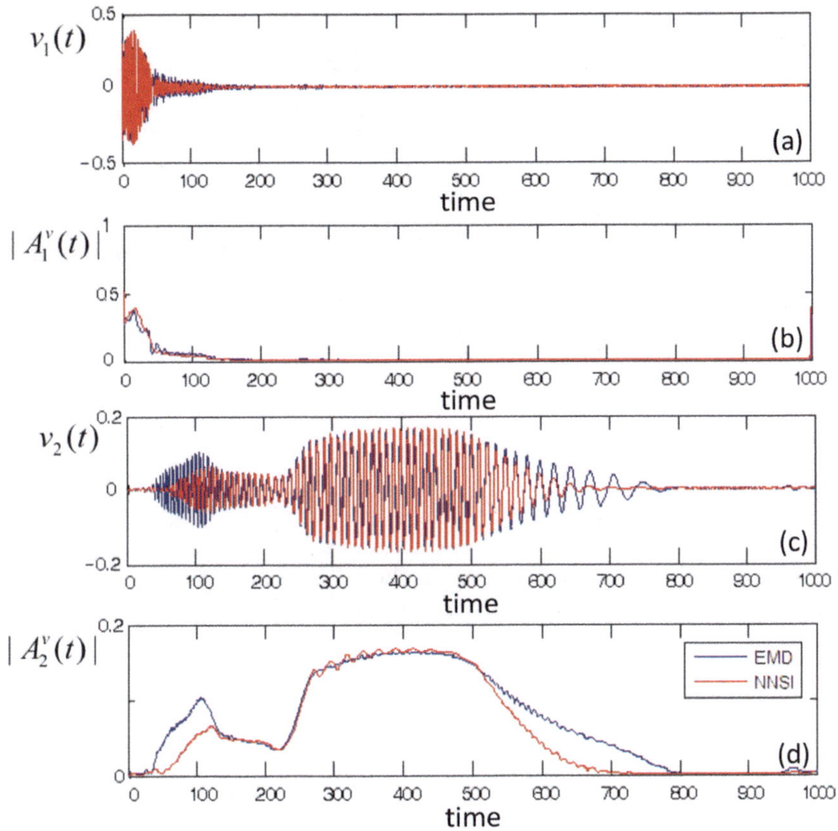

Figure 13. Comparison between the IMOs (22) and the IMFs of the response of the nonlinear attachment: (a,c) first and second dominant IMF compared to the first and second IMO; (b,d) slow amplitudes of (a,c) (22).

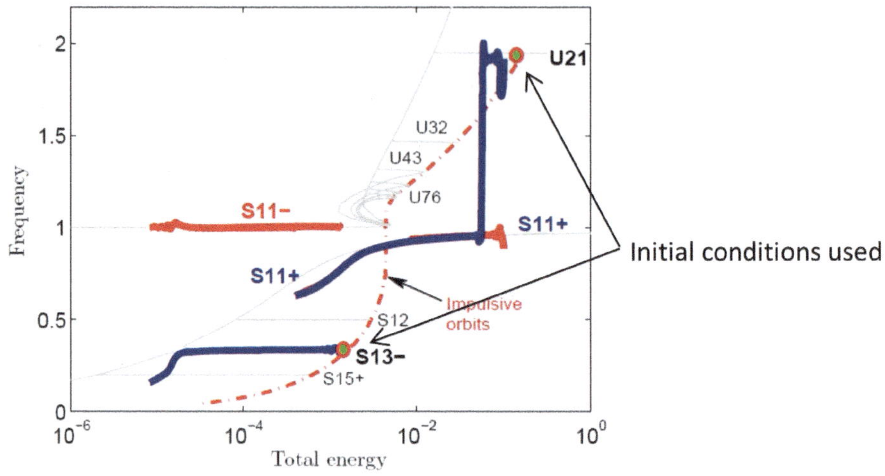

Figure 14. Global identification on the FEP of system (18) based on frequency and energy estimates from identified IMFs (6).

Acoustic mitigation

Bruno Cochelin*

* LMA (Laboratoire de Mécanique et d'Acoustique)
UPR 7051 CNRS, Centrale Marseille
13402 MARSEILLE Cedex 20, France

Abstract These notes correspond to two lectures given by Bruno Cochelin at the CISM course on "Modal Analysis of Nonlinear Mechanical Systems" held at Udine, from 25 to 29 june 2012. The first part is devoted to describe experimental results obtained by Romain Bellet to analyse Target Energy Transfer (T.E.T.) between a thin visco-elastic membrane and an acoustic medium. This study paves the way for a new technic of passive sound control in the low frequency regime, where no effective dissipative material exists. Then, an overview of several extentions and applications of the T.E.T. concept in acoustics is given. All these results have been obtained with the collaboration of R. Bellet, S. Bellizzi, R. Côte, P. Herzog, R. Mariani., P.O. Mattei and J. Shao.

1 INTRODUCTION

For reducing the level of noise and vibrations that arises in many engineering applications, it is often made use of specific devices called dynamic absorber or absorber. In acoustics, classical absorbers are porous materials for high frequency and Helmholtz resonators for low frequency. In mechanics the most popular device is tuned mass-damper system known as the Frahm absorber. Most of these absorbers are linear devices that rely on the anti-resonance concept. In recent theoretical and experimental works Gendelman et al. (2001); Vakakis and Gendelman (2001); McFarland et al. (2005); Gourdon et al. (2007); Nucera et al. (2008), it has been demontrated that the use of a pure nonlinear absorber, i.e. a mass with an essentially nonlinear spring, can be an interesting alternative solution for reducing vibrations. Such absorbers work on the resonance capture phenomenon Vakakis (2001); Kerschen et al. (2005) and they differ radically from the classical linear ones. The principle is to put the absober in a situation where an irreversible transfer of vibrational energy occurs from the primary linear system to the absorber. This energy is finally dissipated by

G. Kerschen (Ed.), *Modal Analysis of Nonlinear Mechanical Systems*, CISM International Centre for Mechanical Sciences DOI 10.1007/ 978-3-7091-1791-0_8 © CISM Udine 2014

damping in the absorber. The result is an efficient cancellation of the vibration in the linear system, since the motion gets localized in the absorber. That phenomenon is called targeted energy transfer in the literature. An interesting feature of such nonlinear aborber is to operate in a given frequency band, instead of a single frequency for a classical Frahm absorber or Helmholtz resonator. Indeed, it is worth noting that, since the spring of the absorber is essentially nonlinear without linear contribution, this system has no natural frequency. A drawback of the nonlinear absorber is that the irreversible transfert of energy occurs only when the primary linear system has reached a certain value of vibrational energy. This can be a limitation for practical applications. More details on the theory and on the pros and cons of such non linear abdorbers can be found in a recent book by Vakakis et al. (2008).

In the following notes, the primary linear system to be protected is an acoustic medium. We investigated the possiblity to design new kind of acoustic passive absorbers that could be efficient for low frequency regime, for loud transient noises or loud broad band noises. The nonlinear absorber is a thin viscoelastic membrane that is connected to the acoustic medium and that performs very large amplitude oscillations, i.e. the amplitude of the excursion is very large as compared to the thickness of the membrane. The principle of the experimental set-up and the associated model will be firstly introduced. Then experimental results will be presented, that is to say the different observed regimes under sinusoidal forced excitation, the transient responses of our system and the different kind of frequency responses. This first experimental verification of energy pumping in acoustics is a first step toward the design of new generation of passive acoustic absorbers. We end these notes by a discussion on several research efforts made to extend the performances and to find applications.

2 EXPERIMENTAL SET-UP AND ASSOCIATED MODELS

2.1 The classical 2 d.o.f. mechanical system

To analyse the targeted energy transfer, many authors have used simple mechanical systems with two degrees of freedom. A mass and a linear spring stand for the linear system that has to be protected. A mass with an essentially nonlinear spring and a damper stand for the dynamic absorber. A coupling spring connects the two oscillators as shown in figure 1. Because the nonlinear oscillator is connected to the ground, this configuration is refered to as the grounded configuration. Target energy transfert typically occurs when the mass of the nonlinear system is of the same order as the mass of the linear oscillator and when the stiffness coupling between the two oscillators is small.

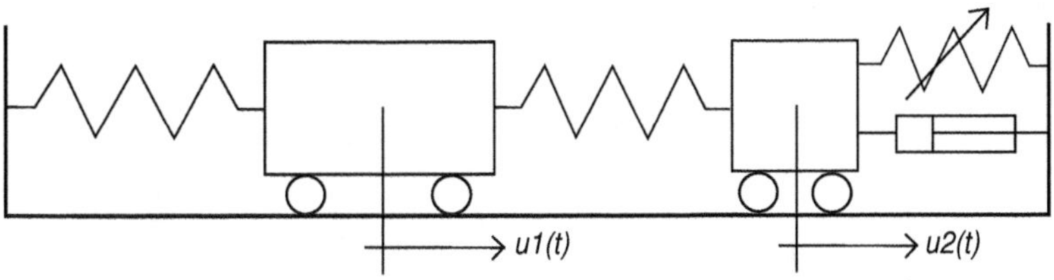

Figure 1. The classical 2 d.o.f. mechanical system used to study the energy pumping phenomenon.

Let $u_1(t)$ and $u_2(t)$ be the displacement of the masses, and assuming a cubic restoring force for the nonlinear spring, the governing equations (nondimentional form) of that grounded configuration are :

$$\begin{aligned} &\ddot{u_1} + a\dot{u_1} + u_1 + b(u_1 u_2) = 0 \\ &c\ddot{u_2} + d\dot{u_2} + eu_2^3 + b(u_2 u_1) = 0 \end{aligned} \tag{1}$$

where b is the small coupling coefficient, c the mass ratio, a and d respectively the damping factors in the tube and in the nonlinear oscillator and e the cubic stiffness coefficient. Despite its simplicity, such a system has a very complex dynamic that has been thouroughly investigated in Lee et al. (2005); Starosvetsky and Gendelman (2008); Vakakis et al. (2003); Gendelman et al. (2005); Gendelman and Lamarque (2005). It should be added that these references deals with the non-grounded configuration, where a small mass is directly connected to the linear mass by the cubic spring. This does not matter since the behaviours of the grounded and the non-grounded configuration are very similar.

2.2 The vibro-acoustic experimental set-up

Our aim is to experimentaly reproduce the pumping phenomenon in a system where the linear oscillator is an acoustic medium. So we replace the linear mechanical oscillator by the air vibrating on the first acoustic mode in a tube, the coupling stiffness by the air in a coupling box and the nonlinear oscillator by a thin visco-elastic membrane as presented on figure 2. The experimental set-up based on these ideas and shown in figure 3 has effectively allowed to observe the targeted energy transfer (energy pumping) phenomena from the acoustic medium to the visco-elastic membrane.

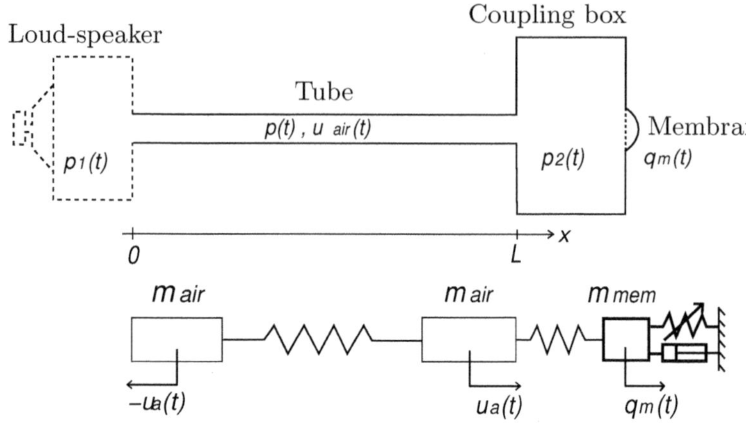

Figure 2. Principle of the set-up.

Practically, the tube (linear system) is an interchangeable U-tube so that its length L can vary between 1.5 and 2.5 m. The first resonance frequencies corresponding to these lengths are then between 75 and 120 Hz. Since the diameter of the tube $d = 0.094$ m is small as compared to L, it does not matter if the tube is straight or U-shape. The volume of the coupling box is $V_2 = 27\ 10^{-3}$ m^3. The device which holds the membrane allows to change the diameter of the working part from 4 to 8 cm. A sliding system permits to apply a constant in-plane pre-stress in the membrane. Once the pre-stress is set, the membrane is clamped to the supporting device. Concerning the material, we use both latex and sillicone, with a Young modulus of about 1.4 MPa, and a density of about 1000 kg.m^3. Finally, various thicknesses h between 0.18 and 1 mm have been tested. For the excitation of the tube (linear system), we use an acoustic source that consists in a loudspeaker and a coupling box which is connected to the other end of the tube. An analyser controls the excitation and collects two measurements: the acoustic pressure at the middle of the tube (microphone) and the velocity of the center of the membrane (laser vibrometer).

2.3 Associated models

A continuous model can be obtained by taking Helmholtz equation for the air contained in the tube, and the nonlinear plate equation of the Von Karman type for the thin membrane. A Kelvin- Voigt model is adopted to account for the viscosity in the membrane : the Piola-Kirchhoff second stress tensor $\mathbf{S}$ is related to the Green-Lagrange strain tensor $\mathbf{E}$ and strain

Figure 3. Photograph of the set-up. The membrane (in green) is supported by a device that allows to change the diameter and the pre-stress.

velocity tensor $\dot{\mathbf{E}}$ by $\mathbf{S} = \mathbf{D} : (\mathbf{E} + \eta\dot{\mathbf{E}})$, where$\mathbf{D}$ is a classical fourth order Hooke elastic tensor and η is the damping factor. As a first approximation, the continuous model can be reduced to a two degrees of freedom system by taking $u_{air}(x,t) = u_a(t)cos(\frac{\pi x}{L})$ for the air in the tube (first undamped mode), and $w(r,t) = q_m(t)(1-(\frac{r}{R})^2)$ for the transversal displacement of the circular membrane (parabolic shape function). The two coordinates $u_a(t)$ and $q_m(t)$ respectively correspond to the displacement at the end of the tube and at the center of the membrane. In the coupling box, the pressure is considered as spatially uniform. It is related to the volume variation by $p_2 = \rho_a c_0^2 \frac{\Delta V_2}{V_2}$ with $\Delta V_2 = u_a(t)St - q_m(t)\frac{Sm}{2}$ Applying a classical Galerkin method and adding a forcing term, we get the following reduced system :

$$\begin{aligned} m_a\ddot{u_a} + c_f\dot{u_a} + k_a u_a + S_t k_b(S_t u_a \frac{Sm}{2} q_m) &= F\cos(\Omega t) \\ m_m\ddot{q_m} + k_3(q_m^3 + 2\eta q_m^2\dot{q_m} + \frac{Sm}{2}k_b(\frac{Sm}{2}q_m - S_t u_a) &= 0 \end{aligned} \tag{2}$$

with

$$\begin{aligned} m_a = \frac{\rho_a S_t L}{2}, \; k_a = \frac{\rho_a S_t c_0^2 \pi^2}{2L}, \; m_m = \frac{\rho_m S_m h}{3}, \\ k_3 = \frac{\pi E h}{3(1-\nu^2)R^2}, \; k_b = \frac{\rho_a c_0^2}{V_2}, \end{aligned} \tag{3}$$

where h, R, Sm and ρ_m are the membrane thickness, radius, section and density, L and St are the tube length and section, V_2 is the volume of the coupling box, E the Young modulus of the membrane, ρ_a the density of the air and c_0 the sound speed. System (2) is then similar to (1) except for the dissipative term which is here nonlinear, because of the geometrical nonlinearity in the membrane. Since the membrane performs large amplitude oscillations, the linear stiffness of the membrane has been neglected.

3 EXPERIMENTAL RESULTS

On figures 4, 5, 6, 7 and 8 , the first channel is the signal of the sinsoidal forcing send to the loudspeaker, the second channel is the acoustic pressure measured at the middle of the tube and the third channel is the velocity of the center of the membrane.

3.1 Different regimes under sinusoidal forcing

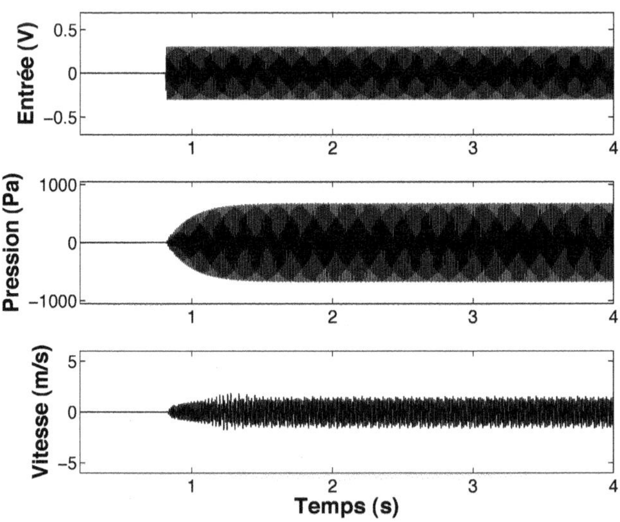

Figure 4. Low excitation amplitude (regime 1): periodic regime localized on the tube. Input voltage : A = 0.29 V.

We are interested in this part in the behaviour of the system under sinusoidal excitation, at the frequency of the first acoustic mode of the tube. Due to the presence of a non-linearity, several behaviors are observed depending on the level of the excitation. In this part, the set-up configuration is : $L = 2$m, $h = 0.4$ mm, $R = 3$ cm, $f1 = 62$ Hz. When the excitation level is below the threshold S1, the observed regime (regime 1) is periodic (figure 4) and vibrational energy is localized on the tube where the noise level is important while the membrane is inactive and has small vibrations. Moreover, on that regime, the displacement of the air at the end of the tube u_a and the displacement of the center of the membrane q_m are simultanous on the same frequency and almost out of phase. If the excitation level is high and more important than a second threshold S2, the regime (regime 2) is also periodic but the energy is now localized on the membrane which vibrates with large amplitudes (figure 6). On that regime, u_a and q_m are still synchronous at the same frequency, but they are now almost in phase. A resonance capture has occured between the tube and the membrane. This is the phenomenon which is at the heart of the energy pumping . These regimes are actually directly connected to two different nonlinear normal modes of the system, as detailled in Bellet et al. (2010). Between these two

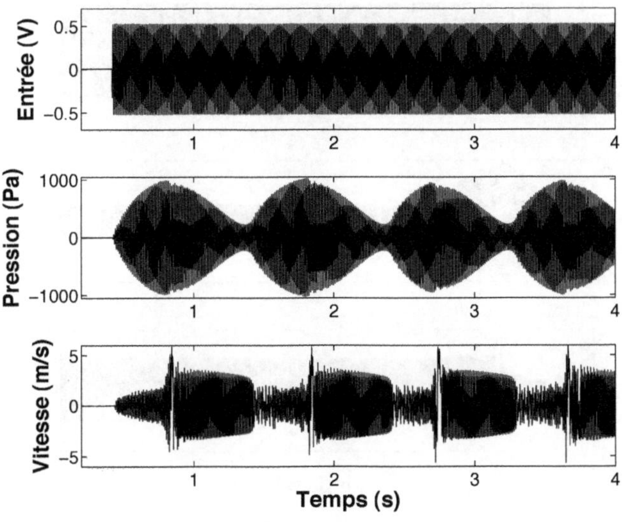

Figure 5. Intermediate excitation amplitude : quasi-periodic regime. Input voltage : A = 0.51 V.

thresholds, the regime is quasi-periodic and the system bifurcates alternatively from one regime to the other (figure 5). Indeed during the sound establishment, the motion is on the regime 1. Then the sound pressure in the tube reaches a certain level corresponding to the threshold S1 which makes the system bifurcating on the regime 2 and the membrane capturing the resonance of the tube. Then the membrane starts vibrating with a great amplitude and the noise level in the tube decreases quickly. The energy is irreversibly transfered from the tube to the membrane and the membrane acts as a nonlinear energy sink (NES). This is during this phase that we talk about energy pumping or targeted energy transfer. At the end of this transfer, the noise level reaches an energy level too low for the regime 2 and the system bifurcates on the regime 1 where the acoustic pressure will grow again and reproduce indefinitely the same cycle. We should add that this figures show an important drawback of the energy pumping: the value of the threshold is very high (depending on the configuration, the membrane is activated if the sound in the tube reaches a threshold between 130 and 160 dB) and represents at the moment an important limitation of the phenomenon.

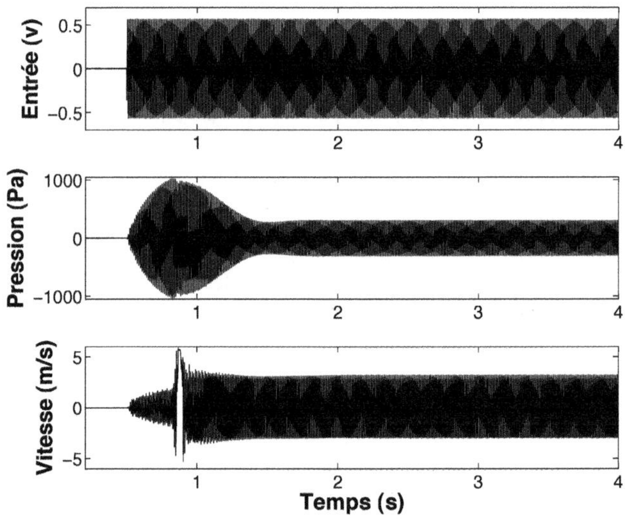

Figure 6. High excitation amplitude (regime 2): periodic regime localized on the membrane. Input voltage : A = 0.59 V .

3.2 Free oscillations

Since the energy pumping is an intrinsically transient phenomenon, this part deals with the behavior of the system under free oscillations with the configuration : $L = 2$m, $h = 0.6$ mm, $R = 3$ cm,$f1 = 57$ Hz. Experimentally, the loudspeaker can not produce a powerful enough pulse to activate the energy pumping. As a transient response, we observe then the free oscillations of the system after a sinusoidal excitation suddenly stopped. If, at that initial instant of the free oscillations, the system vibrates on the regime 1 (for low excitation amplitude), then the sound extinction in the tube follows a natural exponential decrease (figure 7) and the NES is inactive. But, if at the initial instant the system vibrates on the regime 2 (high excitation amplitude), then the sound exctinction in the tube follows a quasi-linear decrease, much faster than the exponential one, during which the membrane still vibrates with a great amplitude until the almost complete cancellation of the sound in the tube (figure 8). During this phase, a targeted energy transfer occurs from the tube to the membrane : the NES quickly captures the resonance of the tube, localizes on it the energy of the acoustic medium and then damps it by viscosity in the membrane and also here, but in a negligible part, by acoustic radiation outside of the set-up.

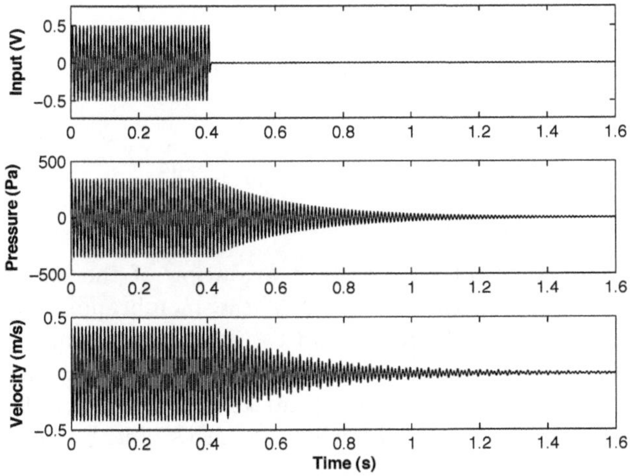

Figure 7. Free oscillations with low energy initial conditions: observation of an exponential decrease for the acoustic pressure. Input voltage: A=0.5 V.

Thanks to the model previously presented, we can define and compute

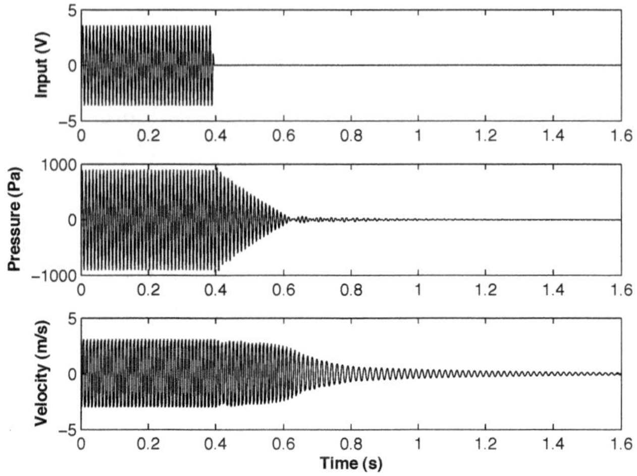

Figure 8. Free oscillations with high energy initial conditions: observation of a quasi-linear decrease for the acoustic pressure. Input voltage: A=3.7 V.

the energy of the different elements of the system :

$$\begin{array}{l} E_{tube} = \frac{1}{2} m_a \dot{u_a}^2 + \frac{1}{2} k_a u_a^2 \\ E_{membrane} = \frac{1}{2} m_m \dot{q_m}^2 + \frac{1}{2} k_3 q_m^4 \\ E_{box} = \frac{1}{2} k_b (S_t u_a \frac{S_m}{2} q_m)^2 \\ E_{total} = E_{tube} + E_{membrane} + E_{box} \end{array} \qquad (4)$$

Then the figures 9 and 10 show the evolution of the energy and the pourcentage of energy in the tube and of the membrane computed with the time series as given in figures 7 and 8 . It appears clearly that in the first case (figure 9) the energy remains always localized in the tube, the membrane never acts. But in the second case (figure 10), the energy is totally transfered to the membrane which localizes quickly almost 100% of the energy of the system.

3.3 Frequency responses

After presenting the temporal behaviour of the system during and after a single harmonic excitation, let us look at the behavior of that acoustic medium coupled to a membrane in the frequency domain. Because of the a non-linearity, we cannot talk about transfer function as the behaviour of the system depends on the amplitude of excitation. Indeed as we see in figure 11,

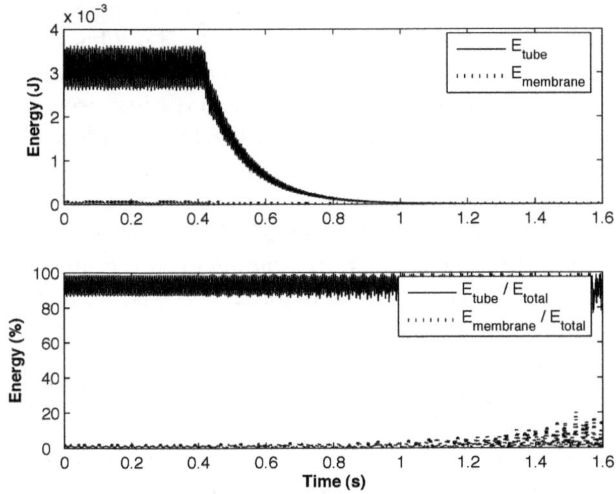

Figure 9. Evolution of the energy in the tube and the energy of the membrane for the time series of figure 7.

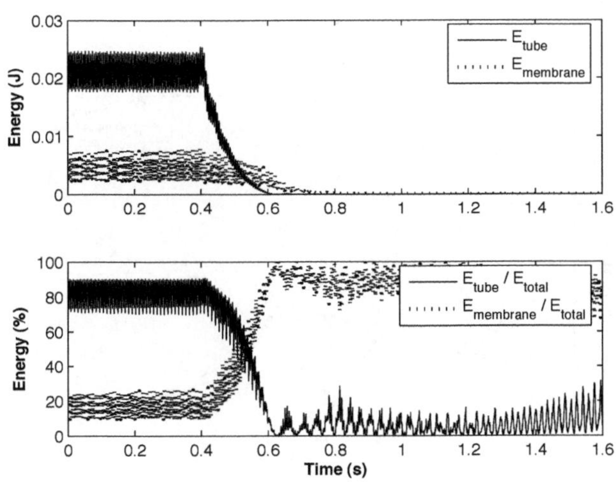

Figure 10. Evolution of the energy in the tube and the energy of the membrane for the time series of figure 8.

there are several kinds of frequency responses. The set-up configuration here is: $L = 2.22$ m,$h = 0.18$ mm, $R = 4$ cm, $f1 = 45$ Hz. This figure shows the experimental normalized values of the amplitude of the acoustic pressure in the tube divided by the amplitude of excitation during sweeps in frequency at constant amplitudes. When the level of excitation is lower than S1, the membrane remains inactive during the entire sweep and the result is simply the resonance peak of the tube. For a level between S1 and S2, a clipping of the peak appears. For frequencies below 85 Hz and above 90 Hz the sound level in the tube is too low to activate the energy pumping and frequency response is identical to that obtained for low levels. But between those frequencies, the level becomes important enough to set the system on the quasi-periodic regime corresponding to the energy pumping and that creates a clipping of the peak. In that frequency range, energy pumping prevents sound pressure from exceeding a certain level. Outside of that range, the presence of the absorber is transparent. When the excitation level is higher than S2, we observe a frequency response with an other resonance peak whom the resonance frequency is smaller than the resonance frequency of the tube and whom maximal amplitude is slightly lower than the amplitude of the resonance peak of the tube. In a noise reduction context, this frequency response is obviously much less interesting than the clipping peak produced by energy pumping, but as we can see, that phenomenon cannot appear with too strong levels.

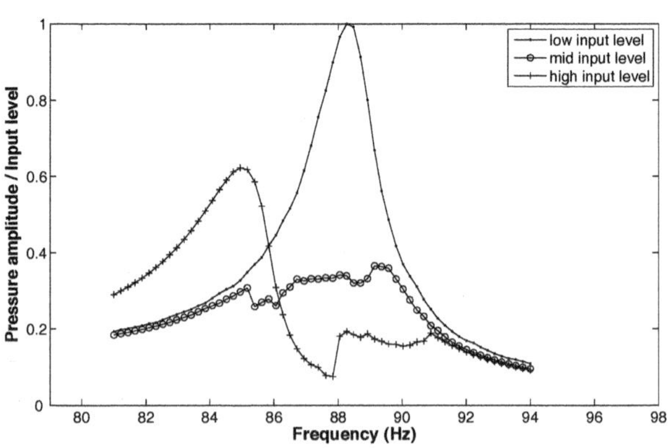

Figure 11. Frequency responses for different amplitudes of sweep sinus forcing.

3.4 Discussion

A more complete description of the set-up, of the model and of the T.E.T. analysis between the visco-elastic membrane and the acoustic medium can be found in Bellet et al. (2010). The link between T.E.T. and the nonlinear modes of the underlying conservative system is discussed, and the important self-tunning property of the membrane absorber is also demonstrated. A further experimental analysis using not only one but several nonlinear membrane absorbers on the coupling box is presented in Bellet et al. (2012). It is shown that using more membranes is a good way to enhance the robustness and the efficiency of the T.E.T. thank to an additive effect of each membrane which are activated in turn and which behaves relativement independentely. This finding allows to think of acoustic aborbing panel containing many small membranes having random characteristics.

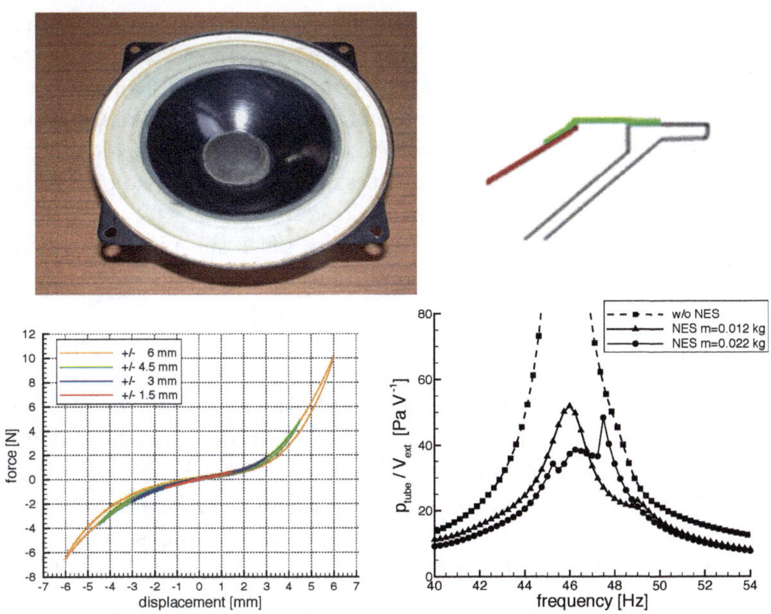

Figure 12. From left to right and top to bottom: The loudspeaker. Surround detail. Force displacement diagramm. Clippling of a resonance peak by adjusting the moving mass

Another experimental study of T.E.T in acoustic has been performed by Mariani et al. (2011). Instead of a membrane, a loudspeaker (suspended piston) working outside its usual linear regime plays the role of the nonlinear absorber. The main advantage of this alternative technology is the possibility to adjust independently the parameters. For example, the mass of the absorbers can be easilly augmented by gluing small masses on the moving part of the loudspeaker and the damping can be also adjusted electronically. Another very important point is the perspective to harvest the transfererd energy instead of dissipating it. On the the three loudspeakers used in Mariani et al. (2011) is presented in figure 12, with its nonlinear (static) force-displacement response and an example of mass adjustement for clipping a resonance peak.

In all these experiments, the acoustic medium was a one dimensionnal tube and a large coupling box was always used between the membrane and the tube. In order to extend the concept to more applications, Shao (2012) has considerer the T.E.T. between a 3D acoustic cavity and several membranes that are mounted directly on the wall of the cavity, without any coupling box. In that case, the weak coupling between the membrane and an acoustic mode is obtained by a correct placement of the membrane with respct to the acoustic modal shape. The design of these acoustic absorbers has been invetigated and tested numerically. Finally, the analysis of a two d.o.f. acoustic tube system coupled to a membrane aborber and sumitted to a two frequency excitation has been investigated by Bellizzi et al. (2013a,b).

We end here this overview of various research studies on acoustic T.E.T. that have started at the LMA in Marseille, after the pionneer results obtained on a simplifed set-up by Cochelin et al. (2006).

Bibliography

R. Bellet, B. Cochelin, P. Herzog, and P.-O. Mattei. Experimental study of target energy transfer from an acoustic system to a nonlinear membrane absorber. *Journal of Sound and Vibration*, 329:2768–2791, 2010.

R. Bellet, B. Cochelin, R. Côte, and P.-O. Mattei. Enhancing the dynamic range of targeted energy transfer in acoustics using several nonlinear membrane absorbers. *Journal of Sound and Vibration*, 331:5657–5668, 2012.

S. Bellizzi, R. Côte, and M. Pachebat. Responses of a two degree-of-freedom system coupled to a nonlinear damper under multi-forcing frequencies. *Journal of Sound and Vibration*, 332:1639–1653, 2013a.

S. Bellizzi, R. Côte, and M. Pachebat. Experimental study of an acoustic system coupled to a nonlinear membrane under multi-frequency forcing. *Proceedings of the ASME 2013 International Design Engineering Technical Conferences Computers and Information in Engineering Conference, IDETC/CIE 2013*, 2013b.

B. Cochelin, P. Herzog, and P.-O. Mattei. Experimental evidence of energy pumping in acoustics. *C. R. Mecanique*, 334(11):639–644, 2006.

O.V. Gendelman and C.H. Lamarque. Dynamics of linear oscillator coupled to strongly nonlinear attachment with multiple states of equilibrium. *Chaos, Solitons and Fractals*, 24:501–509, 2005.

O.V. Gendelman, L.I. Manevitch, A.F. Vakakis, and R. M'Closkey. Energy pumping in nonlinear mechanical oscillators: Part I - Dynamics of the underlying hamiltonian systems. *Journal of Applied Mechanics*, 68:34–41, 2001.

O.V. Gendelman, D.V. Gorlov, L.I. Manevitch, and A.I. Musienko. Dynamics of coupled linear and essentially nonlinear oscillators with substantially different masses. *Journal of Sound and Vibration*, 286:1–19, 2005.

E. Gourdon, N.A. Alexander, C.A. Taylor, C.H. Lamarque, and S. Pernot. Nonlinear energy pumping under transient forcing with strongly nonlinear coupling: Theoretical and experimental results. *Journal of Sound and Vibration*, 300:522–551, 2007.

G. Kerschen, A.F. Vakakis, Y.S. Lee, D.M. McFarland, J.J. Kowtko, and L.A. Bergman. Energy transfers in a system of two coupled oscillators with essential nonlinearity: 1:1 resonance manifold and transient bridging orbits. *Nonlinear Dynamics*, 42:283–303, 2005.

Y.S. Lee, G. Kerschen, A.F. Vakakis, P. Panagopoulos, L.A. Bergman, and D.M. McFarland. Complicated dynamics of a linear oscillator with a light, essentially nonlinear attachment. *Physica D*, 204:41–69, 2005.

R. Mariani, S. Bellizzi, B. Cochelin, P. Herzog, and P.O. Mattei. Toward an adjustable nonlinear low frequency acoustic absorber. *Journal of Sound and Vibration*, 330:5245–5258, 2011.

D.M. McFarland, L.A. Bergman, and A.F. Vakakis. Experimental study of non-linear energy pumping occurring at a single fast frequency. *International Journal of Non-Linear Mechanics*, 40:891–899, 2005.

F. Nucera, F. Lo Iaconoa, D.M. McFarland, L.A. Bergman, and A.F. Vakakis. Application of broadband nonlinear targeted energy transfers for seismic mitigation of a shear frame: Experimental results. *Journal of Sound and Vibration*, 313:57–76, 2008.

J. Shao. *Absorbeurs non linaires de nouvelle gnration bas sur le pompage nergtique: application aux cavits acoustiques 3D*. Phd thesis, Aix-Marseille University, 2012.

Y. Starosvetsky and O.V. Gendelman. Dynamics of a strongly nonlinear vibration absorber coupled to a harmonically excited two-degree-of-freedom system. *Journal of Sound and Vibration*, 312:234–256, 2008.

A.F. Vakakis. Inducing passive nonlinear energy sinks in vibrating systems. *Journal of Vibration and Acoustics*, 123:332, 2001.

A.F. Vakakis and O.V. Gendelman. Energy pumping in nonlinear mechanical oscillators: Part II - Resonance capture. *Journal of Applied Mechanics*, 68:42–48, 2001.

A.F. Vakakis, L.I. Manevitch, O.V. Gendelman, and L.A. Bergman. Dynamics of linear discrete systems connected to local, essentially non-linear attachments. *Journal of Sound and Vibration*, 264:559–577, 2003.

A.F. Vakakis, O.V. Gendelman, L.A. Bergman, D.M. McFarland, G. Kerschen, and Y.S. Lee. *Nonlinear targeted energy transfer in mechanical and structural systems*, volume 156 of *Solid mechanics and its applications*. Springer, 2008.